The Harmony of the World

75 Years of *Mathematics Magazine*

Library of Congress Catalog Card Number 2007926223

ISBN: 978-0-88385-560-7

Printed in the United States of America

Current Printing (last digit):
10 9 8 7 6 5 4 3 2 1

The Harmony of the World

75 Years of *Mathematics Magazine*

Edited by

Gerald L. Alexanderson
Santa Clara University

with the assistance of

Peter Ross
Santa Clara University

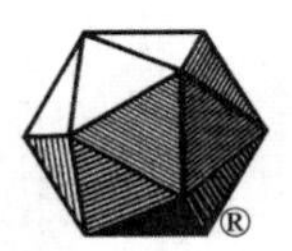

Published and Distributed by
The Mathematical Association of America

SPECTRUM SERIES

The Spectrum Series of the Mathematical Association of America was so named to reflect its purpose: to publish a broad range of books including biographies, accessible expositions of old or new mathematical ideas, reprints and revisions of excellent out-of-print books, popular works, and other monographs of high interest that will appeal to a broad range of readers, including students and teachers of mathematics, mathematical amateurs, and researchers.

777 Mathematical Conversation Starters, by John de Pillis

99 Points of Intersection: Examples—Pictures—Proofs, by Hans Walser. Translated from the original German by Peter Hilton and Jean Pedersen

All the Math That's Fit to Print, by Keith Devlin

Carl Friedrich Gauss: Titan of Science, by G. Waldo Dunnington, with additional material by Jeremy Gray and Fritz-Egbert Dohse

The Changing Space of Geometry, edited by Chris Pritchard

Circles: A Mathematical View, by Dan Pedoe

Complex Numbers and Geometry, by Liang-shin Hahn

Cryptology, by Albrecht Beutelspacher

The Early Mathematics of Leonhard Euler, by C. Edward Sandifer

The Edge of the Universe: Celebrating 10 Years of Math Horizons, edited by Deanna Haunsperger and Stephen Kennedy

Five Hundred Mathematical Challenges, Edward J. Barbeau, Murray S. Klamkin, and William O. J. Moser

The Genius of Euler: Reflections on his Life and Work, edited by William Dunham

The Golden Section, by Hans Walser. Translated from the original German by Peter Hilton, with the assistance of Jean Pedersen.

The Harmony of the World: 75 Years of Mathematics Magazine, edited by Gerald L. Alexanderson with the assistance of Peter Ross

I Want to Be a Mathematician, by Paul R. Halmos

Journey into Geometries, by Marta Sved

JULIA: a life in mathematics, by Constance Reid

The Lighter Side of Mathematics: Proceedings of the Eugène Strens Memorial Conference on Recreational Mathematics & Its History, edited by Richard K. Guy and Robert E. Woodrow

Lure of the Integers, by Joe Roberts

Magic Numbers of the Professor, by Owen O'Shea and Underwood Dudley

Magic Tricks, Card Shuffling, and Dynamic Computer Memories: The Mathematics of the Perfect Shuffle, by S. Brent Morris

Martin Gardner's Mathematical Games: The entire collection of his Scientific American columns

The Math Chat Book, by Frank Morgan

Mathematical Adventures for Students and Amateurs, edited by David Hayes and Tatiana Shubin. With the assistance of Gerald L. Alexanderson and Peter Ross

Mathematical Apocrypha, by Steven G. Krantz

Mathematical Apocrypha Redux, by Steven G. Krantz

Mathematical Carnival, by Martin Gardner

Mathematical Circles Vol I: In Mathematical Circles Quadrants I, II, III, IV, by Howard W. Eves

Mathematical Circles Vol II: Mathematical Circles Revisited and Mathematical Circles Squared, by Howard W. Eves

Mathematical Circles Vol III: Mathematical Circles Adieu and Return to Mathematical Circles, by Howard W. Eves

Mathematical Circus, by Martin Gardner
Mathematical Cranks, by Underwood Dudley
Mathematical Evolutions, edited by Abe Shenitzer and John Stillwell
Mathematical Fallacies, Flaws, and Flimflam, by Edward J. Barbeau
Mathematical Magic Show, by Martin Gardner
Mathematical Reminiscences, by Howard Eves
Mathematical Treks: From Surreal Numbers to Magic Circles, by Ivars Peterson
Mathematics: Queen and Servant of Science, by E.T. Bell
Memorabilia Mathematica, by Robert Edouard Moritz
Musings of the Masters: An Anthology of Mathematical Reflections, edited by Raymond G. Ayoub
New Mathematical Diversions, by Martin Gardner
Non-Euclidean Geometry, by H. S. M. Coxeter
Numerical Methods That Work, by Forman Acton
Numerology or What Pythagoras Wrought, by Underwood Dudley
Out of the Mouths of Mathematicians, by Rosemary Schmalz
Penrose Tiles to Trapdoor Ciphers ... and the Return of Dr. Matrix, by Martin Gardner
Polyominoes, by George Martin
Power Play, by Edward J. Barbeau
The Random Walks of George Pólya, by Gerald L. Alexanderson
Remarkable Mathematicians, from Euler to von Neumann, Ioan James
The Search for E.T. Bell, also known as John Taine, by Constance Reid
Shaping Space, edited by Marjorie Senechal and George Fleck
Sherlock Holmes in Babylon and Other Tales of Mathematical History, edited by Marlow Anderson, Victor Katz, and Robin Wilson
Student Research Projects in Calculus, by Marcus Cohen, Arthur Knoebel, Edward D. Gaughan, Douglas S. Kurtz, and David Pengelley
Symmetry, by Hans Walser. Translated from the original German by Peter Hilton, with the assistance of Jean Pedersen.
The Trisectors, by Underwood Dudley
Twenty Years Before the Blackboard, by Michael Stueben with Diane Sandford
The Words of Mathematics, by Steven Schwartzman

MAA Service Center
P.O. Box 91112
Washington, DC 20090-1112
800-331-1622 FAX 301-206-9789

Introduction

The idea of doing a sampling of the "best" in *Mathematics Magazine* over its 75-year history came to us when we found in Steven G. Krantz's book, *Mathematical Apocrypha*, the story of how it happened that the Stone-Weierstrass theorem, one of the great 20th century theorems in analysis, first appeared in an issue of *Mathematics Magazine* back in 1947/48. A fuller account of how this came about is described in the note accompanying Stone's article included in this volume.

Once we started looking at the contents of the *Magazine* we were surprised and pleased to find so much good material, much of it in the early days of the journal. The lineup of contributors of articles to the *Magazine* is star-studded: Maurice Fréchet, the eminent French analyst; Earle Raymond Hedrick, the first President of the Mathematical Association of America (MAA); Otto Neugebauer, the noted historian of mathematics and founder of *Mathematical Reviews*; E. T. Bell, author of *Men of Mathematics*; D. H. Lehmer, from a family of computational number theorists; H. S. M. Coxeter, largely responsible for the 20th century resurgence of geometry; Dirk Struik, the historian of mathematics; Olga Taussky, the eminent algebraist at Caltech; Ralph P. Boas, Jr., analyst, editor and president of the MAA; Paul Erdős, that peripatetic and fascinating problem solver; Don Knuth, the preeminent computer scientist; Richard W. Hamming, the pioneering expert in coding theory; Ron Graham, combinatorialist and number theorist, and president of both the American Mathematical Society and the MAA; and on and on.

Not all of the above are represented in this collection. There's just too much good material and difficult choices had to be made for balance and accessibility. In an article in *Mathematics Magazine* (78 (2005), 110–123) we were able to give short introductions to far more articles. We rarely include material published since 1990 since that is probably fairly well-known to potential readers and is generally more accessible.

The history of the *Magazine* is complicated and, prior to the publication's being taken over by the MAA, the history sounds like the Perils of Pauline. In fact, the publication did die in 1945 only to be revived in 1947. For details of that we refer the reader to the next section that outlines the history of the publication.

The full archive of the *Magazine* is available on the Web on JSTOR Arts and Sciences II, a database that is unfortunately available only by subscription and many institutions do not yet subscribe.

We wish to acknowledge with thanks the assistance of the recent editor of *Mathematics Magazine*, Frank Farris; Tatiana Shubin who has read material from the various drafts; members of the

Spectrum Editorial Board and the Council on Publications for their suggestions; Leonard Klosinski who provided technical help with the illustrations; Don Albers, Publications Director of the MAA; and the estimable Elaine Pedreira and Beverly Ruedi of the MAA Publications Office.

Gerald L. Alexanderson
Peter Ross
Santa Clara University

Contents

The name of each article is followed by a notation indicating the field of mathematics from which it comes: (Al) algebra, (AM) applied mathematics, (An) analysis, (CG) combinatorics and graph theory, (G) geometry, (H) history, (L) logic, (M) miscellaneous, (NT) number theory, (PS) probability/statistics, and (T) topology.

Introduction vii

A Brief History of Mathematics Magazine xi

Part I: The First Fifteen Years 1

Perfect Numbers, *Zena Garrett* (NT) 3

Rejected Papers of Three Famous Mathematicians, *Arnold Emch* (H) 5

Review of *Men of Mathematics*, *G. Waldo Dunnington* (H) 9

Oslo under the Integral Sign, *G. Waldo Dunnington* (H) 11

Vigeland's Monument to Abel in Oslo, *G. Waldo Dunnington* (H) 19

The History of Mathematics, *Otto Neugebauer* (H) 23

Numerical Notations and Their Influence on Mathematics, *D. H. Lehmer* (NT) 29

Part II: The 1940s 33

The Generalized Weierstrass Approximation Theorem, *Marshall H. Stone* (An) 35

Hypatia of Alexandria, *A. W. Richeson* (H) 45

Gauss and the Early Development of Algebraic Numbers, *E. T. Bell* (Al) 51

Part III: The 1950s 69

The Harmony of the World, *Morris Kline* (M) 71

What Mathematics Has Meant to Me, *E. T. Bell* (M) 79

Mathematics and Mathematicians from Abel to Zermelo, *Einar Hille* (H) 81

Inequalities, *Richard Bellman* (An) 95

A Number System with an Irrational Base, *George Bergman* (NT) 99

Part IV: The 1960s 107

Generalizations of Theorems about Triangles, *Carl B. Allendoerfer* (G) 109

A Radical Suggestion, *Roy J. Dowling* (NT) 115

Topology and Analysis, *R. C. Buck* (An, T) 117

The Sequence $\{\sin n\}$, *C. Stanley Ogilvy* (An) 121

Probability Theory and the Lebesgue Integral, *Truman Botts* (PS) 123

On Round Pegs in Square Holes and Square Pegs in Round Holes, *David Singmaster* (G) 129

π_t: 1832–1879, *Underwood Dudley* (M) 133

Part V: The 1970s ... 135

Trigonometric Identities, *Andy R. Magid* (Al) ... 137
A Property of 70, *Paul Erdős* (NT) ... 139
Hamilton's Discovery of Quaternions, *B. L. van der Waerden* (Al, H) ... 143
Geometric Extremum Problems, *G. D. Chakerian and L. H. Lange* (G) ... 151
Pólya's Enumeration Theorem by Example, *Alan Tucker* (CG) ... 161
Logic from A to G, *Paul R. Halmos* (L) ... 169
Tiling the Plane with Congruent Pentagons, *Doris Schattschneider* (G) ... 175
Unstable Polyhedral Structures, *Michael Goldberg* (G) ... 191

Part VI: The 1980s ... 197

Leonhard Euler, 1707–1783, *J. J. Burckhardt* (H) ... 199
Love Affairs and Differential Equations, *Steven H. Strogatz* (An) ... 211
The Evolution of Group Theory, *Israel Kleiner* (Al) ... 213
Design of an Oscillating Sprinkler, *Bart Braden* (AM) ... 229
The Centrality of Mathematics in the History of Western Thought, *Judith V. Grabiner* (M) ... 237
Geometry Strikes Again, *Branko Grünbaum* (G) ... 247
Why Your Classes Are Larger than "Average", *David Hemenway* (PS) ... 255
The New Polynomial Invariants of Knots and Links,
W. B. R. Lickorish and Kenneth C. Millett (CG, T) ... 257

Briefly Noted ... 273
The Problem Section ... 279
Index ... 281
About the Editors ... 287

A Brief History of *Mathematics Magazine*[1]

The first few issues of what was to become *Mathematics Magazine* were little more than pamphlets soliciting members for the then newly formed Louisiana-Mississippi Section of the Mathematical Association of America (MAA). This was in 1926. The Section worked closely with the National Council of Teachers of Mathematics so the goals of the publication, formally called the *Mathematics News Letter* by the time Volume 2 came out, included the discussion of "the common problems of grade, high school, and college mathematical teaching". By the time Volume 6 appeared, grade school teaching had been stricken from the statement.

Early issues listed as editors the officers of the Section, but in 1928, two editors were actually named, Professor Samuel T. Sanders of Louisiana State University, Baton Rouge, and Mr. Henry Schroeder of Ruston High School in Louisiana. By 1929, Sanders was listed as Editor-in-Chief and he continued as editor until 1945.

The original subscription fees were $.50 for ten issues, described as "the price of two good movie entertainments." During the 1930s, in the middle of the Great Depression, money was tight and it is a wonder the publication survived. Sanders persisted, however, finding contributions where he could while trying to expand the subscription base.

With the 1934/35 issues the name was changed to *National Mathematics Magazine*, to reflect "the journal's expansion." That same year a "Teacher's Department", and a "Notes and News Department" were added to the "Book Review Department" and the "Problem Department", with a "Humanism and History of Mathematics" section added soon after. That same year Louisiana State University began to provide financial support.

In 1941 the *Magazine* nearly became a victim of the notorious political boss and governor of Louisiana, Huey P. Long, when he initiated budget cuts that resulted in the drastic reduction of the contribution from LSU, from $2,700 to $600. The mathematical community rallied around and letters of support were received from leading figures: MAA President Raymond W. Brink, AMS President John R. Kline, C. C. MacDuffee of Wisconsin ("It is with great regret that I learn that the *National Mathematics Magazine* is having financial trouble. I know you have put in years of work and devotion to a journal, which is a credit not only to you but to the University and the South."), Harold M. Bacon of Stanford ("Louisiana State University has rendered a real service to mathematics and education not only in the South but throughout the Nation in supporting the *Magazine*.")

Sanders soldiered on and took on the financial responsibility for the journal himself, in spite of the unfortunate timing—the lack of support from LSU coincided with Sanders' forced retirement from the University. The MAA made a modest contribution for 1942–43 of $400 from its

[1]For a much more detailed history of the first fifty years of the *Magazine*, see "Mathematics Magazine: The First Half Century," by Edwin F. Beckenbach, in Seebach, J. Arthur, Jr., and Lynn Arthur Steen, eds., *Mathematics Magazine: 50 Year Index*, Washington, DC, Mathematical Association of America, 1978. Long out-of-print, the essay can be found at www.maa.org.

Jacob Houck Memorial Fund. Some individuals made modest contributions to the effort and the subscription rate, by then \$2, was increased to \$3. Editor Sanders moved to Mobile, Alabama, to teach at the Mobile Center of the University of Alabama, and later at the University Military School. When he finally retired from teaching in 1957, he was 85.

The MAA continued to provide a modest subsidy until 1945 when, in discussions of whether the MAA should assume responsibility for the continuation of the *Magazine*, the MAA Board decided not to take on a second journal and to cease the subsidy. The result was inevitable: after the November 1945 issue, the *Magazine* ceased publication.

Early in the spring of 1946, Professor Glenn James at UCLA learned of the demise of the *Magazine*. James was best known for the *Mathematics Dictionary* that he had produced in collaboration with his son, Robert C. James. The senior James had contributed articles to the *Magazine* and admired Sanders's work. He proposed that he take over the editorship of a reborn *Magazine* and the necessary arrangements were worked out to transfer the editorship to California. James' plan was to cut expenses by having the production and distribution work done by him and his family in his home. Sanders was reluctant to have the *Magazine* moved from the South since it was the only mathematical journal to originate there, but he had little choice if the *Magazine* was to survive. At the last minute, after an agreement had been reached, the University of Alabama offered to assume responsibility for the *Magazine*, but the offer came too late.

One of the first moves was to drop "National" from the title. E. F. Beckenbach, also on the faculty at UCLA, suggested this since he thought that the *Magazine* by that time had developed international appeal. Beckenbach was asked to join the Executive Committee of the *Magazine*, along with Professor Donald H. Hyers of the University of Southern California and Professor Aristotle D. Michal of Caltech. Beckenbach declined since he was already editing a section of the *Monthly*. The first issue of the newly formed *Mathematics Magazine* was that of September-October, 1947—the subscription rate remained \$3. The production center was in the James family home, with most of the work done on a ping-pong table constructed by their son Raymond. Mrs. James was Circulation Manager and kept all records on cards in file boxes. She addressed by hand thousands of wrappers for copies of issues of the journal. She worked without pay, of course. Composition work was done by sons, daughters, daughters-in-law,. . . very much a family operation.

For twelve years James continued as editor. But with failing eyesight, he realized that he could not continue so he entered into negotiations with Professor Carl B. Allendoerfer, President of the MAA, and this time the Board decided to take over sponsorship of the *Magazine*. Of course, it was easier to negotiate a deal then than earlier since under James' leadership the *Magazine* was operating in the black—largely due to contributed labor by members of his family—but also because the subscription list had increased from a few hundred to a healthy 2200. Thus, in 1959, responsibility for the *Magazine* was transferred to the MAA and a new editor was appointed, Professor Robert E. Horton of Los Angeles City College (and later President of Los Angeles Valley College). In 1960 the role of the *Magazine* was defined as follows: "The mathematical level of the *Magazine* shall be below that of the *Monthly* but above that of the *Mathematics Teacher*." Since this move the MAA has taken on publication of a third journal, the *Two-Year College Mathematics Journal* (now the *College Mathematics Journal*), initiated by the publisher Prindle, Weber and Schmidt in 1970 and taken over by the MAA in 1974.

After the MAA assumed responsibly for the *Magazine* its future has been assured and a succession of editors has been appointed by the President and Board of Governors. The list of editors is (with their terms starting in the years indicated): Roy Dubisch of the University of Washington, Seattle (1964); Stephen A. Jennings of the University of Victoria (B.C.) (1969); G. N. Wollan of Purdue University, West Lafayette (1971); J. Arthur Seebach and Lynn Arthur Steen of St. Olaf College (1976); Doris Schattschneider of Moravian College (1981); Gerald L. Alexanderson of Santa Clara University (1986); Martha J. Siegel of Towson University (1991); Paul Zorn of St. Olaf College (1996); Frank A. Farris of Santa Clara University (2001); and Allan Schwenk of Western Michigan University (2006).

The founding editor, Samuel T. Sanders, died in 1970 at the age of ninety-eight. The editor who saved the *Magazine* from an early demise, Glenn James, died in 1961, one month short of his eightieth birthday.

Let us close with a few little stories about the earlier years of the *Magazine*.

Editors of the *Magazine* even today receive submissions that cause them to do a double take. They have to ask: "Did that manuscript really say that?" Sanders must have had such an experience in 1931 when he looked at an angle-trisection "proof" given by The Reverend J. J. Callahan, President of Duquesne University in Pittsburgh. Fr. Callahan's error was not subtle: he merely showed how with a straightedge and compass one could construct an angle that is three times as large as a given angle! That same year Fr. Callahan published a rather handsome book—it could easily have been confused with a publication of the Oxford University Press at the time—in which he proved in 310 pages the parallel postulate, showed noneuclidean geometries to be "false and baseless" and, further, exposed the theory of relativity, based as it is on Riemann's noneuclidean geometry, to be "from the point of view of physics, mathematics and metaphysics, an utter falsehood on all these counts." Called *Euclid or Einstein: A Proof of the Parallel Theory and a Critique of Metageometry*, the book has not had much impact on the scientific world.

In 1936 the *Magazine* published a review of a book by Johannes Mahrenholz, *Anekdoten aus dem Leben deutscher Mathematiker* (*Anecdotes on the lives of German mathematicians*). The author taught at Adolf Hitler High School in Cottbus, Germany.

In nonmathematical circles, perhaps the best-known *Mathematics Magazine* author was the Metropolitan Opera basso, Jerome Hines. He had been a music, chemistry and mathematics major at UCLA but decided not to go on in science and mathematics, instead pursuing a career as a singer, which he did very well. He was awarded the prestigious Caruso Award at the Met and for that company created the role of Swallow in Benjamin Britten's *Peter Grimes* in 1948. He continued to pursue mathematics as a hobby and occasionally gave seminars on his work. He published a series of articles in the *Magazine* during the 50s, on topics ranging from operator algebras to Stirling numbers. Two of these were reviewed in *Mathematical Reviews*.

Hines was not alone. Other musicians have expressed interest in mathematics. Richard Strauss is reputed to have enjoyed working on mathematical problems. Lou Harrison had a serious interest in Fibonacci numbers and their relevance to the tuning of gamelans. But they didn't write for *Mathematics Magazine*.

Part I

The First Fifteen Years

Perfect Numbers

Zena Garrett

Mathematics News Letter
3 (6) (1928/29), 17–19

Editor's Note: This short pedagogical note appears very early in the history of the *Magazine*. Written by a student, Zena Garrett, at Mississippi Delta State Teachers College, it describes how she worked, under the direction of her teacher, identified only as "Miss Dale,"[2] to find the next perfect number after 6 and 28. (Perfect numbers are positive integers that equal the sum of their proper divisors.) The directions she was given–"we were instructed to work alone and forbidden to use the library"–sound strikingly like the Moore Method, pioneered only somewhat earlier by R. L. Moore at the University of Pennsylvania and the University of Texas. Miss Garrett succeeded in finding 496 and 8128 and then went on to find the formula $2^{n-1}(2^n - 1)$, with $2^n - 1$ a prime, for even perfect numbers. (No odd perfect numbers are known.) This piece appeared at a time when most of the articles in the *Magazine* were concerned with commercial arithmetic, high school algebra and articles on how to teach, so this was one of the early articles to explore a more advanced mathematical question, albeit as an early example of undergraduate research.

Ms. Garrett was 18 when she wrote this article and is still alive and living in Massachusetts. She did not go on in mathematics but instead earned a master's in English at UCLA, became a sculptor, wrote short stories and a book, *The House in the Mulberry Tree*, and eventually moved on to architecture (without degrees in the field), mainly involved in restoration and preservation of old houses in New York and the Hamptons.

[1]Miss Dale was, in fact, Julia Dale, a graduate of Transylvania University, with a PhD from Cornell. She taught for many years at what is now Delta State University, in Cleveland, Mississippi.

Perfect numbers–that is, those equal to the sum of their integral subdivisors–were, in very olden times, a matter of grave consideration and conjecture. The ancients explained many mysteries of Nature by the magic number 6, presumably the first of the perfect numbers, unless 1, or unity, be so considered.

By 500 B.C., we find the Pythagoreans classifying numbers as excessive, perfect, or defective, in comparison with the sum of their integral subdivisors. In the 3rd century B.C., Euclid discussed perfect numbers and concluded by giving a formula for their selection.

In the Middle Ages, theology adopted perfect numbers, and the theologians, led by Alcuin, explained many of the Bible facts by this symbolism. Because of the perfection of the number 6, God created the world in six days, and on the sixth day He created His most perfect work, man. God, who did all things well, created six beings in the first creation. The second origin of man emanated from the number 8, the number of souls said to have been in Noah's Ark. But 8 is a defective number, hence the imperfection of the second creation.

These numbers, though no longer robed with their mystic qualities, were still of interest to mathematicians of later ages. Marin Mersenne, a Frenchman of the 17th century, a Franciscan friar, published in 1764 the first eight perfect numbers. A ninth was discovered in 1885 and a tenth in 1912.

My interest in perfect numbers was first aroused by a discussion in class. Miss Dale explained to us the reason for the name, told us the ancient beliefs concerning them, and gave us the first two—6 and 28. Then she asked if we would be interested in finding the next one, which was between 28 and 500. We were instructed to work alone and forbidden the use of the library.

My first step was to examine closely the structure of the two numbers given us. They are both even numbers. The factors of 6 are 1, 2, and a prime, 3. The factors of 28 are 1, 2, (2^2), 7 (a prime), and multiples of 7 and 2. Neither is divisible by any other number than a prime and powers of 2. I found that numbers divisible by multiples of 3 are excessive when even, defective when odd. Thus were all odd numbers and all numbers divisible by three eliminated as unlikely. Likewise the multiples of 5.

I soon found that perfect numbers did not appear regularly, that is, by the addition of or multiplication by a certain fixed amount; nor could I establish any satisfactory proportion. This made me think the number would probably be near 500, the limit given us.

With these conjectures as a basis, I soon found 496, the next perfect number after 28.

Now curiosity led me to discover the composition. The subdivisors of 496 are also based upon the powers of 2 and a prime–in this case, 31 . The prime for 28 is 7, for 6, it is 3. In each case you will notice that it is a power of 2 minus 1. In 6, 2 is raised to the first power, in 28 to the second, in 496 to the fourth.

I could now write out the formula for 6, 28, and 496, respectively:

Let $x = \frac{1}{2}$ the perfect number N.

$$\therefore 2x = N$$
$$1 + 2 + x = 2x$$
$$x = 3$$

N, or $2x = 6$.

$$1 + 2 + x + (2^2) + \frac{x}{2} = 2x$$
$$2x = 28$$
$$1 + 2 + x + 4 + \frac{x}{2} + 8 + \frac{x}{4} + 16 + \frac{x}{8} = 2x$$
$$x = 248$$

N, or $2x = 496$.

Now I tried my formula for the next one, thus:

$$1 + 2 + x + 4 + \frac{x}{2} + 8 + \frac{x}{4} + 16$$
$$+ \frac{x}{8} + 32 + \frac{x}{16} + 64 + \frac{x}{32} = 2x$$
$$x = 4064$$

N, or $2x = 8128$.

By testing I found this answer correct. I also found that the addition of one more set (e.g., $x + \frac{x}{16}$) to the preceding ones, would not give a perfect number.

But this formula was cumbersome to work with, so I shortened it by the use of geometric progressions:

$$(1 + 2 + 4 + 8 + 16 + \text{etc.})$$
$$+ (x + \frac{x}{2} + \frac{x}{4} + \frac{x}{8} + \text{etc.}) = 2x = N$$
$$\left\{\frac{1 - 2l}{1 - 2}\right\} + \left\{\frac{x - \frac{1}{2}\left(\frac{2x}{l}\right)}{1 - \frac{1}{2}}\right\} = 2x$$
$$(2l - 1) + \left\{\frac{2xl - 2x}{l}\right\} = 2x$$
$$2l^2 - l + 2xl - 2x = 2xl$$
$$2l^2 - l = 2x$$
$$l(2l - 1) = 2x = N$$
$$l = 2^{n-1}$$
$$2^{n-1}(2^n - 1) = N$$

$2^n - 1$ must always be a prime, as it is the prime upon which the perfect number is based. There seems to be no rule for ascertaining what values of n will make $2^n - 1$ prime, except that n will always be either 1 or an even number.

One of the peculiarities of perfect numbers is the fact that they always end in either 6, or 28. Another is the long, irregular leaps between the numbers.

But one thing be regretted about perfect numbers is that nothing in Nature or the Universe has yet been discovered which has any association whatever with the vagary of our number system. Such an association is certainly highly improbable because of the immense rapidity with which the numbers increase. Still we wonder that they could be useless. It would be strange indeed if, in all the Universe, no other thing should follow the path of perfect numbers.

Rejected Papers of Three Famous Mathematicians

Arnold Emch

National Mathematics Magazine
11 (1936/37), 186–89

Editor's Note: Professor Emch here provides some solace to those who have had a good manuscript rejected or ignored by an editor. After stating that merit should be the only criterion for an editor in deciding whether to publish something, he adds that "there is no other science in which this assertion should appear more evident than in mathematics." He then describes three important pieces of work by Ludwig Schläfli, Bernhard Riemann, and Ernest de Jonquières that were mishandled by referees and editors who did not understand the work or failed to see its significance.

Professor Emch, an algebraic geometer, had a PhD from the University of Kansas and spent his career at the University of Illinois, Champaign-Urbana. Shortly prior to his contributing this to *Mathematics Magazine* he had served on the Council of the American Mathematical Society.

1. Introduction

Every scientifically-minded person agrees with those who hold the opinion that in the appraisement or estimate of the value of an intellectual performance, merit, as acknowledged by one or more class or classes of critics, should be the deciding factor. There is no other science in which this assertion should appear more evident than in mathematics. And yet history records some strange exceptions to this rule, due to the fact that mathematicians also are frequently very human, just like the rest of their scientific brethren.

In the following lines I propose to discuss three outstanding examples of extremely meritorious papers which were rejected by supposedly competent critics in editorial or academic committees. This does not mean that these are the only known cases, because there is no doubt that many other instances could be mentioned in support of these curious historic events.

2. Schläfli: Theorie der vielfachen Continuität

This paper on the theory of manifold continuity was written by L. Schläfli, professor of mathematics in the University of Berne, Switzerland, between 1850 and 1852 and was submitted to the Academy of Science in Vienna in July, 1852, but was refused on account of its length. In 1854 Schläfli sent the manuscript to *Crelle's Journal*, which tentatively accepted it for publication, but returned it in 1856. Not until 1901 was this remarkable memoir finally published in the *Denkschriften* (proceedings) of the *Schweizerische Naturforschende Gesellschaft.*

Before giving a short account of Schläfli's accomplishment it is necessary to state that definitions of spaces or manifolds of more than three dimensions were given before. In 1844 Grassmann published *Die lineare Ausdehnungslehre*, a pioneer and classic in the theory of manifolds. A year before Cayley wrote a short paper: "Chapters in the analytic geometry of n dimensions", which appeared in his collected works Vol. IV (1843), pp. 119–27.

The *Comptes Rendus* of the French Academy of Science, Vol. 24 (1847), pp. 885–87, contains an article by Cauchy: Mémoire sur les lieux analytiques, where we find a clear cut statement of an n-dimensional Euclidean geometry as we understand it at the present day. But it indicates rather than establishes an analytic geometry of higher dimensions.

As far as we know, Schläfli was the first to do this explicitly in a masterly fashion. Were it not for the unfortunate fact that his memoir was not published until 1901, historic accounts of the theory of hyperspaces would not ignore but would mention Schläfli's efforts in the first rank.

In this theory of multiple continuity of n dimensions, n variables $x, y, z, \ldots$; $1, 2, 3, \ldots$ equations determine in succession $(n-1), (n-2), \ldots$-fold continua. Distance between two points $x', y', z', \ldots$; $x, y, z, \ldots$ is defined as

$$\sqrt{(x'-x)^2 + (y'-y)^2 + (z'-z)^2 + \cdots},$$

or in oblique coordinate system as

$$\sqrt{(x'-x)^2 + (y'-y)^2 + \cdots + 2k(x'-x)(y'-y) + \cdots}.$$

He studies orthogonal transformations (movements in n-space), defines angles between two lines or two hyperplanes

$$p = ax + by + cz + \cdots + hw,$$
$$p' = a'x + b'y + c'z + \cdots + h'w,$$

as

$$-\cos(\angle pp') = \frac{aa' + bb' + cc' + \cdots + hh'}{\sqrt{a^2 + b^2 + \cdots} \cdot \sqrt{a'^2 + b'^2 + \cdots}}.$$

Schläfli studies the integral

$$S_n = \iiint^n \ldots dx dy dz \ldots$$

with boundary conditions $p_1 > 0$, $p_2 > 0$, ..., $p_n > 0$, $x^2 + y^2 + \cdots + w^2 < 1$. There are at least $n+1$ hyperplanes necessary to bound an n-dimensional solid (polyscheme) whose measure is

$$\iiint \cdots \int dx dy dz \ldots$$

extended over this hypersolid. Higher continua (varieties) are defined by $x = \phi(f_i)$, $y = \psi(f_i)$, $z = \chi(f_i), \ldots$; $i = 1, 2, 3, \ldots, m$; in which ϕ, ψ, $\chi, \ldots$ are not linear functions of $m \leq n$ variables. He studies regular polyhedra in four and more dimensions and their metrical properties. In the second part Schläfli discusses the theory of spherical continua bounded by $x^2 + y^2 + \cdots < 1$ and n linear homogeneous independent polynomials. Part three contains a study of hyperquadrics and their metric properties, including confocal hyperquadrics. Some of these applications are simply generalizations which are included for the sake of completeness. Schläfli himself was very well aware of the intrinsic nonimportance of mere generalizations: *Wenn ich nun auch das Verdienst des Generalisierens nur gering anschlage.*

3. Riemann: Commentatio mathematica qua respondere tentatur quaestioni ab Illma Academia Parisiensi propositae

In the year 1858 the Academy of Science in Paris proposed a prize-question, which in the original (to be very accurate) is as follows: *Trouver quel doit être l'état calorifique d'un corps solide homogène indéfinie pour qu'un système de courbes isothermes après un temps quelconque, de telle sorte que la température d'un point puisse s'exprimer en fonction du temps et de deux autres variables indépendantes.*

Riemann headed his memoir with the motto: *Et his principiis via sternitare ad majora.* Nothing was done! It was handed in July 1, 1861. The prize was refused, because the methods by which the results were obtained were not fully explained. The explicit treatment of the problem was prevented by Riemann's ill health. The memoir, which was in Latin, was published by H. Weber on pp. 391–423 in the collected works of Bernhard Riemann in 1892. In this memoir Riemann presents the foundations of manifolds of multiple dimensions and makes an ingenious application of this theory. The original manuscript was returned to the Göttingen Academy of Science on request, by the secretary of the French Academy.

The ideas involved find expression in Riemann's classic "*Probevorlesung*" at Göttingen, June 10, 1854, for habilitation as a lecturer at the university: *Ueber die Hypothesen, welche der Geometrie zu Grunde liegen.* This was published separately by H. Weyl in 1919, including valuable comments. In this Riemann explains n-dimensional manifolds and their various geometric interpretations and applications.

Again to preserve absolute accuracy we quote the concluding statement of Riemann in the author's original version:

> *Es muss also entweder das dem Raume zugrunde liegende Wirkliche eine diskrete Manigfaltigkeit bilden, oder der Grund der Massverhältnisse ausserhalb, in daraufwirkenden bindenden Kräften gesucht werden .... Es führt dies hinüber in das Gebiet einer andern Wissenschaft, in das Gebiet der Physik.*

In his comments Weyl remarks (translation): "Thus Riemann's idea which tore down the barrier between geometry and physics has found its most brilliant confirmation by Einstein's theory.

4. De Jonquières: De la transformation géométrique des figures planes

The *Nouvelles Annales de Mathématiques*, Vol. 3 (1864), pp. 97–111 contains a paper by De Jonquières under this title in which he states that he had sent a memoir on this subject to the Institute of France. The *Comptes Rendus* of 1859, Vol. 49, p. 542 acknowledges the receipt of such a memoir. But there never was a report made, so that it rested for nearly a quarter of a century among the Archives of the Academy.

In the first part of the memoir is established a plane transformation in which to lines in one plane correspond in another plane curves of order n which all have a point of multiplicity $n-1$ in common. De Jonquières thus considered birational transformations between two planes before Cremona published his important results in 1863.

In a conversation with G. B. Gucia, De Jonquières mentioned the results of these investigations, whereupon Gucia expressed the desire to read an authentic copy of the memoir with a view to publishing it in a scientific journal of Italy. This was eventually done in the *Giornale di Matematiche*, Vol. 23, 1885, pp. 48–75. The title of this paper is *Mémoire sur les figures isographiques et sur un uniforme de génération des courbes à double courbure d'un ordre quelconque au moyen de deux faisceaux correspondants de droites.*

In it De Jonquières studies the net of curves of order m through $2(m-1)$ simple points and an $(n-1)$-fold point. Then he considers three arbitrary lines $L=0$, $L'=0$, $L''=0$ so that every other line can be represented by

$$L+\lambda L'+\mu L''=0.$$

With these are associated three curves $c=0$, $c'=0$, $c''=0$, so that to

$$c+\alpha\lambda c'+\beta\mu c''=0$$

corresponds $L+\lambda L'+\mu L''=0$, with α, β as constants. To the intersection of two lines (L) and (L_1) corresponds uniquely the one variable intersection of the corresponding curves (C) and (C_1). In this birational transformation, which is now universally known as a De Jonquières transformation, the fundamental points are the $2m-2$ simple points and the $(m-1)$-fold point. De Jonquières, who was "capitaine" in the French navy, wrote the memoir on board the frégate le d'Assas, November 25, 1859.

Review of *Men of Mathematics* by E. T. Bell

G. Waldo Dunnington

National Mathematics Magazine
11 (1936/37), 406–7

Editors' Note: E. T. Bell's *Men of Mathematics* is one of the all-time best sellers among books on mathematics or mathematicians. We now recognize that the title appears to exclude the possibility of women being mathematicians. Sonja Kovalevsky, the only woman with a biography in the text, is given only half a chapter. The title was not Bell's, however; it was forced on him by the publishers who wanted a counterpart to their popular *Men of Art*. In recent years it has become fashionable to "discover" errors in Bell's accounts in *Men of Mathematics*, a serious problem with the book (see, for example, Tony Rothman: Genius and Biographers: The Fictionalization of Évariste Galois, *Amer. Math. Monthly* 89 (2) (1982), 84–106).

Readers of *Mathematics Magazine* at the time *Men of Mathematics* first appeared, however, would not have been surprised to learn that Bell often made exaggerated statements about his subjects, sometimes seemingly making up the details as he went along. Dunnington would seem to refute that charge, but he clearly alerts readers of the *Magazine* to the shortcomings in Bell's narrative. He also points out that the book is remarkably entertaining. As Dunnington predicted, the book developed a large and loyal following and it has influenced countless numbers of students and amateurs to become interested in mathematics. The book has not been out of print since its appearance in 1937.

The references to Gauss in the review are not surprising: Dunnington, a professor of German who was very interested in mathematics and edited the Humanism and History of Mathematics Section of the *Magazine*, wrote the first biography of Gauss in English. That biography has been reissued by the MAA in its Spectrum Series (*Gauss: Titan of Science*, G. Waldo Dunnington, MAA, 2004).

There exists today a strong tendency to pay homage and attention to the heroes of science, to humanize its essential creators, and this book is good evidence of the fact. A. Macfarlane (1916), H. W. Turnbull (1929), and G. Prasad (2 Vols., 1933–1934) have issued collections of mathematical biography, but this one is done on a much larger and more pretentious scale. A parade of the following mathematicians passes in rapid review: Zeno, Eudoxus, Archimedes, Descartes, Fermat, Pascal, Newton, Leibniz, the Bernoulli family, Euler, Lagrange, Laplace, Monge, Fourier, Poncelet, Gauss, Cauchy, Lobachevsky, Abel, Jacobi, Hamilton, Galois, Sylvester, Cayley, Weierstrass, Sonja Kovalevsky, Boole, Hermite, Kronecker, Riemann, Kummer, Dedekind, Poincaré, and G. Cantor. Of course much incidental information about other important mathematicians is included. From the above, it will readily be seen that the volume is devoted almost exclusively to the modern period; also the first three names are compressed into one chapter. Gauss receives 52 pages, considerably more than the amount of space devoted to any other mathematician. The sketch of Gauss is based largely on Sartorius' monograph. Albrecht Dürer's *Melencolia* is reproduced as the frontispiece. The portrait illustrations, taken primarily from the David Eugene Smith Library, add much to the value of the volume.

Footnotes and bibliography are in general excluded and sources or authority for statements are usually not indicated. The author has purposely done this, since the aim of the book is popularization, or an appeal to the general reader. He specifically states that it is not intended to be a history of mathematics, or any section of it. He has based this colorful readable book on printed sources, as far as the present reviewer can ascertain. Most of them are obscure and not very accessible to the general reader. Dr. Bell is a seasoned, skillful writer with a fluent style; he writes with a realistic, curt, potent wit and stark, frank humor which does not stop short of vigorous, rollicking slang (regarded by some as "undignified"—but this is a matter of taste). Doubtless this will amuse many casual, rapid readers, but it also leads him to a certain type of exaggeration. Hence the historian will be largely interested in the *mode* of treatment, provided most of the facts are already familiar to him. Without attempting to do full justice to the skill of presentation within the limits of so brief a review, it may be said that the circumspect, careful reader will find a considerable number of examples of such "exaggeration." The author states that an ordinary high school course in mathematics plus interest and an undistracted head is sufficient to understand everything presented.

Dr. Bell frequently recounts the charming legends, interesting traditions, and melodramatic fiction concerning the various mathematicians, but fortunately he usually discounts them properly by elucidating that they may be apocryphal anecdotes, rather than fact.

Occasionally slips occur, such as on p. 130, where we read: "Today over three hundred years after his death, Leibniz's reputation as a mathematician is higher ... ." 1839 was not the only occasion on which Jacobi visited Gauss (p. 331). Herbart died in 1841, about five years before Riemann went to Göttingen (p. 491). The reader also finds certain errors of others presented as fact, such as Landau's statement that Dedekind was "the last pupil of Gauss" (p. 516). Also, Dedekind was professor in the "Institute of Technology", rather than a technical "high school" and this should not be thought inferior to a university, if we stop to consider the German system, (p. 518). He went there in 1862, probably glad to return to his home city after being in a foreign country (Zurich). The ties which grew with the years must have held him there; an alumnus often returns to his alma mater at a certain sacrifice, in order to teach there, and remains stuck there for the remainder of his life.

It is worth noting that no American was of the proper calibre to be included in this volume; J. J. Sylvester is the closest approach, and mention of his first sojourn in America is exceedingly brief. Any writer is always puzzled to know just what spelling to adopt for many foreign geographical and proper names, since often there is no standard.

Dr. Bell allows his imagination to play; conjecture, personal opinion, and speculation are abundant throughout these chapters as to what the historical development would have been if conditions had been different, or some mathematicians had lived longer, behaved differently, or if mortals were constituted otherwise.

To write such a work requires a huge amount of labor, but the author will probably be rewarded by finding a large circle of readers. Its laudable merit of popularizing, if this succeeds, will justify its publication as well as the toil required to produce it. The paper, printing, binding, etc., are all that one could desire. Misprints are extremely rare.

Oslo under the Integral Sign

G. Waldo Dunnington

National Mathematics Magazine
11 (1936/37), 84–93

Editors' Note: Among International Mathematical Congresses, the one held in Oslo in 1936 holds a special place. Of strictly mathematical interest is the fact that the first Fields Medals were awarded there—to Lars Ahlfors and Jesse Douglas. Of wider interest is the political climate in Europe at the time. The Nazis were already in control in Germany and it would not be long before they would take over countries throughout Europe. Even by the time of the 1936 Congress, Italian mathematicians (all but two women) were prevented from coming. Germans attended, but by 1936 many German mathematicians had fled to other countries, mainly the United States. Attendance was low, only 487, whereas the two previous congresses in Bologna and Zurich had had 836 and 667 attend, respectively.

Here Dunnington describes the Congress and its setting. The mathematical community in those days was very different from what it is today. It was relatively small; people seemed to know each other. It was certainly more formal and more genteel, like the society around it; note that the King and Queen of Norway held a tea for the participants at the Royal Palace. What Dunnington is too polite to say is that not all the mathematicians behaved in a way suitable for a Palace tea. From some reports, Queen Maud was shocked to see the supply of food disappear so quickly. The mathematicians descended on the tables of food (like locusts?). It was, after all, the middle of the Depression and some may have been grateful for free food.

Another indication of changed times is a description of the closing. (We have omitted the last part of the article concerning reports on the talks in the sessions for specialists.) But there was special mention at the end of the "excellent work of the Committee on the Entertainment of Ladies." We probably won't see a committee of that sort at congresses again. The assumption that all the "Ladies" would be visiting art museums, sightseeing and attending teas while "the mathematicians" would be doing serious work, is not supported, however, by the list of delegates, which included Professors M. I. Logsdan (Chicago), I. Harris (Richmond), and C. C. Krieger (Toronto), all women. Further, we know that Dame Mary Cartwright attended, though not as a delegate. (We have a picture of her taken at the Congress.)

Dean Luther P. Eisenhart of Princeton extended on behalf of the American mathematical organizations an invitation to hold the next Congress in 1940 in New York "or a smaller city on the Atlantic seaboard." That did not happen, of course, and the next Congress was held in 1950 in Cambridge, Massachusetts.

The integral sign in the title derives from the fact that members of the Congress wore badges in the form of integral signs so they could ride the streetcars and busses in Oslo free.

Norway and Switzerland are not large countries, but their significance in the history of mathematics is not difficult to assess. One need merely mention for the former, Abel and Sophus Lie; for the latter, Euler and the Bernoulli family. The choice of Oslo as a meeting place for the International Congress of Mathematicians, July 13–18, 1936, proved to be a most happy and satisfactory one. Harald the Hard Hearted, one of the Viking kings, founded Oslo in 1045, and Haakon V built about 1300 a typical medieval fortress or castle called Akershus on a rocky cliff on the Oslo fjord. Less than 300 years ago King Christian IV founded a new city on a rocky plateau below Akershus and called it Christiania. In 1925 the name Oslo was restored, since it had been thus known from 1047 to 1624. Greater Oslo has today a population of about 400,000 inhabitants, a pleasing climate, is clean, and as a result of all this, is visited by many tourists. The museums are at once ideal and unique. Probably the Viking ships interested the mathematicians more than any other exhibit. The writer of these lines cannot resist recording here a delightful dinner and an evening spent with Dr. Dietrich Hildisch, the Consul General, and Mrs. Hildisch, at their lovely island estate in the Oslo fjord.

The meetings of the Congress were held in the buildings of the University of Oslo, which were erected during the years 1839–1854. On Monday evening, July 13, Prof. Sem Saeland, rector of the University, gave a reception for visiting mathematicians and their families in the "Aula." He was represented by Prof. Poul Heegaard, dean of the fac-

Figure 1. Opening session, International Congress of Mathematicians, Aula, University of Oslo. July 14, 1936. King Haakon VII is seated in the center aisle in a chair to the left.

ulty of mathematics and natural sciences, who spoke warm words of welcome in French, German, and English. This was an evening of social pleasure and reunion for the mathematicians, in the beautiful Aula which is adorned with the colossal murals of Edvard Munch. In front of the Aula are monuments of Schweigaard and P. A. Munch, two important historians. On the campus of the University are several rune stones, one of which is about 1500 years old. Several other important, interesting buildings which space forbids us to describe here, are: The Storting (parliament), the National Theater, the Palace of Justice, the Nobel Institute, the University library with 800,000 volumes, the Stock Exchange, and the Navigation School. Oslo has been called the "City of Frescoes," since it probably surpasses other European cities in this respect.

The formal opening session of the Congress occurred on Tuesday morning, July 14, in the Aula, King Haakon VII being present. Prof. Carl Størmer, astrophysicist at Oslo University and chairman of the Committee on Organization, officially opened the Congress. Official greetings were extended to the Congress by the Norwegian Minister of Foreign Affairs. Prof. Størmer was then elected President of the Congress. This was a well deserved honor, since he seems to have done more than any other individual to make it a success. In this connection it should be stated that Prof. Alf Guldberg, who had served as chairman of the above committee, died on February 15, 1936. The Fields Medals, established in 1932 at the Zurich Congress, were conferred on Prof. Lars V. Ahlfors, of Helsingfors, Finland, and Prof. Jesse Douglas, of Cambridge, Mass., for their work in function theory. Prof. Elie Joseph Cartan, University of Paris, officiated at this event.

Prof. Størmer lectured at the above session on a "*Program for the Quantitative Discussion of Electron Orbits in the Field of a magnetic Dipole, with Application to Cosmic Rays and Kindred Phenomena.*" This lecture was unusually well illustrated on the screen. Following this, Prof. R. Fueter (Zurich) gave a paper entitled "*Die Theorie der regularen Funktionen einer Quarternionenvariablen.*" At the conclusion of this morning session the delegates and participants were photographed in front of the Aula.

The afternoon of July 14th was given over to sectional meetings, which will be touched on later. At 5:30 p. m. the Congress members were received by King Haakon and Queen Maud at tea in the Royal Castle. This castle was built in 1848 and although somewhat larger, reminds one in external appearance of the White House. It is situated on a hill at the end of Karl Johann-gate, Oslo's main street, just beyond the University, and commands a splendid view of the city. The castle is surrounded by a stately park; in front of it is an equestrian monument of King Karl Johann. In the park are a portrait bust of Camilla Collett, the pioneer Norwegian feminist, and a monument of Abel, both by the great Norwegian sculptor Gustav Vigeland. The writer plans to publish some brief discussion of this Abel monument in the next issue of the *National Mathematics Magazine*. Nearby is Vigeland's statue of the composer Richard Nordraak, who wrote Norway's national hymn "Ja, vi elsker." The rather democratic royal family in Norway is extremely popular. Besides the royal couple, it consists of Crown Prince Olaf, who acted as honorary president of the Congress, Crown Princess Martha, and their two little daughters, the Princesses Ragnhild and Astrid.

From an historical viewpoint the general session on Wednesday morning, July 15, was probably the high watermark of the Congress. Mathematicians from 38 nations were assembled in the Aula for the unveiling of Dyre Vaas' bust of Sophus Lie. According to the program, Director J. Sejersted Bødtker was to speak and present the bust to the University of Oslo. However, he sent a telegram stating that he was unavoidably detained by a late train, and thus Georg Lous, an Oslo lawyer, spoke as follows:

"In the name of the friends of the Sophus Lie family and also of the admirers of his mathematical genius, I have the honor of presenting to our university a bust of this our great mathematician. Beside Niels Henrik Abel's name that of Sophus Lie is one of those which shines brightest in our mathematical firmament, indeed in all Norwegian science. As far as I know this will not be the first work of art by one of our distinguished sculptors to ornament our University on Blindern (hill). In asking that the veil be dropped, I desire to express the hope that the University will cherish this work of art, and that in the future on Blindern (hill) many busts of great Norwegian scientists will be erected, who shall have cast as much splendor over science and their fatherland as that potent Norwegian genius, Sophus Lie."

Dean Poul Heegaard accepted the gift in the name of the University. He expressed the hope that the mathematical work in the University of Oslo would always be worthy of this great model, now embodied and symbolized by the bust. He felt it a good omen that this hope may be fulfilled by the fact that Prof. Cartan was present to give the historical lec-

Figure 2. Georg Lous, Oslo lawyer, presenting the bust of Sophus Lie to the University of Oslo, July 15th, 1936, at the session of the International Congress of Mathematicians.

ture (interspersed with personal reminiscences) entitled: "*The role of Sophus Lie's theory of groups in the development of modern geometry.*" (Incidentally, it may be mentioned that Prof. Cartan early in September gave a paper at the Harvard Tercentenary Conference on "*The Extension of Tensor Analysis to Non-Affine Geometries.*")

Dean Heegaard said: "I am glad to be able to state on this occasion that the printing of Lie's Collected Works, with the exception of a fascicle of notes, is now completed. For this we are grateful to all institutions and persons who have helped us; especially to the Norwegian Mathematicians Association and above all to the untiring efforts of Prof. Friedrich Engel (University of Giessen). The Works have become not only a monument to Sophus Lie, but also to him."

Prof. Engel, a pupil of Lie, was prevented at the last moment from attending the Congress, which sent him a telegram of greetings and admiration. Following this, two lectures were given simultaneously: Prof. Carl Ludwig Siegel (Frankfurt a. M.) on "*Analytic Theory of Quadratic Forms*" concerning which we plan to report more in detail in a later issue, and Prof. Oswald Veblen (Princeton) on "*Spinors and Projective Geometry.*" Prof. Jakob Nielsen (Copenhagen) then gave a paper on "*Topologie der Flachenabbildungen.*" Unfortunately an important paper "*Theorie des nombres transcendants*" by Prof. A. O. Gelfond (Moscow) had to be omitted since he and the other Russian delegates did not appear. No one seemed to know the exact reason for this. The writer was told that Italy was not represented since Norway had belonged to the sanctions.

The afternoon was devoted to sectional meetings and a session of the International Commission on the Teaching of Mathematics. This session was presided over by Prof. Henri Fehr (Geneva); unfortunately the reports read from the various countries were rather dry and perfunctory, and there was little enthusiasm or discussion. However, the report of Dr. Walter Lietzmann, professor in Göttingen and vice-chairman of the Commission, was exceedingly interesting and enlightening as to mathematics in the schools of present-day Germany. Much praise is also due the Japanese delegates who illustrated their report with actual textbooks used there, and distributed to those present a well printed summary in pamphlet form.

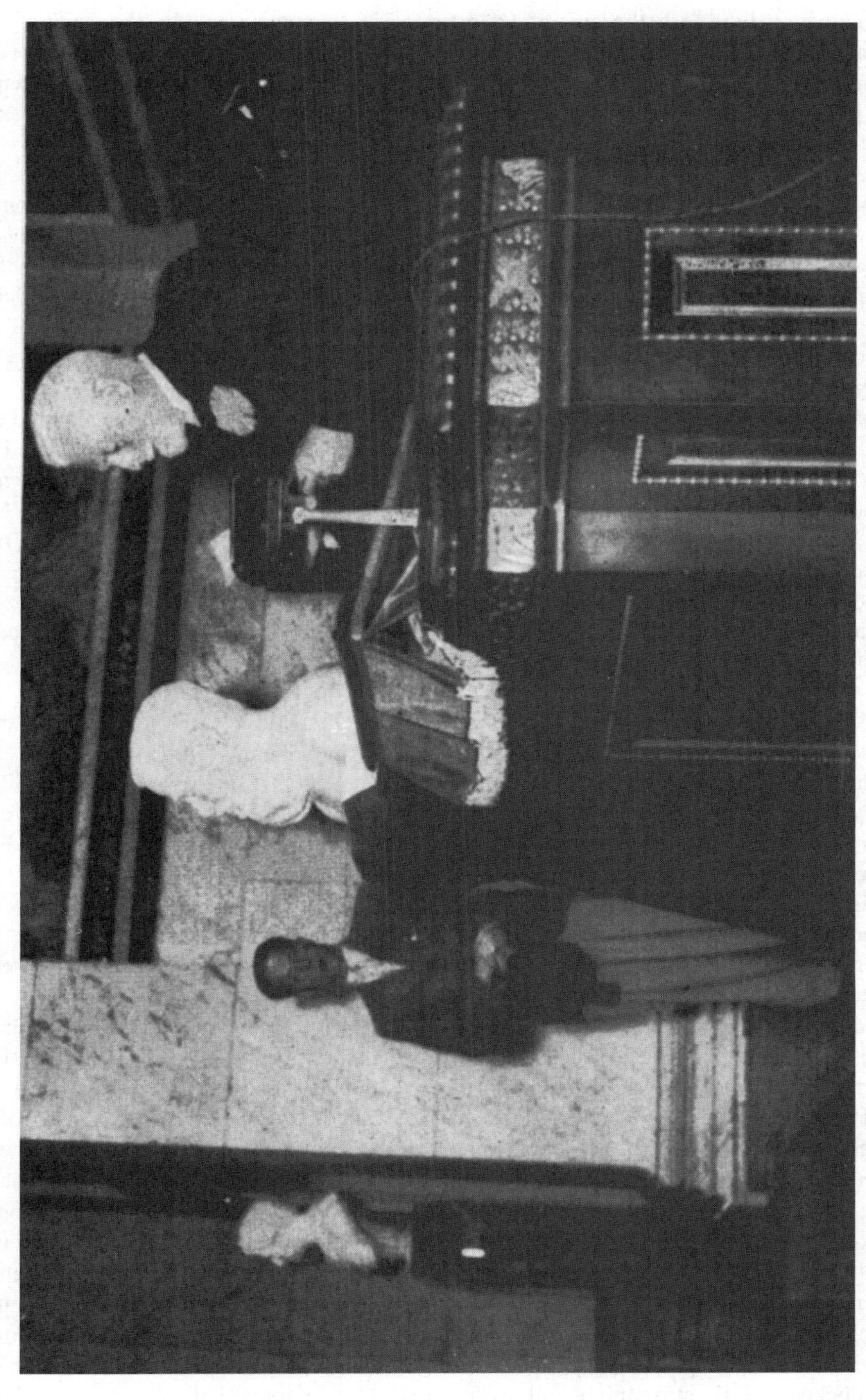

Figure 3. Dean Poul Heegaard, University of Oslo, accepting the Sophus Lie bust in the name of the University. Aula, July 15, 1936. International Congress of Mathematicians.

We hope to review this in our Teachers' Department in an early issue.

The evening of July 15th the city of Oslo tendered a banquet at the Hotel Bristol to the members of the Congress. Several prominent city officials and mathematicians present spoke briefly after the dinner, followed by a social evening in the upper parlors of the hotel.

A genuine master in his field, Prof. Ernst Hecke, University of Hamburg, started off the Thursday morning session (July 16th) with a paper "*New Progress in Theory of Elliptic Modular Functions*." Dr. Otto Neugebauer (Copenhagen) furnished a highlight on the program with his well illustrated lecture "*On pre-Grecian Mathematics and its Comparison with the Grecian*," stressing always the algebraic nature of the former and the geometric nature of the latter. Following this, Prof. C. W. Oseen (Stockholm) lectured on "*Problems in Geometric Optics*" and Prof. V. Bjerknes (Astrophysical Institute, Oslo; president of the Norwegian Academy of Science and Letters, son of C. A. Bjerknes, Abel's biographer) gave a paper on "*New Lines in Hydrodynamics*." Prof. Helmut Hasse, head of the mathematics department in the University of Gottingen, closed the morning session with a brilliant lecture "*Über die Riemannsche Vermutung in Funktionenkörpern*."

Thursday afternoon and evening, July 16th, were devoted to a trip through the Oslo fjord on the SS. *Stavangerfjord*, the largest vessel of the Norwegian American line. The Crown Prince and Princess participated in this excursion, and at 6 o'clock a banquet was served in four dining halls of the ship. Following the dinner Prof. Størmer gave the opening address; the secretary-general of the Congress, Prof. Edgar B. Schieldrop (Oslo) welcomed the guests. The last named deserves special praise for the splendid and efficient manner in which he handled the entire secretariat of the Congress, in the Domus Academica of the University. The response to the above welcome was given by Prof. Emile Borel (Paris). These brief addresses were transmitted to all the dining halls by loudspeaker, and were interspersed with music of Grieg, Sibelius, Strauss, and a march dedicated to Crown Prince Olaf. In the evening there was dancing and cardplaying, happy conversation and reminiscing, with restaurant and bar on board in full swing. The ship docked in Oslo at midnight and taxis were waiting to take the guests to their hotels.

Dean G. D. Birkhoff (Harvard) opened the Friday morning session with a lecture "*On the Foundations of Quantum Mechanics*," going back to and elaborating certain ideas he had worked out as early as 1907. He plans to publish his final results soon in a memoir in the *Proceedings of the National Academy of Arts and Sciences*. He was specially honored by the presence of Crown Prince Olaf at his lecture, and had the further pleasure of seeing his son, Garrett Birkhoff, give papers in three sections of the Oslo Congress on "*Generalized Convergence*," "*Order and the Inclusion Relation*," and "*Product Integration of Non-linear Differential Equations*." Mrs. Birkhoff accompanied him; they attended the 550th anniversary Jubilee of Heidelberg University, the Centenary of the University of London, and returned home to prepare for the Harvard Tercentenary Mathematical Colloquium.

Four additional lectures featured the Friday morning general session as follows: Prof. S. Banach (Lwow, Poland) on "*Le rôle de la théorie des operations dans l'analyse*"; Prof. L. J. Mordell (Manchester, England) on "*Minkowski's Theorems and Hypotheses on Linear Forms*"; Prof. Lars V. Ahlfors (Helsingfors) on the "*Geometry of Riemann Surfaces*"; and Prof. J. G. van der Corput (Groningen) on "*Diophantine Approximation*." Sectional meetings occupied the afternoon.

On Saturday morning (July 18th) the lecture of Prof. A. Khintchine (Moscow) on "*Main Features of the Modern Theory of Probability*" had to be omitted, due to his absence. There were three remaining lectures, as follows: Prof. Maurice Fréchet (Paris) on "*Mélanges mathématiques*"; Prof. Norbert Wiener (Cambridge, Mass.) on "*Tauberian Gap Theorems*"; and Prof. Øystein Øre (Yale) on "*The Structure of Algebraic Systems*." There were sectional meetings during the afternoon.

The closing session of the Congress convened in the Aula at 5:15 p. m. At this session a telegram of thanks (replying to greetings sent him) from Prof. Friedrich Engel was read. Letters of thanks from the Queen and Crown Princess for flowers the Congress had sent them, were read. The chairman read telegrams of greeting and admiration sent to David Hilbert in Göttingen, Emile Picard in Paris, and Vito Volterra in Rome. Prof. Henri Fehr (Geneva) then reported on the session of the International Commission on the Teaching of Mathematics. Prof. Gaston Julia (Versailles) spoke about the International Mathematical Union.

Dean Luther P. Eisenhart (Princeton) presented the invitation of the American mathematical organizations to hold the next International Congress of Mathematicians (1940) in New York or a smaller city on the Atlantic seaboard. This invitation was accepted with loud, prolonged applause. Prof. J. A. Schouten (Delft) gave the farewell address for the guests. He thanked most heartily Norway, the city of Oslo, the Committee on Organization, and the Committee for Entertainment of the Ladies. Prof. Størmer then expressed congratulations to those present, best wishes for the coming years, and closed the Congress.

. . .

In closing this report special mention should be made of the excellent work of the Committee on the Entertainment of Ladies, which included the widow of Prof. Guldberg, and the wives of Professors Heegaard, Schieldrop, Skolem, and Størmer. This entertainment included sightseeing by bus, side trips to Frognerseteren and Skaret, visiting the museums and art galleries, and a tea given by Mrs. Henny Løvenskiold on her estate at Baerum's Verk.

The information office at the Congress secretariat in the Domus Academica and the writing room there were well managed. The mail of the guests was well handled, and the students of the University were very helpful to the guests in various incidental matters. Delegates to the Congress wore a badge in the form of an integral sign, which entitled them to ride free on the street cars and busses in Oslo and vicinity. The book stores featured special exhibits of mathematical books and materials during the week. Steamship lines and railroads serving Norway allowed special rates to the visiting mathematicians, many of whom took advantage of this to remain a week after the Congress, in order to see beautiful scenery for which Norway is famous.

Vigeland's Monument to Abel in Oslo

G. Waldo Dunnington

National Mathematics Magazine
11 (1936/37), 145–46

Editors' Note: Continuing his report of the Oslo Congress of 1936, Dunnington describes here the Abel monument in the Royal Park in Oslo, something that should be on any list of sights for mathematicians visiting the Norwegian capital. The sculptor, Gustav Vigeland, is best known for his many works in the gardens at Frogner Park in Oslo, but that collection contains no sculptures of mathematicians. For other versions of the statue of Abel and for maquettes produced in preparation for the piece in the Royal Park, one should visit Vigeland's house and studio, also in Oslo.

In 1908 Norway honored the memory of Niels Henrik Abel by erecting a monument of him in the park in front of the Royal Castle in Oslo. Gustav Vigeland, the leading modern Norwegian sculptor, created this monument. The late Felix Klein in his *Vorlesungen über die Entwicklung der Mathematik im 19. Jahrhundert* (vol. 1, p. 108) compares Abel and Mozart, speaking in this connection of the beautiful monument to Mozart in Vienna. But of the Abel monument he writes:

"I cannot avoid referring on this occasion to the entirely different sort of monument which has been erected to Abel in Oslo and which must severely disappoint everyone who knows his nature. On a lofty steep block of granite a youthful athlete of the Byronic type is striding upward, over two atrocious victims. In any case, if one can interpret the hero as a symbol of the human mind, then one questions in vain about the deeper meaning of this monstrosity. Are they conquered equations of the fifth degree or elliptic functions? The base of the monument bears in huge letters the inscription ABEL."[1]

Thorvald Holmboe, an engineer in Oslo and editor of a technical journal there, sent a letter in reply to Klein's criticism to Professor Wilhelm Lorey, to whom I am much indebted for calling my attention to it:

"Felix Klein's judgment probably is based on an ignorance of the independent artistic value of the Abel monument. The inspired Abel monument is not primarily to be interpreted as a portrait of the man Niels Henrik Abel; it is rather the homage of the artist to the genius of the mind, of thought—embodied in the greatest whom we have had. Therefore, also the simple vigorous inscription ABEL, purposely without any commentary.

"If Felix Klein had known the character of the Norwegian people better, that would certainly have been clear to him. The two allegorical figures, on whose backs the main figure is riding, are the embodiment of the genii ("guardian angels") of unconfined human thought, who carry the thought of a genius through the universe. The fact that these figures are represented somewhat grotesquely—and in execution they are quite naturalistic—agrees well with the idea of the severe and groping struggle of the human mind for truth.

"The masculine strength of the monument corresponds to the task before the artist much better

[1] We are told that Vigeland felt that every Norwegian must know who Abel was, hence no more would be necessary.

Figure 1. Vigeland's Monument to Abel

than the trivial, mawkish figures which otherwise adorn the monuments of the large cities. A comparison with the Mozart monument in Vienna is irrelevant: those works of art originate in entirely different epochs and interpretations of art."

Albert Dresdner in a book on Swedish and Norwegian art since the Renaissance (1924) writes of the Abel monument:

"Vigeland is not so sure of himself in executing monuments. As a monument for the mathematician Abel he has created a splendid figure of a youth, winged with ideal buoyancy, but the relationship of this figure to those carrying it (which probably are to symbolize forces of nature) remains obscure."

On April 11, 1929, Vigeland's 60th birthday, Dresdner wrote in the *Deutsch-Nordisches Jahrbuch für Kulturaustausch und Volkskunde*, Jena, 1929, (p. 127) concerning the Abel monument:

"Vigeland's most excellent accomplishment in the field of monuments (as far as I know) is the monument for the mathematician Abel erected in 1908 on the Castle Hill in Oslo. On a mighty monolith rises abruptly a slender naked figure of a youth, borne by figures in swaying motion, who may symbolize mental powers or thoughts. A base for the figures is missing, the allegorical figures reach out over the edges of the monolith. The lower part of the group is full of unrest, rich in cross-cuttings, space-consuming openings and opposing motifs; but the body of the youth finally grasps and holds together the plastic form. The structure is naturalistic and illusionistic, the architectural principle is completely excluded. I couldn't say that this work appears to me thoroughly determined in every respect, but in the end one's impression is dominated by the nobility, power, and vibrancy of the youth's form rising in the open against the heavens, and it must not be overlooked that Vigeland has presented here the human being free and victorious through the mind."

The History of Mathematics

Otto Neugebauer
Translated by G. Waldo Dunnington

National Mathematics Magazine
11 (1936/37), 17–23

Editor's Note: The author, Otto Neugebauer, is remembered for various contributions to mathematics, but none greater than his convincing the publisher Springer-Verlag, in 1931, to put out an abstracting journal, *Zentralblatt für Mathematik und ihre Grenzgebiete*, that would cover all of mathematics. This journal ended up supplanting the earlier *Jahrbuch über die Fortschritte der Mathematik*, which was published between 1868 and 1942, and with the increase in the number of published papers had, by the time of its demise, fallen hopelessly behind schedule and was no longer very useful to research mathematicians. The database for the *Jahrbuch* has subsequently been incorporated into *Zentralblatt*.

Neugebauer became the first editor of the new journal. With the rise of the Nazis in Germany, Springer had to follow the party line and restrict the reviewing of work by Jewish mathematicians, so Neugebauer, on the invitation of Harald Bohr, moved to Copenhagen in 1933, taking the offices of *Zentralblatt* with him. With much of Europe being taken over by the Germans, Neugebauer had to move to the United States in 1939—but not before destroying the *Zentralblatt* records (except for the cumulative index). In the U.S. he took a position at Brown University. During the Second World War, with *Zentralblatt* only sporadically available in the Allied countries and the content of the journal no longer determined by mathematical criteria alone, Neugebauer proposed that the American Mathematical Society publish an abstracting journal in the United States, *Mathematical Reviews*. It began publishing in 1940 and, again, Neugebauer was appointed editor. Today both *Mathematical Reviews* and *Zentralblatt* survive.

Neugebauer received his doctorate at Göttingen and published a joint paper on periodic functions with Harald Bohr. But that was his first and last purely mathematical paper. After that he worked exclusively in the history of mathematics, mainly the history of ancient times. He was an admirable man, respected for his integrity and good sense. Ralph Boas told a story about Neugebauer when he was editing *Mathematical Reviews*. A mathematician (who was not German) wrote to complain that Neugebauer had written him in English, not in his "Muttersprache" (Mother tongue). Neugebauer replied that it was not a question of his mother's language but of his secretary's. At Brown he established a highly prestigious department of the history of mathematics.

Born in 1899 in Austria, he died in 1990.

Here Neugebauer reflects on the unfortunate fact that many mathematicians appear to be uninterested in the history of their subject, a phenomenon perhaps more common in his day than now. He gives us some reasons why history is important. The article was translated by G. Waldo Dunnington, presumably from German. Still, when this was published Neugebauer was living in Copenhagen, so it could have been written in Danish originally. Boas wrote of Neugebauer that when he and his family moved to Denmark they spoke only Danish at home, and when they moved on to the United States they spoke only English.

The historiography of a science usually does not enjoy any too high an esteem among its productive representatives. The reasons for this are several and they are not difficult to recognize, especially in a science like mathematics, which can distinguish with such precision between secured possession and unsolved problem. Mathematical problems and methods are indeed, like every other element of existence, historically conditioned; but that portion of the way already traversed—which for their continuation one must know and as such survey—is a relatively short one. Probably long centuries worked only in closest connection with antiquity, but the great rise of modern science begins with the appearance on the scene of entirely *new* ideas whose import lies in the opening up of hitherto unknown questions rather than in the settlement of old ones. In all natural sciences definitive results may have an absolute value—in mathematics a settled theory (the technical term for this is a "classical theory") is something dead, which cannot captivate fresh forces.

With this set-up the non-historical character of mathematics is not to be wondered at. In addition however, or rather connected with the above is the fact that the existing historical presentations of this science cannot interest the professional because their authors, as outsiders, with all their formal knowledge of subject matter, do not touch the real essence and interest of the problems. Instead one finds things treated such as the subject "Geometric forms that were in existence before the advent of life on the planet" (in a "History of Mathematics" appearing in 1923), which have nothing at all to do with mathematics. And only too often an anecdote collection, or even worse, an endless chain of priority questions must replace *history.*

It would not be worthwhile to write about such things, if one had to regard them as irrevocably united with the substance of the history of mathematics. Indeed I do not believe that the above mentioned purely objective relationship between progressing research and its history can be changed; but I regard it as an absolutely attainable goal, so to re-cast the history of mathematics *in itself,* that, like the history of philosophy, it will become an integral member in the series of modern sciences and not lead to an entirely meaningless existence untouched by mathematical as well as historical spirit. Then one will again dare to hope that even the purely professional mathematician, from this broader viewpoint, can profit by occupation with it.

I have already touched on the point where according to my opinion the chief deficiency in the present condition lies: in the lack of an *historical-problem* attitude. I should like to explain somewhat more in detail by a very special question the way I desire this to be understood: What is the treasure of mathematical knowledge which the Greeks took over from the Orient?

Immediately an objection: What interest at all does such a question have? Indeed it is quite immaterial to know whether some Egyptian or this or that Greek possessed a formula for the volume of a truncated cone or not. And such an objection really exists quite properly, as long as one contemplates such knowledge only on account of its absolute content, but not as points of demarcation on a scale which one needs in order to be able to draw any sort of historical comparisons at all. However immaterial it may be in itself, whether these points of the scale are constituted by propositions of "elementary mathematics" or by propositions of an optional brand of modern analysis—that we have to deal with the one or the other is historically accidental—nevertheless the gaining of reliable factual material becomes important as a *preliminary* labor in order to have safe ground under our feet, and not to go to seed by merely attitudinizing esthetically.

Here I must interpolate a purely methodical remark which refers to the securing of the foundation for answering the question asked above. Our entire tradition from the ancient Orient exhibits one great advantage: we scarcely know one name of an artist or scholar. The entire cultural evolution of those periods is from the very beginning most closely united with the national unit, its picture is presented in much more tranquil lines before a larger background, than in an epoch when the struggle for "master or school" or for "genuine or false" distorts the perspectives. The chronological arrangement necessary for every historical comprehension must of itself ensue in a much broader framework: *in the framework of general history.* Thus with all the scantiness of our knowledge the history of Egyptian art, religion, and indeed of linguistic history forms with the "pragmatic" history a much more closely knit unit than is the case for analogous Grecian conditions. Directly therefrom arises however the demand to connect the utterances of mathematical thought with these general viewpoints. Not until the investigation of purely objective questions takes place strictly on the basis of the history of civilization can one expect to attain a relatively correct evaluation of the separate problems. Then too such an investigation obtains, on the other hand, a much more general meaning because it is able to bring to light in a quite precisely comprehensible manner very charactertistic features of a people.

The disappearance of the individual in the history of Egyptian civilization has not always been regarded as an advantage. Thus the poor scribe "Ahmes" who immortalized himself as the copyist of the mathematical Rhind papyrus (the most important monument of Egyptian mathematics), has had to assume all possible titles from "king" or "teacher in an agricultural school" down to unskilled "pupil." Or however one has taken refuge in an intentional concealment of the individual behind the very popular "priest castes," although they do not exist, at least in the genesis period of all Egyptian sciences, thus it is maintained: "Mathematics as a science was in Egypt the exclusive possession of the priest caste and was carried on in the priesthood as an occult science and concealed from the people,"— with such success indeed, that not the slightest vestige of this "occult science" has been handed down to us!

To the demand for arrangement of historico-mathematical investigations in the scheme of general history of civilization is joined a sphere of questions much more difficult of access, as soon as it has to do with ancient history: the connection with philology. Such a fundamental investigation as Sethe's[1]

[1]The distinguished Egyptologist, Prof. Kurt Sethe (1866–1934),

book "*On Numbers and Numeral Words among the ancient Egyptians*" shows how much can still be summoned from these things for the beginnings of mathematical thought. One must not carelessly pass by these things, as soon as one asks questions about the historical development of the most important categories of thought, especially in a period of mathematical research like the present, in which the question about the logical foundations of mathematics assumes a central position. Here is really one of the points where the most uncognate sciences encroach quite directly on each other, so that it is not pertinent to grant philosophical speculation the only decision.

If one turns to the actual content of pre-Grecian mathematics, the first impression is that it has an exclusively "elementary" character, and is in itself rather homogeneous: simple problems of calculation, executed in part with the help of numerical tables, or problems involving the calculation of areas and volumes of geometrical figures such as those demanded by agriculture or at the most stonemasonry. But if one observes the role which these things played in their own cultures, this picture is quite essentially changed and offers problems which are by all means worthy of historical investigation. The deep-reaching distinction between the two great cultural units Egypt and Babylonia, even in simple numerical notation and application of the first calculation exercises, asserts itself here quite essentially. The beginnings are indeed in both districts the same: Hieroglyphics with special signs for the most important numerical values 1, 10, 1/2, 1/3 etc. and a purely additive counting foundation of all calculation. Now however the individual development sets in. The fundamental problem of Egyptian mathematics (which bears a purely "arithmetical" character) can be bluntly formulated: to provide those oldest *additive* methods with as sizable a domain of operations as possible. That which we today call "multiplication" is in Egypt repeated addition (by continued doubling and suitable collecting); indeed the entire fraction calculation, which in this form extended its influence over all of later antiquity far into our Middle Ages, owes its externally very intricate methodology only to the consistent execution of the same fundamental principle. We have here in abstract pure culture so to speak the same tenacious adherence to old traditional forms which characterizes all other portions of Egyptian life, which has made its theological systems a chaos so difficult to unravel. The inevitable transformation of religious concepts does not ensue by the formation of clear, new systems but by artificial interpretation of the oldest texts, whereby their simple meaning is distorted and (from *our* viewpoint) the most contradictory statements are entangled, rather than give up the fiction that everything has been thus from olden times.

Quite different in Mesopotamia with its much more stirring history. Even the further development of the hieroglyphic system created by the Sumerians ensues in an essentially different direction. While the hieroglyphic system in Egypt, at least as a system for inscriptions, was preserved to the latest period and while the hieratic writing represents only a levelling-off of it, nevertheless in Babylonia the hieroglyphic symbols, already of a marked linear style, were finally replaced by a number of pure "cuneiform symbols" which give up every conscious connection with the old hieroglyphics. Of course this is also tied up with external influences such as the inconvenience of clay as a writing material and the rise of new elements in the population. Consequently for numerical notation a system of figures very deficient in symbols is formed, which approximate closely our present notation by "local value."[2] However the latter is again conditioned by a very early creation of independent multiplication (as is shown by multiplication tables and tables of squares preserved from a very ancient period) and a strong influence on the entire system of notation due to the standardizing of weights and measures. The details of this process are of course much too complicated for me to discuss here. As to general history it is stimulating to remark that the Babylonian system of *writing* ran into a cul de sac by preventing the gradual transition from hieroglyphics to "uniconsonantal symbols" and finally to "letters," that mathematics however from the beginning on pointed in a direction which in its consistent expansion by means of the Hindu local value system (with which our present notation by digits is identical) has furnished one of the most important supports for modern further development.

I hope one will be able to see even from these hasty references that the elementary, indeed fre-

liked to observe the phenomena of Egyptian culture and civilization from the viewpoint of general history and by comparison with other cultures, to assign them their place in the evolution of civilization. His work *Über Zahlen und Zahlworte bei den alten Ägyptern und was für andere Völker und Sprachen daraus zu lernen ist* (1916), as well as a number of his other papers, definitely enriched the history of mathematics. —G. W. D.

[2]"Principle of position", or value due to position of the digits. —G. W. D.

quently primitive character of the mathematical problems and methods of this early period, which has never been calmly admitted as such and efforts made to conceal it by phrases like "primeval Egyptian wisdom," results in quite the opposite by granting us an especially clear insight into the historical beginnings of mathematical thought. To be sure, one must waive the desire to build up a linear chain of mathematical knowledge from earliest antiquity to our time, but one must learn to place independent cultures like separate personalities beside each other. Then it will be recognized that every people and every epoch seeks in its own manner to meet the problems presented to it and the *comparison* of these various phases receives a new meaning. Not until Oriental mathematics in its singularities is really known, will one know how to evaluate properly *those* of Grecian mathematics and look for those points where specifically Grecian problem-attitudes set in.

But historical processes do not originate only by the encampment of various cultural types beside each other, for beside this "horizontal" articulating of cultures a "vertical" stratification plays an ever recurring role. The desire to regard such a complex structure as Grecian civilization as a unity would be a very fundamental mistake. Between the influences from outside enter the great differences between groups of the most varied intellectual tendencies. Pythagoreans, the Academy and the Sophists cannot be brought into one line of development, but stand in their mutual influence as independent simultaneous factors beside each other. Thus one cannot co-ordinate the attitude of all these tendencies on mathematical problems, even though the external result of occupation with mathematical questions is a permanent increase of objective knowledge. The really essential thing however is not the question whether one multiplies the square of the radius by 3.16 like the Egyptians or by 3.14 in order to determine the area of a circle, but to fathom the collective attitude on the problem of measuring the area of figures bounded by curves and the meaning of infinite processes to which this leads. And in the answering of *these* questions one will again have groups quite separated by principle to differentiate, groups whose transformations are to be pursued in detail, in order to gain a true-to-life picture of the entire process. And when the Arabs or West European civilization later tie on to the Grecian acquisitions, this occurs every time in a special manner and with preference for quite a definite method among these various currents, in spite of all continuity with reference to mere content. Thus the history of mathematics suddenly reaches out far beyond its narrower framework and offers no end of the most interesting questions, which reward the effort to review steps long archaic in mathematical thought.

And beside this general historical significance here comes to the front yet another which refers to mathematics in the narrower sense, but on that very account must not be forgotten. I mean, viz., that only *historical* thinking can possibly form a balance to the much deplored specialization. The last phase of our science inaugurated in grand style at the passage of the eighteenth into the nineteenth century, of whose universal character the great French scholars and the Humboldt brothers may serve as an example, has not been able to maintain this initial level. It is clear that a rigorous establishment of the newly unlocked sciences is to be accomplished only by the greatest division of labor in careful separate investigation. A consequence thereof was however not only a separation of the single sciences from each other, but also a crumbling of these disciplines themselves into divisions scarcely understandable or interesting to each other. There is no doubt that a serious reaction must be set in against this condition and in part already has set in in a very perceptible manner. The question about the whence and whither of a science, about its place in the broader sphere of our entire civilization, is being asked more and more decidedly. In all fields it is being shown that only in the *synthesis* of modern research methods with the less hampered perspectives of a deeper intellectual content can a guarantee for restoration of the unity of all sciences be found. The work[3] by Felix Klein

[3]Professors Neugebauer, Richard Courant, and Erich Bessel-Hagen (Bonn) prepared this volume for publication. It is entitled *Vorlesungen über die Entwicklung der Mathematik im 19. Jahrhundert* and appeared with the imprimatur of the publishing firm Julius Springer in Berlin, 1926, a year after Klein's death. These lectures cover a period of approximately 1914–1919 and were delivered by Klein to a small circle of students in his home. His death in 1925 brought to nought his plan to issue a more voluminous work on the subject, although this presentation fills nearly 400 pages. These lectures are especially charming because they are published just as he gave them, and never received literary "finishing touches." The book is a history of mathematical ideas rather than mathematicians, and an amazing example of his power of presentation; he portrays the train of thought leading to some discovery and as Professor Neugebauer has so well said above, he views isolated pieces of research within a larger framework. Klein's main thesis, running through the entire book, is that regardless of this or that genius a fixed and definite evolution of mathematical ideas exists. The science moves and must move ahead continually on this stream of development. The appearance

"*Lectures on the Development of Mathematics in the Nineteenth Century*" shows as does none other what the historical view in this sense can mean for mathematics. Truly historical thinking united with the most intimate research activity speaks to us here, reminding each one to understand and evaluate his own research tendency as an element of a great historical process.

It will not be vouchsafed to many to write the history of a science in this sense. However every single historical investigation can count as a usable *preliminary* performance toward further synthesis only if it is guided by two viewpoints: to see the history of mathematics in the framework of general history and to understand mathematics itself not as a collection of formulas to be continually increased, but as a living unity.

---

of a genius merely hastens the current. —G. W. D.

Numerical Notations and Their Influence on Mathematics

D. H. Lehmer

Mathematics News Letter
7 (10) (1932/33), 8–12

Editor's Note: Derrick Henry Lehmer was a distinguished number theorist at the University of California, Berkeley, where his father, Derrick Norman Lehmer, had also been a number theorist known for his pioneering use of computing devices in the field, though to call them computers would be an exaggeration by modern standards. Two of the elder Lehmer's publications were *Factor Table for the First Ten Millions Containing the Smallest Factor of Every Number Not Divisible by 2, 3, 5, or 7 between the Limits 0 and 10,017,000* (1956) and *List of Prime Numbers from 1 to 10,006,721* (1956). D. H. Lehmer's wife, Emma Lehmer, is also widely known as a number theorist and the three Lehmers shared a common interest in computation as an aid to doing number theory. D. H. Lehmer's name is forever attached to the Lucas-Lehmer primality test which remains the means of finding larger and larger Mersenne primes, a search that still goes on.

Derrick H. Lehmer graduated from Berkeley and went on to work with L. E. Dickson at the University of Chicago before transferring to Brown University to complete his PhD there under J. D. Tamarkin. He returned to teach at Berkeley after postdoctoral appointments at the California Institute of Technology, Stanford University and a longer appointment at Lehigh. This article on mathematical notation was written while he was at Stanford.

Lehmer was one of the most prominent Berkeley faculty to be caught up in the red-baiting that went on in the 1950s, and that resulted in the infamous loyalty oath required of faculty at the University of California. Lehmer refused to sign the oath and was dismissed. He became Director of the National Bureau of Standards' Institute for Numerical Analysis, so it didn't turn out too badly and he could carry on his work in computation in his new position. When the McCarthy era waned, Lehmer returned to Berkeley where he remained until his death in 1991. He had been born in Berkeley in 1905.

Known for his dry wit, during the long period after the Second World War, when Communist China was out of favor with the United States government, Lehmer referred to the Chinese Remainder Theorem, one of the topics touched on in this paper, as the Taiwanese Remainder Theorem.

After primitive man had learned to count, his next task was to invent ways of representing and recording whole numbers. This can be done in such a vast number of ways that even today the possibilities have not been exhausted and new ways of expressing numbers are being devised each year. Any good history of mathematics gives in detail the story of man's early attempts at writing numbers. It is not my intention to dwell on this subject from a historical point of view. I merely wish to discuss three types of notation which are in use today.

In any system for writing numbers each number must have one and only one representation. But something more is needed in order that the system have more than historical importance: there must be some operation with numbers to which the system is particularly well adapted. Now the operations most frequently met with are addition and multiplication. Hence the more familiar systems of notation are additive and multiplicative.

The simplest additive notation is the tally. Every number can be written uniquely as a sum of units. Addition is performed by merely placing together the numbers to be added. In this system the fact that six plus seven is thirteen is written

$$111111 + 1111111 = 1111111111111$$

This operation is so simple that it becomes laborious to apply. It is really the basis of all mechanical devices for adding.

The most prevalent additive system today is the decimal notation by which numbers are written to the base ten. In fact we have become such slaves to this method that we have confused the concept of a number itself with what it looks like when written in the decimal system. We seldom stop to realize that Archimedes, were he alive today, would not understand that $23 + 59 = 82$ without some explanation of what these symbols really mean. It never occurs to the numerologist of today that it is possible to destroy utterly some of the mystic properties of numbers by merely writing them to a different base. I recall reading an explanation of the number π which the numerologist wrote 3.1416. The number 3 stands for the trinity while 1 is the unity of the Trinity; 6 is the perfect number of days in which the earth was created.[1] If we choose to write π to the perfect base 6 however we obtain 3.05025. Mathematics itself has not escaped from the influence of the number 10. Many theorems have been announced and problems proposed that depend not on numbers but on the digits of numbers. Dickson devotes a chapter of his *History of the Theory of Numbers* to results of this nature. The curious example

$$29 \cdot 83 = 2407 \qquad 23 \cdot 89 = 2047$$

is typical. Such facts and problems are not of paramount importance but belong to the field of mathematical recreations. They have always had a certain fascination however especially among non-professional mathematicians. A prize has been offered recently for the largest list of squares with 9 distinct digits.

The base ten has been forced upon us by physiology. Many practical men have urged the adoption of the base 12 with its many divisors, while mathematicians would prefer a prime number for a base. At any rate it is clear that 10 is by no means the best choice. The smaller the base the smaller are the tables of addition and multiplication which must be committed to memory. On the other hand a smaller base requires more digits to represent a given number. An attempt to compromise these conflicting desiderata leads us to the base 6 or 7 according as we are interested in arithmetic from a practical or a theoretical standpoint. The base two with its simple rules : $1 + 1 = 10$, $1 \cdot 1 = 1$ recommends itself in a great many ways. It is a useful tool in almost every branch of mathematics. However the base ten has become such an integral part of our civilization that any universal change in base is impractical.

An entirely different method of writing numbers, and one which is designed to facilitate the investigation of the multiplicative properties of numbers, arises from the so-called fundamental theorem of arithmetic: Every whole number may be decomposed into the product of powers of prime numbers in one and only one way. Thus we may write

$$65520 = 2^4 \cdot 3^2 \cdot 5 \cdot 7 \cdot 13 \qquad (1)$$

We may even go further and suppress the actual primes leaving them to be implied by a positional notation and record only the exponents of the various primes just as we record only the coefficients of the various powers of ten in writing numbers in the decimal system. Thus

$$65520 = 4, 2, 1, 1, 0, 1 \qquad (2)$$

In this system the nth prime is supposed to be "known" in much the same way that the nth power of ten is "known". The systems illustrated by (1) or (2) must be regarded as more fundamental than the decimal system in which ten is so arbitrary. In either case the primes or their exponents may become very large so we must fall back on the decimal notation again as we did in writing the prime 13 in (1). Of course it is possible in (2) to write the exponents in terms of their prime factors by merely repeating the notation of (2) perhaps several times if necessary. This has never been done however.

Multiplication in the system reduces itself to the addition of the exponents. If instead of adding each pair of exponents we select the larger (or the smaller) we obtain the exponents for the least common multiple (or the greatest common divisor) instead of the product. Thus the product, L.C.M. and G.C.D. of the two numbers 4,2,1,1,0,1 and 2,0,3,5,0,0,0,2 are respectively 6,2,4,6,0,1,0,2; 4,2,3,5,0,1,0,2 and 2,0,1,1. A number of other functions important to the theory of numbers may be easily dealt with in either system (1) or (2) and the majority of theorems presuppose that numbers are expressed in terms of their prime factors. Of course addition in this system is out of the question because we can say little or nothing about the prime factors of a sum. It is easy to derive the left side of (1) from the right. But to go from left to right is notoriously difficult. This one problem occupies a central position in the theory of numbers.

In conclusion I wish to call attention to still another way of expressing numbers which, in a sense, is both additive and multiplicative. This method seems to have originated in China and was considered by one Sun Tsu as early as the first century. The story is told of a wise Chinese general who wished to count the exact number of men in his huge army. He knew that this number was somewhere between 100,000 and 150,000. It would have been too risky and tedious to count the men in the ordinary fashion. So the general commanded his men to arrange themselves in groups of 4. Two men were left over. Groups of 5 were next in order and in this case no men were left over. Grouping by 7, 9, 11, and 13 was tried and in these respective cases 4, 6, 5, and

[1] At the present time I fail to recollect the true meaning of the number 4. There is also the matter of the decimal point. Perhaps some reader can complete these details.

3 men were left over. With these results the general made a few rapid calculations and announced the number of men to be precisely 145,590. Whether or not we see exactly how this answer was found, it is clear that numbers may be described by the remainders which they leave on division by a selected set of divisors. Furthermore, if these divisors are relatively prime (and it serves no useful purpose to have them otherwise) it is easy to see that this method of description is unique for all numbers less than the product of the divisors. In contradistinction to the prime factor method, it is easier to pass from the decimal system to some Chinese system than to go in the opposite direction although this latter step is by no means formidable.

Those of us who are true worshippers of the decimal system will ask: "Would it not have been better to arrange the army in groups of ten and to observe the number of men left over. This would give the last digit in the answer. By arranging these groups in groups of ten, the next to the last digit would be obtained, and so on until there were no groups left." The answer is no. By using such a practical method the general would not have won for China the distinction of being the source of what was later to become the most effective method for solving certain problems in the theory of numbers. In fact, information about the answers to many interesting problems comes to us in the same way that our general received his. Unfortunately there is often more than one army to count; in some cases thousands or even millions of armies are concerned. These must be arranged according to size. The smallest army is the one we are looking for. I shall not attempt to describe the large class of problems to which this Chinese method is applicable. One of these problems is that of representing a number as a product of primes. As a recent example of this I may cite the factorization:

$$2^{79} - 1 = 2687 \cdot 202029703 \cdot 1113491139767.$$

This result was obtained by sorting armies according to size at the rate of 5,000 armies a second.

It is a familiar fact to the student of algebra or geometry that many a seemingly difficult problem may often become remarkably simple when one makes the right change in variable or the appropriate choice of coordinates. In the same way a suitable system for representing numbers will sometimes facilitate and simplify problems in higher arithmetic. Conversely, he who devises a new numerical notation is sure to discover new properties of numbers and to realize more fully the difference between a symbol for a number and the number itself.

Part II

The 1940s

The Generalized Weierstrass Approximation Theorem, Part I (Abridged)

Marshall H. Stone

Mathematics Magazine
21 (1947/48), 167–84

Editors' Note: Marshall Harvey Stone was one of the leading American mathematicians of the mid-twentieth century. Educated at Harvard—his PhD advisor was G. D. Birkhoff—he joined the faculty there after short appointments at Columbia and Yale. In 1946 he was brought to the University of Chicago to revitalize the Mathematics Department there. Thus he ushered in the often-cited "Stone Age" at Chicago, attracting such senior faculty as S. S. Chern, Saunders Mac Lane, André Weil, and Antoni Zygmund, as well as a younger group that included Paul R. Halmos, Irving E. Segal, and Edwin H. Spanier, who joined existing faculty Irving Kaplansky and A. A. Albert.

Stone was the son of the U. S. Supreme Court Chief Justice (1941–46), Harlan Fiske Stone,who is reported to have said, "I am puzzled but happy that my son has written a book of which I understand nothing at all."

The classical Weierstrass approximation theorem asserts that any continuous function on a closed interval $[a,b]$ can be uniformly approximated there by a polynomial function. This theorem can be reformulated in terms of the algebra $C([a,b])$ of all continuous functions on $[a,b]$, which contains as a subalgebra the family P of all polynomials in a single variable x. $C([a,b])$ is a complete metric space under the so-called supremum norm, where the distance between two continuous functions f and g is $\max_{x\in[a,b]} |f(x)-g(x)|$. Weierstrass's approximation theorem then asserts that the uniform closure of P is $C([a,b])$ itself or, equivalently, that P is dense in $C([a,b])$.

Stone first published a generalization of this result in a paper in 1937, but the theorem was tucked away at the end of the paper and not even mentioned in the introduction. The work was prompted by a conversation with John von Neumann who, according to Stone, asked the "right" question. After a decade of improving on the original proof, modifying and extending the theorem, and finding "many interesting applications to classical problems of analysis," Stone chose *Mathematics Magazine* for "collecting relevant material in an expository article where everything could be presented in the light of our most recent knowledge." Steven G. Krantz reports in his book, *Mathematical Apocrypha* (MAA, 2003), that Stone sent it to the *Magazine* "because he had promised them a paper to help them get off to a good start," referring to the revival of the *Magazine* in the late 40s by Glenn James.

Introducing a result as important as the Stone-Weierstrass Theorem, one of the cornerstones of 20th century analysis, was certainly an achievement for *Mathematics Magazine*. Here we include only the first three sections of Part I. Section 4 on the characterization of closed ideals and Part II, which appeared in *Mathematics Magazine* 21 (1947/48), 237–54, are more technical and the exposition is probably more challenging to readers than other material included in this collection.

1. Introduction

Some years ago the writer discovered a generalization of the Weierstrass approximation theorem suggested by an inquiry into certain algebraic properties of the continuous real functions on a topological space [1]. This generalization has since shown itself to be very useful in a variety of similar situations. Interest in it has stimulated several improvements in the proof originally given and has also led to some modifications and extensions of the theorem itself. At the same time many interesting applications to classical problems of analysis have been observed by those working with the generalized approximation theorem. The writer, for instance, has noted a number of such applications in his lectures of 1942–1945, dealing with this and other subjects. Since the proofs thus obtained for several important classical theorems are remarkably simple, there would seem to be some advantage in collecting the relevant material in an expository article where everything could be presented in the light of our most recent knowledge. To offer such an article is our present purpose.

2. Lattice Formulations of the Generalized Theorem

The Weierstrass approximation theorem states, of course, that any continuous real function defined on a bounded closed interval of real numbers can be uniformly approximated by polynomials. The generalization with which we shall be concerned here seeks in the first instance to lighten the restrictions imposed on the domain over which the given functions are defined. The difficulty which has to be turned at the very outset in formulating such a generalization is that there are no polynomials on a general domain. It is rather easy, however, to circumvent this difficulty by orienting our inquiry towards the solution of the following question: what functions can be built from the functions of a prescribed family by the application of the algebraic operations (addition, multiplication, and multiplication by real numbers) and of uniform passages to the limit? In the classical case settled by the Weierstrass approximation theorem, the prescribed family consists of just two functions, f_1 and f_2, where $f_1(x) = 1$ and $f_2(x) = x$ for all x in the basic interval. In this, as in other cases which will be noted below, the answer is especially interesting because a very small prescribed family suffices to generate a very much more inclusive family. In his first discussion of the general problem posed above, the author focussed attention on the role played in approximation theory by the operations of forming the maximum and the minimum of a pair of functions. The reason why these operations are technically appropriate to the end in view can be seen even in the classical case of Weierstrass. There it is geometrically evident that a given continuous real function can be uniformly approximated by continuous piecewise linear functions, since to obtain such approximations one has only to inscribe polygons in the graph of the given function; and each piecewise linear function can be obtained from linear functions by means of the operations in question. The approximation of piecewise linear functions by polynomials then becomes the issue. The parts of the author's proof which involve these operations have since been much improved by Kakutani, with the aid of suggestions made by Chevalley, and the results given explicit formulation as a theorem about lattices of continuous functions [2]. Further modifications will be indicated below in the course of our present discussion.

In accordance with the preceding remarks, we shall start with an arbitrary topological space X, the family $\mathfrak{X}$ of all continuous real functions on X, and a prescribed subfamily $\mathfrak{X}_0$ of $\mathfrak{X}$. Our object is to determine the family $\mathfrak{U}(\mathfrak{X}_0)$ of all those functions which can be built from functions in $\mathfrak{X}_0$ by the application of specified algebraic operations and uniform passage to the limit. We shall consider first the case where the specified operations are the lattice operations $\cup$ and $\cap$ defined as follows:

$$f \cup g = \max(f, g) \quad \text{and} \quad f \cap g = \min(f, g)$$

are the functions h and k respectively, where

$$h(x) = \max\big(f(x), g(x)\big)$$

and

$$k(x) = \min\big(f(x), g(x)\big)$$

for all x in X. Later we shall take up other cases. In general we shall require of X that it be a compact space or even a compact Hausdorff space; but in the course of our preliminary remarks no such restriction will be necessary.

In all the cases we shall consider, $\mathfrak{U}(\mathfrak{X}_0)$ is a part of $\mathfrak{X}$ closed under uniform passage to the limit—in symbols,

$$\mathfrak{U}(\mathfrak{X}_0) \subset \mathfrak{X}, \qquad \mathfrak{U}\big(\mathfrak{U}(\mathfrak{X}_0)\big) = \mathfrak{U}(\mathfrak{X}_0).$$

Let us discuss these statements briefly in the case of the lattice operations. Since $f \cup g$ and $f \cap g$ are continuous whenever f and g are continuous (the mapping of X into the plane given by $x \to \big(f(x), g(x)\big)$ is continuous and the mappings of the plane into the real number system given by

$$(\xi, \eta) \to \max(\xi, \eta) \quad \text{and} \quad (\xi, \eta) \to \min(\xi, \eta)$$

respectively are both continuous, so that the composite mappings

$$x \to \max\big(f(x), g(x)\big)$$

and

$$x \to \min\big(f(x), g(x)\big)$$

are continuous also) and since the uniform limit of continuous functions is a continuous function, we see that the operations applied in the construction of $\mathfrak{U}(\mathfrak{X}_0)$ work entirely within $\mathfrak{X}$ and hence that $\mathfrak{U}(\mathfrak{X}_0) \subset \mathfrak{X}$. We now observe that $\mathfrak{U}(\mathfrak{X}_0)$ can be constructed in two steps: we first form all the functions obtainable by applying the algebraic operations alone to members of $\mathfrak{X}_0$ and we then form all the functions obtainable from these by uniform passage to the limit. For convenience let us designate the family of functions obtained in the first step by $\mathfrak{U}_1(\mathfrak{X}_0)$ and the family obtained in the second step by $\mathfrak{U}_2(\mathfrak{X}_0)$. It is evident that

$$\mathfrak{X}_0 \subset \mathfrak{U}_1(\mathfrak{X}_0) \subset \mathfrak{U}_2(\mathfrak{X}_0) \subset \mathfrak{U}(\mathfrak{X}_0).$$

We shall show that $\mathfrak{U}_2(\mathfrak{X}_0)$ is closed under the operations allowed and hence that $\mathfrak{U}(\mathfrak{X}_0) = \mathfrak{U}_2(\mathfrak{X}_0)$. It is then trivial that $\mathfrak{U}(\mathfrak{X}_0)$ is also closed under those operations. It is easy to see that any function f which is a uniform limit of functions f_n in $\mathfrak{U}_2(\mathfrak{X}_0)$ is itself a member of $\mathfrak{U}_2(\mathfrak{X}_0)$: in fact, each f_n can be uniformly approximated by functions in $\mathfrak{U}_1(\mathfrak{X}_0)$ so that, if ϵ is any positive number, f_n and a corresponding function g_n in $\mathfrak{U}_1(\mathfrak{X}_0)$ can be found satisfying the inequalities

$$\left|f(x) - f_n(x)\right| < \epsilon/2, \quad \left|f_n(x) - g_n(x)\right| < \epsilon/2,$$

and hence the inequality

$$\left|f(x) - g_n(x)\right| < \epsilon$$

for all x in X. It is also fairly easy to see that whenever f and g are in $\mathfrak{U}_2(\mathfrak{X}_0)$ so also are $f \cup g$ and $f \cap g$. For this it is sufficient to observe that, when f and g are uniform limits of the respective sequences f_n and g_n in $\mathfrak{U}_1(\mathfrak{X}_0)$, then $f \cup g$ and $f \cap g$ are uniform limits of the respective sequences $f_n \cup g_n$, and $f_n \cap g_n$—which are obviously in $\mathfrak{U}_1(\mathfrak{X}_0)$ too. The validity of this observation depends upon the inequalities

$$\left|\max(\xi, \eta) - \max(\xi', \eta')\right| \le |\xi - \xi'| + |\eta - \eta'|,$$
$$\left|\min(\xi, \eta) - \min(\xi', \eta')\right| \le |\xi - \xi'| + |\eta - \eta'|,$$

for which formal proofs based on the equations

$$\begin{aligned} \max(\xi, \eta) &= \tfrac{1}{2}(\xi + \eta + |\xi - \eta|) \\ \min(\xi, \eta) &= \tfrac{1}{2}(\xi + \eta - |\xi - \eta|) \end{aligned} \tag{1}$$

are easily given. Using these inequalities and choosing n so that

$$\left|f(x) - f_n(x)\right| < \epsilon/2, \quad \left|g(x) - g_n(x)\right| < \epsilon/2$$

for all x in X, we find directly that

$$\left|\max\left(f(x), g(x)\right) - \max\left(f_n(x), g_n(x)\right)\right| < \epsilon,$$
$$\left|\min\left(f(x), g(x)\right) - \min\left(f_n(x), g_n(x)\right)\right| < \epsilon,$$

for x in X. In case we assume X to be compact, every function in X is automatically bounded. By virtue of this assumption, or by virtue of a direct restriction to the bounded continuous functions on X in the general case, we put ourselves in a position to summarize the preceding. remarks in a particularly brief form. In fact, if we restrict $\mathfrak{X}$ to consist of the bounded continuous functions on X and define the distance between two bounded functions f and g to be $\sup_{x \in X} \left|f(x) - g(x)\right|$, we thereby make $\mathfrak{X}$ into a complete metric space in which metric convergence is equivalent to uniform convergence. The lattice operations are continuous with respect to this metric. As before, when $\mathfrak{X}_0 \subset \mathfrak{X}$ the relations

$$\mathfrak{X}_0 \subset \mathfrak{U}(\mathfrak{X}_0) \subset \mathfrak{X}, \quad \mathfrak{U}(\mathfrak{X}_0) = \mathfrak{U}\big(\mathfrak{U}(\mathfrak{X}_0)\big)$$

are valid. The first states that the uniform limit of bounded continuous functions is a bounded continuous function, the second that $\mathfrak{U}(\mathfrak{X}_0)$ is metrically and algebraically closed. The proof of the latter fact runs as before; but it can be more briefly stated as follows: if $\mathfrak{U}_1(\mathfrak{X}_0)$ is the family of all "lattice polynomials" formed from $\mathfrak{X}_0$ and $\mathfrak{U}_2(\mathfrak{X}_0)$ is its metric closure, then $\mathfrak{U}_2(\mathfrak{X}_0)$ is obviously metrically closed and the fact that it is algebraically closed with respect to the lattice operations is a simple, direct consequence of their metric continuity.

We are now ready to determine, in the important case where X is compact, what functions belong to $\mathfrak{U}(\mathfrak{X}_0)$.

Theorem 1 *Let X be a compact space, $\mathfrak{X}$ the family of all continuous (necessarily bounded) real functions on X, $\mathfrak{X}_0$ an arbitrary subfamily of $\mathfrak{X}$, and $\mathfrak{U}(\mathfrak{X}_0)$ the family of all functions (necessarily continuous) generated from $\mathfrak{X}_0$ by the lattice operations and uniform passage to the limit. Then a necessary and sufficient condition for a function f in $\mathfrak{X}$ to be in $\mathfrak{U}(\mathfrak{X}_0)$ is that, whatever the points x, y in X and whatever the positive number ϵ, there exists a function f_{xy} obtained by applying the lattice operations alone to $\mathfrak{X}_0$ and such that*

$$\left|f(x) - f_{xy}(x)\right| < \epsilon, \quad \left|f(y) - f_{xy}(y)\right| < \epsilon.$$

Proof. The necessity of the stated condition is trivial. It is the sufficiency which requires discussion. Starting with the functions f_{xy} in $\mathfrak{U}_1(\mathfrak{X}_0)$, we shall construct an approximant for f. Let G_y designate the open set $\{z : f(z) - f_{xy}(z) < \epsilon\}$, where x is fixed. By hypothesis x and y are in G_y, so that the union of all the sets G_y is the entire space X. The compactness of X implies the existence of points $y_1, \ldots, y_n$ such that the union of the sets

$G_{y_1}, \ldots, G_{y_n}$ is still the entire space X. Setting

$$g_x = f_{xy_1} \cup \cdots \cup f_{xy_n} = \max(f_{xy_1}, \ldots, f_{xy_n}),$$

we see that for any z in X we have $z \in G_{y_k}$ for a suitable choice of k and hence

$$g_x(z) \geq f_{xy_k}(z) > f(z) - \epsilon.$$

On the other hand, the fact that

$$f_{xy}(x) < f(x) + \epsilon$$

implies that $g_x(x) < f(x) + \epsilon$. We can now work in a similar manner with the functions g_x. Let H_x designate the open set

$$\{z : g_x(z) < f(z) + \epsilon\}.$$

Evidently x is in H_x, so that the union of all the sets H_x is the entire space X. The compactness of X implies the existence of points $x_1, \ldots, x_m$ such that the union of the sets $H_{x_1}, \ldots, H_{x_m}$ is still the entire space X. Setting

$$h = g_{x_1} \cap \cdots \cap g_{x_m} = \min(g_{x_1}, \ldots, g_{x_m}),$$

we see that for any z in X we have $z \in H_{x_k}$ for a suitable choice of k and hence

$$h(z) \leq g_{x_k}(z) < f(z) + \epsilon.$$

On the other hand, the fact that

$$g_x(z) > f(z) - \epsilon$$

for all z and all x implies that

$$h(z) > f(z) - \epsilon$$

for all z. Thus we have $|f(z) - h(z)| < \epsilon$ for all z in X. To complete the proof we note that, since only the lattice operations have been used in constructing the functions g_x and h from the functions f_{xy}, these functions are all in $\mathfrak{U}_1(\mathfrak{X}_0)$, as desired.

We may note two simple corollaries, as follows:

Corollary 1.1 *If $\mathfrak{X}_0$ has the property that, whatever the points, x, y, $x \neq y$, in X and whatever the real numbers α and β, there exists a function f_0 in $\mathfrak{X}_0$ for which $f_0(x) = \alpha$ and $f_0(y) = \beta$, then $\mathfrak{U}(\mathfrak{X}_0) = \mathfrak{X}$—in other words, any continuous function on X can be uniformly approximated by lattice polynomials in functions belonging to the prescribed family $\mathfrak{X}_0$.*

Corollary 1.2 *If a continuous real function on a compact space X is the limit of a monotonic sequence f_n of continuous functions, then the sequence converges uniformly to f.* (Professor André Weil remarks that the extension to monotonic sets is immediate.)

Proof. We take $\mathfrak{X}_0$ to be the totality of functions occurring in the sequence f_n. Then $\mathfrak{U}_1(\mathfrak{X}_0) = \mathfrak{X}_0$ since monotonicity implies that $f_m \cup f_n$ coincides with one of the two functions f_m and f_n while $f_m \cap f_n$ coincides with the other. The assumption that

$$\lim_{n \to \infty} f_n(x) = f(x)$$

for every x now shows that the condition of Theorem 1 is satisfied. Hence f is in $\mathfrak{U}(\mathfrak{X}_0)$; and f is therefore the uniform limit of functions occurring in $\mathfrak{X}_0$. Since $|f(x) - f_n(x)|$ decreases as n increases and since

$$|f(x) - f_N(x)| < \epsilon$$

for all x and a suitable choice of N, we see that

$$|f(x) - f_n(x)| < \epsilon$$

for all $n > N$, as was to be proved.

Theorem 1 tells us that the question, "Can a given function f be approximated in terms of the prescribed family $\mathfrak{X}_0$?", has an answer depending only on the way in which f and $\mathfrak{X}_0$ behave on pairs of points in X. The contraction of a function obtained by suppressing all points of X except the two points x, y of a pair is a function of very simple kind—it is completely described by the ordered pair (α, β) of those real numbers which are its values at x and at y respectively. If $\mathfrak{X}_0(x, y)$ designates the family of functions obtained by contracting every function in $\mathfrak{X}_0$ in this manner, and if $\mathfrak{X}(x, y)$ has a corresponding significance, then everything depends on an examination (for all different pairs x, y) of the question, "Can a given element of $\mathfrak{X}(x, y)$ be approximated in terms of $\mathfrak{X}_0(x, y)$?" This question is that special case of our original problem in which X is a two-element space! When X has just two elements, the approximation problem can be described in slightly different language, as follows. We have to deal with all ordered pairs (α, β) of real numbers—that is, with the cartesian plane. On two such pairs we can perform the operations $\cup$ and $\cap$ defined by the equations

$$(\alpha, \beta) \cup (\gamma, \delta) = \big(\max(\alpha, \gamma), \max(\beta, \delta)\big)$$

$$(\alpha, \beta) \cap (\gamma, \delta) = \big(\min(\alpha, \gamma), \min(\beta, \delta)\big).$$

Geometrically these operations produce the upper right vertex and lower left vertex respectively of a rectangle with its sides parallel to the coordinate axes and one pair of opposite vertices falling on the points (α, β), (γ, δ). For any given subset S of the plane the problem to be solved is that of finding what points can be generated from it by the above operations and passage to the limit. From what has been said above, it is clear that the points so generated constitute a closed subset S^* of the plane which contains with (α, β) and (γ, δ) the two points described above. It is also clear that this subset is the smallest set enjoying these properties and containing the given subset S. Reverting now to the interpretation of Theorem 1, we see that it can be restated in the following form: if $f \in \mathfrak{X}$, then $f \in \mathfrak{U}(\mathfrak{X}_0)$ if and only if

$$\big(f(x), f(y)\big) \in \mathfrak{X}_0(x, y)^*$$

for every pair of distinct points x, y in X. We have not asserted that the conditions corresponding to various pairs x, y are independent of one another, nor have we asserted that every point (α, β) in $\mathfrak{X}_0(x, y)^*$ can be expressed in the form $\alpha = f(x)$, $\beta = f(y)$ for some f in $\mathfrak{U}(\mathfrak{X}_0)$. Indeed, even in the case where $\mathfrak{X}_0 = \mathfrak{U}(\mathfrak{X}_0)$ we know only that $\mathfrak{X}_0(x, y)^*$ is the closure of $\mathfrak{X}_0(x, y)$.

It is convenient to express some of the results sketched in the preceding paragraph as a formal theorem. This we do as follows.

Theorem 2 *Let X be a compact space, $\mathfrak{X}$ the family of continuous real functions on X, and $\mathfrak{X}_0$ a subfamily of $\mathfrak{X}$ which is closed under the lattice operations and uniform passage to the limit. Then $\mathfrak{X}_0$ is completely characterized by the system of planar sets $\mathfrak{X}_0(x, y)^* = \mathfrak{X}_0(x, y)$.*

Proof. Our hypothesis that $\mathfrak{X}_0 = \mathfrak{U}(\mathfrak{X}_0)$ shows that $\mathfrak{X}_0(x, y)$ has $\mathfrak{X}_0(x, y)^*$ as its closure, as we remarked above. Let us suppose that

$$\mathfrak{Y}_0 = \mathfrak{U}(\mathfrak{Y}_0) \subset \mathfrak{X}$$

and that $\mathfrak{X}_0(x, y)^* = \mathfrak{Y}_0(x, y)^*$ for all pairs of points x, y in X. Then the conditions for f in $\mathfrak{X}$ to belong to $\mathfrak{X}_0$ are identical to those for it to belong to $\mathfrak{Y}_0$. Hence $\mathfrak{X}_0$ and $\mathfrak{Y}_0$ coincide.

We pass now to the modifications of Theorems 1 and 2 which result when we take into consideration the operations of linear algebra as well as the lattice operations. The newly admitted operations are, more precisely, addition and multiplication by real numbers. In view of the equations (1), which express the lattice operations in terms of the linear operations and the single operation of forming the absolute value, we may take the specified algebraic operations to be simply addition, multiplication by real numbers, and formation of absolute values. The remarks preliminary to Theorem 1 apply, *mutatis mutandis*, to the present situation. The family $\mathfrak{U}(\mathfrak{X}_0)$ of all functions which can be constructed from $\mathfrak{X}_0 \subset \mathfrak{X}$ by application of the linear lattice operations and uniform passage to the limit is again seen to be obtainable in two steps, the first being algebraic and the second consisting in the adjunction of uniform limits. This family is closed under the operations used to generate it. We now have the following analogue of the results contained in Theorems 1 and 2.

Theorem 3 [2] *Let X be a compact space, $\mathfrak{X}$ the family of all continuous (necessarily bounded) real functions on X, $\mathfrak{X}_0$ an arbitrary subfamily of $\mathfrak{X}$, and $\mathfrak{U}(\mathfrak{X}_0)$ the family of all functions (necessarily continuous) generated from $\mathfrak{X}_0$ by the linear lattice operations and uniform passage to the limit. Then a necessary and sufficient condition for a function f in $\mathfrak{X}$ to be in $\mathfrak{U}(\mathfrak{X}_0)$ is that f satisfy every linear relation of the form $\alpha g(x) = \beta g(y)$, $\alpha\beta \geq 0$, which is satisfied by all functions in $\mathfrak{X}_0$. If $\mathfrak{X}_0$ is a closed linear sublattice of $\mathfrak{X}$—that is, if $\mathfrak{X}_0 = \mathfrak{U}(\mathfrak{X}_0)$—then $\mathfrak{X}_0$ is characterized by the system of all the linear relations of this form which are satisfied by every function belonging to it. The linear relations associated with an arbitrary pair of points x, y in X must be equivalent to one of the following distinct types:*

1. *$g(x) = 0$ and $g(y) = 0$;*
2. *$g(x) = 0$ and $g(y)$ unrestricted, or vice versa;*
3. *$g(x) = g(y)$ without restriction on the common value;*
4. *$g(x) = \lambda g(y)$ or $g(y) = \lambda g(x)$ for a unique value λ, $0 < \lambda < 1$.*

Proof. Since $\mathfrak{Y}_0 = \mathfrak{U}(\mathfrak{X}_0)$ is closed under the lattice operations and uniform passage to the limit, Theorem 2 can be applied to $\mathfrak{Y}_0$. However, the fact that $\mathfrak{Y}_0$ is also closed under the linear operations can be expected to produce effective simplifications. Indeed we see that the planar set $\mathfrak{Y}_0(x, y)$, where x and y are arbitrary points in X, must be the entire plane, a straight line passing through the origin, or the one-point set consisting of the origin alone. This appears at once when we observe that if

$$(\alpha, \beta) \in \mathfrak{Y}_0(x, y)$$

then

$$(\lambda\alpha, \lambda\beta) \in \mathfrak{Y}_0(x, y)$$

for every λ, and that if (α, β) and (γ, δ) are in $\mathfrak{Y}_0(x, y)$, then

$$(\alpha + \gamma, \beta + \delta) \in \mathfrak{Y}_0(x, y).$$

Since $\mathfrak{Y}_0(x, y)$ is obviously a closed subset of the plane, we have

$$\mathfrak{Y}_0(x, y)^* = \mathfrak{Y}_0(x, y).$$

When $\mathfrak{Y}_0(x, y)$ is a straight line through the origin we write its equation as $\alpha\xi = \beta\eta$ and observe that

$$(\beta, \alpha) \in \mathfrak{Y}_0(x, y).$$

Since $\mathfrak{Y}_0$ is closed under the operation of forming absolute values, we see that

$$(|\beta|, |\alpha|) \in \mathfrak{Y}_0(x, y).$$

Hence $\alpha|\beta| = |\alpha|\beta$ so that $\alpha\beta|\beta| = |\alpha|\beta^2 \geq 0$ and $\alpha\beta \geq 0$. When $\mathfrak{Y}_0(x, y)$ consists of the origin alone, we have the case enumerated as (1) in the statement of the theorem. When $\mathfrak{Y}_0(x, y)$ is a straight line through the origin we have case (2) if it coincides with one of the coordinate axes, case (3) if it coincides with the bisector of the angle between the positive coordinate axes, and case (4) otherwise. When $\mathfrak{Y}_0(x, y)$ is the entire plane there is no corresponding linear relation, of course. Theorem 2 shows that $\mathfrak{Y}_0$ is characterized by the sets

$$\mathfrak{Y}_0(x, y) = \mathfrak{Y}_0(x, y)^*$$

—in other words, that f in X belongs to $\mathfrak{Y}_0 = \mathfrak{U}(\mathfrak{X}_0)$ if and only if

$$(f(x), f(y)) \in \mathfrak{Y}_0(x, y).$$

Since $\mathfrak{X}_0 \subset \mathfrak{Y}_0$, it is clear that the conditions thus imposed on the functions in $\mathfrak{U}(\mathfrak{X}_0)$ are satisfied by the functions in $\mathfrak{X}_0$. On the other hand if all the functions in $\mathfrak{X}_0$ satisfy relations of the kind enumerated in (1)–(4) it is clear that every function in $\mathfrak{U}(\mathfrak{X}_0)$ must do likewise: for the sums, constant multiples, absolute values, and uniform limits of functions which satisfy a condition of any one of these types must satisfy the same condition. Thus the linear relations of the form $\alpha g(x) = \beta g(y)$, $\alpha\beta \geq 0$, satisfied by the functions in $\mathfrak{X}_0$ are identical with those satisfied by the functions in $\mathfrak{U}(\mathfrak{X}_0)$ and serve to characterize the latter family completely.

We may note some simple corollaries to the theorem just proved.

Corollary 3.1 *In order that $\mathfrak{U}(\mathfrak{X}_0)$ contain a nonvanishing constant function, it is necessary and sufficient that the only linear relations of the form $\alpha g(x) = \beta g(y)$, $\alpha\beta > 0$, satisfied by every function in $\mathfrak{X}_0$ be those reducible to the form $g(x) = g(y)$.*

Proof. It is obvious that of conditions (1)–(4) in Theorem 3, only condition (3) can be satisfied by a nonvanishing constant function.

Corollary 3.2 *In order that $\mathfrak{U}(\mathfrak{X}_0) = \mathfrak{X}$, it is sufficient that the functions in $\mathfrak{X}_0$ satisfy no linear relation of the form (1)–(4) of Theorem 3.*

In order to state a further corollary, we first introduce a convenient definition.

Definition 1 A family of arbitrary functions on a domain X is said to be a separating family (for that domain) if, whenever x and y are distinct points of X, there is some function f in the family with distinct values $f(x)$, $f(y)$ at these points.

In terms of this definition we have the following result.

Corollary 3.3 *If X is compact and if $\mathfrak{X}_0$ is a separating family for X and contains a nonvanishing constant function, then $\mathfrak{U}(\mathfrak{X}_0) = \mathfrak{X}$.*

Proof. Since $\mathfrak{X}_0$ contains a nonvanishing constant function, the only one of conditions (1)–(4) satisfied by every function in $\mathfrak{X}_0$ are those of the form (3). Since $\mathfrak{X}_0$ is a separating family, no linear relation of the form $g(x) = g(y)$, where $x \neq y$, is satisfied by every function in $\mathfrak{X}_0$. Hence Corollary 3.2 yields the desired result.

Corollary 3.4 *If $\mathfrak{X}_0$ is a separating family, then so is $\mathfrak{X}$. If $\mathfrak{X}$ is a separating family and $\mathfrak{U}(\mathfrak{X}_0) = \mathfrak{X}$, then $\mathfrak{X}_0$ is also a separating family.*

Proof. The first statement is trivial. The second statement follows at once from the fact that $\mathfrak{X}_0$ is subject to no linear relation of the form $g(x) = g(y)$ which is not also satisfied by every function in $\mathfrak{U}(\mathfrak{X}_0) = \mathfrak{X}$.

It should be remarked that in general the family $\mathfrak{X}$ of *all* continuous functions on a compact space X need not be a separating family. In case X is a compact *Hausdorff* space, however, it is well known that $\mathfrak{X}$ is a separating family: if $x \neq y$, there exists a continuous function f on X such that $f(x) = 0$, $f(y) = 1$.

3. Linear Ring Formulations of the Generalized Theorem

We are now ready to discuss the approximation problem when the specified algebraic operations used in the construction of approximants are the linear operations and multiplication. Since the product of two continuous functions is continuous we see that the family $\mathfrak{X}$ of all continuous functions on a topological space X is a commutative ring with respect to the two operations of addition and multiplication, and a commutative linear associative algebra or linear ring with respect to the operations of addition, multiplication, and multiplication by real numbers. Hence the formally stated results of this section constitute what may be called the *linear-ring* formulation of the generalized Weierstrass approximation theorem.

If we now designate by $\mathfrak{U}(\mathfrak{X}_0)$ the family of all functions generated from $\mathfrak{X}_0 \subset \mathfrak{X}$ by means of the linear-ring operations and uniform passage of the limit, we have to note a slight modification which must be made in the general statements made in the lattice case. If f and g are uniform limits of the sequences f_n and g_n respectively, the product fg is not in general the uniform limit of the sequence $f_n g_n$—consider, for example, the case where $g_n = g$ is a nonbounded function and f_n is the constant $1/n$. *We shall therefore suppose that $\mathfrak{X}$ consists of all bounded continuous functions on a topological space X,* this boundedness restriction being automatically satisfied when X is compact. By virtue of this restriction we can apply the inequality

$$|fg - f_n g_n| \leq |f||g - g_n| + |g||f - f_n| + |f - f_n||g - g_n|$$

to show that when f_n and g_n are uniformly convergent sequences in $\mathfrak{X}$ their respective limits f and g are in $\mathfrak{X}$ and that the sequence $f_n g_n$ converges uniformly to the product fg, in $\mathfrak{X}$. When $\mathfrak{X}_0 \subset \mathfrak{X}$ we see as before that $\mathfrak{U}(\mathfrak{X}_0) \subset \mathfrak{X}$, $\mathfrak{U}(\mathfrak{U}(\mathfrak{X}_0)) = \mathfrak{U}(\mathfrak{X}_0)$. It is easy to see that $\mathfrak{U}(\mathfrak{X}_0)$ consists of all those functions, necessarily in $\mathfrak{X}$, which are uniform limits of polynomials in members of $\mathfrak{X}_0$—in other words, $f \in \mathfrak{X}$ is in $\mathfrak{U}(\mathfrak{X}_0)$ if and only if, whatever the positive number ϵ, there exist functions $f_1, \ldots, f_n$ and a polynomial function $p(\xi_1, \ldots, \xi_n)$ of the real variables $\xi_1, \ldots, \xi_n$ with $p(0, \ldots, 0) = 0$ such that

$$\left|f(x) - p(f_1(x), \ldots, f_n(x))\right| < \epsilon$$

for every x in X.

Now in order to prove our principal theorem we shall establish a very special case of the classical Weierstrass approximation theorem, using for this purpose direct and elementary methods which do not depend on any general theory. The result we need is the following proposition.

Theorem 4 *If ϵ is any positive number and $\alpha \leq \xi \leq \beta$ any real interval, then there exists a polynomial $p(\xi)$ in the real variable ξ with $p(0) = 0$ such that $\left||\xi| - p(\xi)\right| < \epsilon$ for $\alpha \leq \xi \leq \beta$.*

Proof. Unless the point $\xi = 0$ is inside the given interval (α, β), we can obviously take $p(\xi) = \pm\xi$. Thus there is no loss of generality in confining our attention to intervals of the form $(-\gamma, \gamma)$ where $\gamma > 0$, since the given interval (α, β) can be included in an interval of this form. Moreover it is obviously sufficient to study the case of the interval $(-1, 1)$ since, if $q(\eta)$, $q(0) = 0$, is a polynomial such that

$$\left||\eta| - q(\eta)\right| < \epsilon/\gamma$$

for $-1 \leq \eta \leq 1$, then

$$p(\xi) = \gamma q(\xi/\gamma), \qquad p(0) = 0,$$

is a polynomial such that $\left||\xi| - p(\xi)\right| < \epsilon$ for $-\gamma \leq \xi \leq \gamma$. We shall obtain the desired polynomial q for the interval $-1 \leq \eta \leq 1$ as a partial sum of the power series development for $\sqrt{1-\zeta}$ where $\zeta = 1 - \eta^2$. The validity of the development has to be established directly.

We commence by defining a sequence of constants α_k recursively from the relations

$$\alpha_1 = \frac{1}{2}$$

$$\alpha_k = \frac{1}{2} \sum_{m+n=k} \alpha_m \alpha_n = \frac{1}{2}(\alpha_1 \alpha_{k-1} + \alpha_2 \alpha_{k-2} + \cdots + \alpha_{k-1}\alpha_1)$$

It is obvious that $\alpha_k > 0$. Putting

$$\sigma_n = \sum_{k=1}^{n} \alpha_k,$$

we can show inductively that $\sigma_n < 1$. In fact we have

$$\sigma_1 = \alpha_1 = \frac{1}{2} < 1$$

and note that $\sigma_n < 1$ implies

$$\begin{aligned}\sigma_{n+1} &= \alpha_1 + \sum_{k=2}^{n+1} \alpha_k \\ &= \frac{1}{2} + \frac{1}{2}\sum_{k=2}^{n+1}\sum_{i,j\geq 1}^{i+j=k} \alpha_i\alpha_j \\ &\leq \frac{1}{2} + \frac{1}{2}\sum_{i,j=1}^{n} \alpha_i\alpha_j \leq \frac{1}{2}(1+\sigma_n^2) < 1.\end{aligned}$$

Accordingly the positive term series

$$\sum_{k=1}^{\infty} \alpha_k$$

converges to a sum σ satisfying the inequality $\sigma \leq 1$; and the power series

$$\sum_{k=1}^{\infty} \alpha_k \zeta^k$$

converges uniformly for $|\zeta| \leq 1$ to a continuous function $\sigma(\zeta)$. It is now comparatively easy to identify this function with the function $1 - \sqrt{1-\zeta}$. To do so we prove that

$$\sigma(\zeta)\big(2 - \sigma(\zeta)\big) = \zeta.$$

Looking at the partial sums of the power series for $\sigma(\zeta)$, we observe that

$$\begin{aligned}&\left(\sum_{i=1}^{n} \alpha_i \zeta^i\right)\left(2 - \sum_{j=1}^{n} \alpha_j \zeta^j\right) \\ &= 2\sum_{k=1}^{n} \alpha_k\zeta^k - \sum_{i,j=1}^{n} \alpha_i\alpha_j\zeta^{i+j} \\ &= 2\sum_{k=1}^{n} \alpha_k\zeta^k - 2\sum_{k=2}^{n} \alpha_k\zeta^k - \sum_{1\leq i,j\leq n}^{i+j\geq n+1} \alpha_i\alpha_j\zeta^{i+j} \\ &= \zeta - \sum_{1\leq i,j\leq n}^{i+j\geq n+1} \alpha_i\alpha_j\zeta^{i+j}\end{aligned}$$

in accordance with the definition of the coefficients α_k. The final term here can now be estimated as follows:

$$\begin{aligned}\left|\sum_{1\leq i,j\leq n}^{i+j\geq n+1} \alpha_i\alpha_j\zeta^{i+j}\right| &\leq \sum_{1\leq i,j\leq n}^{i+j\geq n+1} \alpha_i\alpha_j \\ &\leq \sum_{k=n+1}^{\infty}\sum_{i,j\geq 1}^{i+j=k} \alpha_i\alpha_j \\ &\leq 2\sum_{k=n+1}^{\infty} \alpha_k.\end{aligned}$$

When n becomes infinite, therefore, this term tends to zero; and passage to the limit in the identity above accordingly yields the relation

$$\sigma(\zeta)\big(2 - \sigma(\zeta)\big) = \zeta.$$

For each ζ such that $-1 \leq \zeta \leq 1$ we have

$$\sigma(\zeta) = 1 \pm \sqrt{1-\zeta}.$$

Here we decide upon the choice of sign by showing that $\sigma(\zeta) \leq 1$, an inequality incompatible with the upper sign. It is evident that $\sigma(1) = 1$, independently of the choice of sign, and hence that

$$\sum_{k=1}^{\infty} \alpha_k = \sigma(1) = 1.$$

Inasmuch as α_k is positive it follows that

$$\sigma(\zeta) \leq \sigma\big(|\zeta|\big) \leq \sigma(1) = 1,$$

as we intended to show. It is now clear that the power series for $\sqrt{1-\zeta}$ is given by

$$\begin{aligned}\sqrt{1-\zeta} &= 1 - \sigma(\zeta) \\ &= 1 - \sum_{k=1}^{\infty} \alpha_k\zeta^k = \sum_{k=1}^{\infty} \alpha_k(1-\zeta^k).\end{aligned}$$

Taking η so that $-1 \leq \eta \leq 1$, we have $0 \leq 1 - \eta^2 \leq 1$ and hence

$$\begin{aligned}|\eta| = \sqrt{\eta^2} &= 1 - \sigma(1-\eta^2) \\ &= \sum_{k=1}^{\infty} \alpha_k\big(1 - (1-\eta^2)^k\big),\end{aligned}$$

the series being uniformly convergent. The general term of this series is a polynomial in η which vanishes for $\eta = 0$. Hence we can take a suitable one of its partial sums as the required polynomial $q(\eta)$, thus completing our discussion.

We are now ready to give our principal results concerning the generalization of the Weierstrass theorem for the linear-ring operations.

Theorem 5 *Let X be a compact space, $\mathfrak{X}$ the family of all continuous real functions on X, $\mathfrak{X}_0$ an arbitrary subfamily of $\mathfrak{X}$, and $\mathfrak{U}(\mathfrak{X}_0)$ the family of all functions (necessarily continuous) generated from $\mathfrak{X}_0$ by the linear-ring operations and uniform passage to the limit. Then a necessary and sufficient condition for a function f in $\mathfrak{X}$ to be in $\mathfrak{U}(\mathfrak{X}_0)$ is that f satisfy every linear relation of the form $g(x) = 0$ or $g(x) = g(y)$ which is satisfied by all functions in $\mathfrak{X}_0$. If $\mathfrak{X}_0$ is a closed linear subring of $\mathfrak{X}$—that is, if $\mathfrak{X}_0 = \mathfrak{U}(\mathfrak{X}_0)$—then $\mathfrak{X}_0$ is characterized by the system of all the linear relations of this kind which are satisfied by every function belonging to it. In other words, $\mathfrak{X}_0$ is characterized by the partition of X into mutually disjoint closed subsets on each of which every function in $\mathfrak{X}_0$ is constant and by the specification of that one, if any, of these subsets on which every function in $\mathfrak{X}_0$ vanishes.*

Proof. By virtue of Theorem 4, we see that if f is in $\mathfrak{U}(\mathfrak{X}_0)$, then $|f|$ is also in $\mathfrak{U}(\mathfrak{X}_0)$. Indeed, since X is compact, the function f is bounded. Assuming accordingly that

$$\alpha \leq f(x) \leq \beta$$

for all x, we can find a polynomial $p_n(\xi)$ such that

$$\big||\xi| - p_n(\xi)\big| < 1/n$$

for $\alpha \leq \xi \leq \beta$, while $p_n(0) = 0$. It is clear that $p_n(f)$ is in $\mathfrak{U}(\mathfrak{X}_0)$ and that

$$\big||f(x)| - p_n(f(x))\big| < 1/n$$

for all x in X. Hence $|f|$ is the uniform limit of functions—namely, the functions $p_n(f)$—in $\mathfrak{U}(\mathfrak{X}_0)$. Thus $|f|$ is in $\mathfrak{U}(\mathfrak{X}_0)$, as we wished to prove. By virtue of the formulas (1) connecting the operations $\cup$ and $\cap$ with the operation of forming the absolute value, we now see that whenever f and g are in $\mathfrak{U}(\mathfrak{X}_0)$, then so also are $f \cup g$ and $f \cap g$—in other words, $\mathfrak{U}(\mathfrak{X}_0)$ is closed under the linear lattice operations, as well as under the ring operations and uniform passage to the limit. The characterization of closed linear sublattices of $\mathfrak{X}$ given in Theorem 3 applies, naturally, to $\mathfrak{U}(\mathfrak{X}_0)$. It is easy to see that none of the characteristic linear relations can be of the type (4) described there. In fact, if every function in $\mathfrak{U}(\mathfrak{X}_0)$ were to satisfy a linear relation of the form $g(x) = \lambda g(y)$, we would find for every f in $\mathfrak{U}(\mathfrak{X}_0)$ that, f^2 being also in $\mathfrak{U}(\mathfrak{X}_0)$, the relations

$$f(x) = \lambda f(y) \qquad f^2(x) = \lambda f^2(y),$$

$$\lambda^2 f^2(y) = \lambda f^2(y)$$

would hold; and we would conclude that $f(y) = 0$ for every f in $\mathfrak{U}(\mathfrak{X}_0)$ or that $\lambda = 0, 1$. Thus we conclude that f is in $\mathfrak{U}(\mathfrak{X}_0)$ if and only if it satisfies all the linear relations $g(x) = 0$ or $g(x) = g(y)$ satisfied by every function in $\mathfrak{X}_0$.

The first characterization of the closed linear subrings of $\mathfrak{X}$ given in the statement of the theorem follows immediately. As to the second characterization, we remark first that the relation $\equiv$ defined by putting $x \equiv y$ if and only if $f(x) = f(y)$ for all f in $\mathfrak{X}_0$ is obviously an equivalence relation *stronger* than the natural equality in X: $x = y$ implies $x \equiv y$; $x \equiv y$ implies $y \equiv x$; $x \equiv y$ and $y \equiv z$ imply $x \equiv z$. Consequently, X is partitioned by this equivalence relation into mutually disjoint subsets, each a maximal set of mutually equivalent elements. The set of all points y such that $x \equiv y$ is just that partition class which contains x. Since this set is the intersection or common part of all the sets

$$X_f = \{y : f(x) = f(y)\}$$

for the various functions f in $\mathfrak{X}_0$ and since each set X_f is closed by virtue of the continuity of f, we see that the partition class containing x is closed. If x and y are in distinct partition classes, then there exists a function f in $\mathfrak{X}_0$ such that $f(x) \neq f(y)$, since otherwise we would have $x \equiv y$ and the two given partition classes could not be distinct. If a partition class contains a single point x such that $f(x) = 0$ for every f in $\mathfrak{X}_0$, then all its points obviously have this property. On the other hand, at most one partition class can contain such a point since, if x and y are points such that $f(x) = 0$, $f(y) = 0$ for every f in $\mathfrak{X}_0$, then $f(x) = f(y)$ for every f in $\mathfrak{X}_0$, $x \equiv y$, and x and y are in the same partition class.

We cannot expect that an arbitrary partition of X into mutually disjoint closed subsets can be derived in the manner just described from some closed linear subring $\mathfrak{X}_0$ of $\mathfrak{X}$. However, partitions obtained from *distinct* closed linear subrings are necessarily *distinct*—except in the case where one subring consists of all the functions in $\mathfrak{X}$ which are constant on each partition class and the other consists of all those functions which are in the first subring and in addition vanish on one specified partition class. Thus we see that a closed linear subring is specified by the partition of X into the closed subsets on each of which all its members are constant and the specification of that particular partition class, if any, on which all its members vanish.

We have at once a pair of useful corollaries.

Corollary 5.1 *In order that* $\mathfrak{U}(\mathfrak{X}_0)$ *contain a nonvanishing constant function it is necessary and sufficient that for every* x *in* X *there exist some* f *in* $\mathfrak{X}_0$ *such that* $f(x) \neq 0$.

Corollary 5.2 *If* $\mathfrak{X}_0$ *is a separating family for* X, *then* $\mathfrak{U}(\mathfrak{X}_0)$ *either coincides with* $\mathfrak{X}$ *or is, for a uniquely determined point* x_0, *the family of all functions* f *in* $\mathfrak{X}$ *such that* $f(x_0) = 0$. *If, conversely,* $\mathfrak{X}$ *is a separating family for* X *and* $\mathfrak{U}(\mathfrak{X}_0)$ *either coincides with* $\mathfrak{X}$ *or is the family of all those* f *in* $\mathfrak{X}$ *which vanish at some fixed point* x_0 *in* X, *then* $\mathfrak{X}_0$ *is a separating family.*

Proof. If $\mathfrak{X}_0$ is a separating family, so also are $\mathfrak{U}(\mathfrak{X}_0)$ and $\mathfrak{X}$. Hence the partition classes associated with $\mathfrak{U}(\mathfrak{X}_0)$ must each consist of a single point. It follows that $\mathfrak{U}(\mathfrak{X}_0)$ must be as indicated. Conversely, when $\mathfrak{X}$ is a separating family and $\mathfrak{U}(\mathfrak{X}_0)$ is as stated, then $\mathfrak{U}(\mathfrak{X}_0)$ is a separating family. If it were not, every f in $\mathfrak{U}(\mathfrak{X}_0)$ vanishes at some point x_0; and there would exist distinct points x and y in X such that $f_0(x) = f_0(y)$ for every f_0 in $\mathfrak{U}(\mathfrak{X}_0)$. Consider now an arbitrary function f in $\mathfrak{X}$. Clearly, the function f_0 defined by putting

$$f_0(z) = f(z) - f(x_0)$$

is continuous and vanishes at x_0. Thus f_0 is in $\mathfrak{U}(\mathfrak{X}_0)$, the equation $f_0(x) = f_0(y)$ is verified, and in consequence $f(x) = f(y)$. Thus we find that $f(x) = f(y)$ for every f in $\mathfrak{X}$, against hypothesis. Since $\mathfrak{U}(\mathfrak{X}_0)$ is a separating family, $\mathfrak{X}_0$ must be also. Otherwise, of course, there would exist distinct points x, y in X such that $f_0(x) = f_0(y)$ for every f_0 in $\mathfrak{X}_0$; and then the equation $f(x) = f(y)$ would hold for every f in $\mathfrak{U}(\mathfrak{X}_0)$, contrary to what was just established.

References

1. Stone, M. H., *Transactions of the American Mathematical Society*, 41 (1937), pp. 375–481, especially pp. 453–481.
2. Kakutani, Shizuo, *Annals of Mathematics*, (2) 42 (1941), pp. 994–1024, especially pp. 1004–1005.
3. Silov, G., *Comptes Rendus* (Doklady), Akademia Nauk, U.S.S.R., 22 (1939), pp. 7–10.
4. Čech, E., *Annals of Mathematics*, (2) 38 (1937), pp. 823–844.
5. Alexandroff, P., and H. Hopf, “Topologie I,” (Berlin, 1935), pp. 73–78.
6. Dieudonné, J., *Comptes Rendus* (Paris), 205 (1937), p. 593.
7. Weierstrass, Karl, *Mathematische Werke*, Band 3, Abhandlungen III, pp. 1–37, especially p. 5 (= Sitzungsberichte, Kon. Preussischen Akademie der Wissenschaften, July 9 and July 30, 1885).
8. Peter, F., and H. Weyl, *Mathematische Annalen*, 97 (1927), pp. 737–755, especially p. 753.
9. Wiener, Norbert, “The Fourier Integral,” (Cambridge, England, 1933) especially Chapter II, §14.

Hypatia of Alexandria

A. W. Richeson

National Mathematics Magazine
15 (1940/41), 74–82

Editors' Note: Today one can easily find a good deal of information about Hypatia, even a full-length biography, *Hypatia of Alexandria*, by Maria Dzielska (Harvard, 1995). Of course, little is known of her mathematical contributions even now since in most cases we have to rely on comments of others on works that are now lost. An attempt to tell what has come to light since the Richeson article is Michael A. B. Deakin's article "Hypatia and Her Mathematics" in the *American Mathematical Monthly*, 101 (1994), 234–243. Much recent work on Hypatia was done by the late Wilbur Knorr, a distinguished historian of mathematics at Stanford University. This work was not available to Richeson in 1940, of course. The article here is strikingly early in the literature in English devoted to Hypatia. So with this article *Mathematics Magazine* was publishing a groundbreaking work.

Richeson, a professor at the University of Maryland, received his PhD at Johns Hopkins under the direction of Frank Morley, best known for Morley's theorem in geometry, Morley was president of the American Mathematical Society, 1919–20. Richeson wrote on the history of mathematics and astronomy, and for the *Magazine* he contributed various articles including one on Laplace's work in pure mathematics.

The first woman mathematician regarding whom we have positive knowledge is the celebrated mathematician-philosopher Hypatia. The exact date of her birth is not known, but recent studies indicate that she was born about A. D. 370 in Alexandria. This would make her about 45 years of age at her death. Hypatia, it seems, was known by two different names, or at least by two different spellings of the same name; the one, Hypatia; the other, Hyptachia. According to Meyer,[1] there were two women with the same name living at about this time; Hypatia, the daughter of Theon of Alexandria; the other, the daughter of Erythrios. Hypatia's father was the well-known mathematician and astronomer Theon, a contemporary of Pappus, who lived at Alexandria during the reign of Emperor Theodosius I. Theon, the director of the Museum or University at Alexandria, is usually considered as a philosopher by his biographers.

Hypatia's biographers have given us but little of her early personal history. We know that she was reared in close touch with the Museum in Alexandria, and we are probably safe in assuming that she received the greater part of her early education from her father. If we are to judge from the records which the historians have left us, we would conclude that her early life was uneventful. It would seem that she spent the greater part of her time in study and reading with her father in the Museum.

Suidas[2] and Socrates,[3] as well as others who lived at the same time, lead us to believe that Hypatia possessed a body of rare beauty and grace. They attest not only to her beauty of form and coloring, but each and every one speaks just as highly of the beauty of her character. In the absence of a life painting of Hypatia we must depend upon the conception of others for a picture of the philosopher. In the introduction to his edition of Theon's Commentary[4] Halma has given us a short biography of Hypatia. On the title-page there is a medallion which gives his conception of the philosopher. Meyer feels that this drawing is unfortunate, as he does not believe it gives a true impression of the woman Hypatia. Charles Kingsley, on the other hand, in his novel *Hypatia* has written a

[1]Meyer, Wolfgang Alexander, *Hypatia von Alexandria*, Heidelberg, 1886, p. 52.

[2]Suidæ, *Lexicon, Lexicographi Graeci*, Vol. I, Pars IV, ed. Ada Adler, Lipsiæ, 1935.

[3]Socrates, *The Ecclesiastical History*, Trans. by Henry Bohn, London, 1853.

[4]*Theon d'Alexandrie, Commentaire sur le livre III de l'Almageste de Ptolemee*, ed. Halma, Paris, 1882.

Figure 1. Gasparo's portrait of Hypatia

vivid description of his impression of the philosopher.

If we are to believe the historians as to her beauty, we would expect that she was eagerly sought after in marriage. This apparently was the case: her suitors included not only outsiders, but many of her students as well. The question of her marriage, however, leads us to one of the controversial points of her life. Suidas states she was the wife of the philosopher Isidorus; then 25 lines later, he states she died a virgin. This apparent contradiction has been explained in several ways by later writers.

Toland[5] believes she was engaged to Isidorus before she was murdered, but was never married. Hoche[6] is of the opinion that the mistake arose from Suidas' abstract of the works of Damascius, a conclusion which Meyer does not believe to be true, pointing out that he found on the margin of one of Photius' works the statement, "Hypatia, Isidore uxor." Since Photius transcribed Hesychius' works, it is possible that the error arose in this manner. The evidence against such a marriage is further substantiated by the fact that Damascius states that Isidorus was married to a woman named Danna and had a child by this wife. Another fact which should be taken into consideration is that Proclus was much older than Isidorus: it has been pretty definitely established that Proclus was born about 412, and, since Hypatia's death occurred in the year 415, it would be impossible for Hypatia to have been the wife of Isidorus. The present writer is inclined to agree with Meyer that the mistake arose in Photius' transcription of Hesychius' work and that Hypatia was not married at any time in her life.

The second controversial point is the question of her death. In studying the statements made by many of the historians in regard to her death it seems desirable to review the murder in relation to the events which had happened previously. It is necessary for us to investigate not only Hypatia's relation to paganism, but also the relation between Cyril, the Christian bishop at Alexandria at this time, Orestes, the Roman Governor at Alexandria, and Hypatia. In view of this triangular relationship, we shall recall briefly some of the important events just prior to and during the episcopate of Cyril and their relationship to the authority of the Roman Governor.

On October 12, 412, Theophilus, the Bishop at Alexandria, died, and six days later his nephew Cyril was elevated to the episcopate of Alexandria. From the outset the new bishop began to enforce with zeal the edicts of Theodosius I, the Roman Emperor, against the pagans, along with restrictions which he himself promulgated against the Jews and unorthodox Christians. He further began to encroach upon the jurisdiction which belonged to the civil authorities; that is, to the Roman Governor. It must be remembered that the population of the city of Alexandria in the fourth and fifth centuries of the Christian era consisted of a conglomeration of nationalities, creeds, and opinions, and that nowhere in the Empire did the Romans find a city so difficult to rule as Alexandria. The people were quick-witted and quick-tempered, and we read of numerous clashes, street fights, and tumults, not only between the citizenry and the soldiers, but also between the different classes of citizens themselves. There were frequent riots between the Jews and the Christians on the one hand and the pagans and the Christians on the other. The Christian population did little or nothing to quiet these people, but even added one more controversial topic for them to quarrel about. Consequently we find that the edicts and promulgations of Cyril not only caused strife among the people but aroused the anger of the Roman Governor, Orestes, the one person who stood in the way of the complete usurpation of the civil authority by Bishop Cyril. Friction continued between these two until there was a definite break in their relations.

Because of her intimacy with Orestes, many of the Christians charged that Hypatia was to blame, at least in part, for the lack of a reconciliation between Orestes and Cyril. Socrates states that some of

---

[5]Toland, John, *Tetradymus*, London, 1720, pp. 101–136.

[6]Hoche, Richard, *Hypatia die Tochter Theons*, Philologus, Fünfzehnter Jahrgang, Göttingen, 1869, pp. 435–474.

them, whose ringleader was named Peter, a reader, driven on by a fierce and bigoted zeal, entered into a conspiracy against her. They followed her as she was returning home, dragged her from her carriage, and carried her to the church Caesareum, where they stripped her and then murdered her with shells. They tore her body to pieces, took the mangled limbs to a place called Cinaron, and burned them with rice straws. This brutal murder happened, he says, under the tenth consulate of Honorius and the sixth of Theodosius in the month of March during Lent, so that the year of her death may be set as 415.

Socrates' report of Hypatia's death is corroborated not only by Suidas, but also by other historians such as Callistus,[7] the ecclesiastical historian, Philistorgus,[8] Hesychius[9] the Illustrious, and Malalus.[10] Damascius says that Cyril had vowed Hypatia's destruction, while Hesychius states that his envy was caused by her extraordinary wisdom and skill in astronomy. Damascius also relates that at one time Cyril, passing by the house of Hypatia, saw a great multitude, both men and women, some coming, some going, while others stayed. When he was told that this was Hypatia's house and that the purpose of the crowd of persons was to pay their respects to her, he vowed her destruction.

When we compare these statements, it would seem that Hypatia's death, or at least the occasion of it, was due to her friendship with Orestes. This friendship enraged the Christian populace because they felt that she prevented a reconciliation between Cyril and Orestes. We are also led to believe that the more sober-minded of the Christians yearned for a reconciliation between these two and that no doubt her death was ordered by Cyril.

Among the later writers on the subject there is a divergence of opinion. Toland lays the death of Hypatia directly at the feet of Cyril. Wolf,[11] on the other hand, is inclined to believe that Cyril knew beforehand that the murder was being plotted but did nothing to prevent it. As to the causes of the murder, Wolf mentions her belief in paganism and her teaching of Neoplatonism, along with the practice of treating the mentally diseased with music, all of which might be considered as coming under the pale of the edicts of Theodosius I regarding pagan worship.

The present writer is inclined to follow Meyer part of the way in the interpretation of these events; that is, Hypatia was used as a sacrifice for a political or personal vengeance, possibly a political vengeance. Cyril and Orestes were at odds; both had made various reports to the Emperor, each one attempting to show that his actions were justified. On the other hand, Orestes was the one person who stood in the way of the complete assumption of the civil power by Cyril, and naturally Cyril was eager to use every incident which would embarrass Orestes. In the case of Hypatia's death it would seem that its underlying cause was not so much a struggle for the assumption of the civil authority, but rather a struggle of the Christian church against the pagan society of Alexandria. It must be remembered that although Orestes professed Christianity, the fact still remained that his profession was more one of policy than of faith. In all justice it would certainly seem that Cyril should be held at least indirectly responsible for her death. Certainly he could have prevented the mob's violence, if he had made the slightest effort.

Meyer feels the relation between Cyril and Synesius should be considered in investigating Hypatia's death. He is of the opinion that possibly there was an old difference between these two, and that her death was brought about by Cyril in order to settle this difference with Synesius. Meyer bases his conclusions on the contents of Epistle 12[12] of Synesius, in which he exhorts Cyril to go back to the Mother Church, from which he had been separated for a period of time for the expiation of sin. The present writer is of the opinion that Meyer has no justification for this assumption. Although we do not know the exact date of Synesius' death, it was probably between 412 and 414, and it must be remembered Cyril was not raised to the bishopric until late in the year 412. It is very probable that Epistle 12 was written before Cyril was made Bishop at Alexandria, though as a matter of fact we have no convincing evidence that the letter was written to Saint Cyril. Furthermore, there is no evidence to support the belief there ever existed any difference between Cyril and Synesius.

It has been stated above that little is known concerning Hypatia's early life. Consequently there is

[7] *Nicephori Kallisti historia ecclesiastica Migne, Patrologiae Graecae*, Tome 147, Paris, 1856.

[8] *Ex ecclesiastici Philostorgii historia epitome confecta a Photio patriarcha*, H. Valesio interprete, Parisis, 1873.

[9] *Hesyychii Milesii Onomatologie que supersunt cum prolegomenis*, ed. J. Flacch, Lipsiae, 1882.

[10] Malalae, Johannus, *Chronographia ex recensione Ludovici Dindorfii*, Bonnae, 1831.

[11] Wolf, Stephan, *Hypatia, die Philosophin von Alexandrien*, Vienna, 1879.

[12] Synesii, *Opera quae extant omnia, Patrologiæ, Graecæ*, Tomus LXVI, Paris, 1864.

little on which to base our conclusions regarding her early education. It goes without saying that her father taught her in mathematics, astronomy, and science. Beyond this we do not know who her teachers were, but we may rest assured that, with an intellect as fertile as hers, she was not long satisfied with the narrow training in mathematics and astronomy. In order to understand the possible trend of her education it is necessary to take a look at the working of the Museum at Alexandria. The Museum had its origin in the efforts of Ptolemy Soter about 300 B. C., when he brought to the city of Alexandria all the philosophers and writers it was possible for him to obtain. To these he gave every encouragement possible, not only financial aid, but also in books and manuscripts from Greece. The later rulers of Egypt continued their support until the country came under Roman authority in 30 B. C. This ended the first period of intellectual activity, which is characterized as purely literary and scientific in nature. With the conquest of the country by the Romans, intellectual activity was again in the ascendency and Roman, Greek, and Jewish scholars were again attracted to the city. This second school of thought was somewhat different from the first. We have an intermingling of nationalities with their varying philosophies and personalities all of which developed into the speculative philosophy of the Neoplatonist, the religious philosophy of the early Christian fathers, and the gnosticism of the Oriental philosophers. This second period of intellectual activity continued until about 642, when the city was destroyed by the Arabs. Considered as a whole, the Alexandrian School stood for learning and cosmopolitanism, for erudition rather than originality, and for a marked interest in all literary and scientific techniques. It was at the Museum that these philosophers, writers, and scientists gathered to lecture to their students and to converse with one another. Theon, Hypatia's father was director or fellow in the Museum, and it is reasonable to infer that Hypatia came into close contact with the leading educators and philosophers of Alexandria.

The question is frequently asked whether or not Hypatia studied at Athens. Here again we come to a point which has not been definitely decided. Suidas says she obtained part of her education there, or at least the passage has been so interpreted, for both Meyer and Hoche are of the opinion that Suidas has been misinterpreted on this point. Wolf states that Hypatia studied at Athens under Plutarch but Meyer again points out that this was highly improbable, as at the time Plutarch was lecturing at Athens, Hypatia was probably 30 years of age and was herself lecturing at Alexandria. Suidas also makes mention of the fact that she studied under another philosopher at Alexandria, but he does not identify this philosopher except to say that it was not Theon. Meyer thinks it might have been Plotinus. Regardless of how or where she received her education, we do know that she received a thorough training in arts, literature, science, and philosophy under the most competent teachers of the time.

It was with this training that she succeeded to the leadership of the Neoplatonic School at Alexandria. The exact date at which she assumed control of the school is not known, but Suidas informs us that she flourished under Arcadius, who was Emperor of the Eastern Roman Empire from 395 to 408. We are naturally led to ask two questions regarding her teaching: first, what was her ability as a teacher? second, what was the nature of her teaching? The first question is much simpler than the second, although there are sufficient facts relating to the nature of her teaching to enable us to draw a fairly definite conclusion.

All the contemporary and later writers on this period testify to the high reputation of her work as a teacher. Each one attributes an extraordinary eloquence and an agreeable discourse to her lectures. Suidas speaks highly of her teaching methods, while Synesius in one letter praises her voice and in another mentions that her philosophy was carried to other lands. Socrates and Philistorgius tell us that not only the Egyptians, but students from other quarters of Europe, Asia, and Africa came to her classes until there was in reality a friendly traffic in intellectual subjects. Suidas states that, on account of her ability as a teacher and her personality, Orestes sought out her house to be trained in the art of public manners. Damascius states she far surpassed Isidorus as a philosopher, and it should be remembered that Damascius was a friend and pupil of Isidorus.

Among her disciples there are many well-known men other than Synesius. The names of these include Troillius, the teacher of the ecclesiastical historian Socrates, Euoptius, the brother of Synesius and probably the Bishop of Tolemais after the death of Synesius, Herculianus, Olympius, Hesychius, and finally Herocles the successor of Hypatia in the Platonic School at Alexandria.

From her teaching position she expounded the philosophy of the Neoplatonic School and her fame rests primarily upon the manner in which she conducted this school. In her teaching she no doubt lectured not only on philosophy as we know it today, but also included the scientific subjects of mathematics, astronomy, and the subject of physics as known at the time. She was apparently well versed in astronomy, since Suidas tells us that she excelled her father in this field. We may also assume that she taught the rudiments of mechanics, since there is a reference in one of Synesius' letters to an astrolabe which she constructed, and in another letter Synesius requests Hypatia to make a hydroscope for him.

Neoplatonism, as a philosophic system of thought, had its inception during the second century of the Christian era. It was built up from the remains of many of the systems of philosophies of ancient Greece and became a religion for many of the heathens, who could no longer believe in the old gods of Olympus. The Neoplatonist believed in a supreme being or power, which was the Absolute or One of the system. This supreme power was mystic, remote, and unapproachable in a direct fashion by finite beings. Hence there existed between man and the Absolute lesser gods or agencies. The first in this series was Nous or Thought, which was emanated by the Absolute as an image of itself. Below Nous there existed the triad of Souls, which pervaded all of the material universe, and all of those beings with which it is peopled are a direct emanation from the triad of Souls. Matter or material things were thought of as belonging to an evil category, while the triad of Souls belonged to a pure category. Man, a mixture of the material and the spiritual, has the power by indulging in self-discipline and subjugation of the senses, to lift himself to a level where he may receive from the Absolute a revelation of divine realities. Once man has caught a glimpse of this vision, he is able to free himself entirely from the thralldom of matter.

It should be noted that the development was from a higher to a lower or descending series. Since each series participated in the one above it, there was also a turning back, where the soul by an ascending process was able to return to the Absolute. The object of life, when the soul was perfectly free, was to rise by the practice of virtue from the category of matter to the higher category of intelligible realities. There were purifying virtues, which disciplined the soul till it became capable of union with the Absolute.

We have no writings of Hypatia, but we may rest assured that she at least subscribed to the general principles of Neoplatonism. Plotinus' works show that he succeeded in contempt of bodily cares and needs, and we find the same thing to be true with Hypatia. No doubt Hypatia's use of logic, mathematics, and the exact sciences gave her a discipline which kept her and her pupils from going too far in the superstitions and speculations of some members of this group of thinkers. Synesius in his speech before the Arcadians, acknowledges the purely subjective character of the different attributes which are conceived of by man as belonging to the divine nature. He also felt a wholesome reticence in his attempts to reach towards the Incomprehensible. He believed in the Trinity of Plotinus, but did not assign to the World-soul the creating or animating of the entire universe. He thought occasional supernatural communications between God and the human soul were possible, and he also believed that man was able to purify his soul to such an extent that he would be able to elevate the imagination to a point where it would be possible for him to share in the ecstacy of the upper light. He believed that the final goal aimed at in life was a pure and tranquil state of mind, undistracted by fierce passions, gross appetites, or the demands of worldly affairs. It would be reasonable to assume that these tenets of Synesius' faith were inculcated in him by his beloved teacher Hypatia.

In considering the writings of Hypatia we have but little information to fall back on. Suidas is the only historian to give us any information concerning her writings. He gives us the names of three: a commentary on the *Arithmetica* of Diophantus of Alexandria, a commentary on the *Conics* of Apollonius of Pergassus, and a commentary on the *Astronomical Canon* of Ptolemy. None of these are extant at the present time.

We are naturally led to the question why Hypatia, a student of philosophy, a teacher of renown, and the leader of the Neoplatonic School at Alexandria, left only three works and those three purely mathematical or astronomical. The answer is probably that Suidas quoted the writings of Hypatia as given by Hesychius, who for some reason gives an account only of the Astro-Mathematical works of Hypatia. It is rather difficult for us to believe that with approximately twenty years of teaching she would produce not more than three works, and those three commentaries. So we are led to the conclusion that Hypatia did leave other writings, which were probably lost in the destruction of the library at Alexandria, and that these works were principally philosophic in nature. It is true that both Halma and Montucla[13] make

[13]Montucla, J. F., *Histoire des Mathematiques*, Tome I, Part I, Liv. V, Paris, 1799.

mention of other works of Hypatia; Halma in particular says she left behind "beaucoup d'ecrits." At the present time it is impossible to determine from what source Halma obtained this information, and it is more than probable this is only a conjecture on his part.

With the passing of Hypatia we have no other woman mathematician of importance until late in the Middle Ages. Although we have no definite information to indicate that she exerted any great influence on the development of mathematics or science in general, nevertheless she certainly passed on to her scholars and followers a discipline and restraint which were carried over to a later period. It is possible that the effects of her teachings have been lost sight of, since any works she might have left behind were certainly lost when the Arabs destroyed the Library at Alexandria in 640.

Gauss and the Early Development of Algebraic Numbers

E. T. Bell

National Mathematics Magazine
18 (1943/44), 188–204, 219–33

Editors' Note: Eric Temple Bell is largely known today for his popular works—*Men of Mathematics, Mathematics: Queen and Servant of Science, The Last Problem, Numerology, The Search for Truth*, among others–and perhaps even for his science fiction, published under the pseudonym of John Taine. He was nevertheless a serious mathematician. Constance Reid, his biographer, notes that Bell's bibliography, including his mathematical papers, general articles, and reviews, runs to over three hundred items.

Bell took his PhD at Columbia, after receiving a master's at the University of Washington. He spent only twelve months at Columbia working on his doctorate and wrote a dissertation on cyclotomy, extending some work of Eisenstein. The identity of his dissertation advisor at Columbia remains a mystery, even after extensive inquiries and interviews with those who knew Bell. The best conjecture is that his advisor was Frank Nelson Cole, the number theorist at Columbia for whom the Cole Prize of the American Mathematical Society is named. Others conjecture that it was the department head, C. J. Keyser. Unfortunately the dissertation was not signed by an advisor and documents at Columbia have the relevant lines left blank. It's one of many "mysteries" in Bell's life—one still remaining after the extraordinary sleuthing done by Reid when she wrote his biography (*The Search for E. T. Bell*, MAA, 1993).

In 1940 Bell wrote a more scholarly and detailed history than *Men of Mathematics*, his *Development of Mathematics*, still a valuable reference. The present article on algebraic numbers may have arisen from some of the work he did for this volume. In the vocabulary of working mathematicians in combinatorics, Bell's name survives in the study of Bell numbers and Bell polynomials, concepts related to Stirling numbers and partition polynomials.

1. From arithmetic to abstract algebra

An unexpected turn in twentieth-century mathematics was the abrupt change in the motivation and objectives of algebra. The change became evident by 1925 at the latest, and in about ten years made some of the algebra of the nineteenth and early twentieth centuries seem rococo and strangely antiquated to algebraists of the younger generation.

The transition from individually developed theories, overloaded with masses of intricate theorems—often the seemingly fortuitous outcome of elaborate calculations carried through with consummate manipulative skill—to the deliberate search for unifying abstract principles was sudden. It did not occur without preparation, of course; but the passage from calculation to preoccupation with fundamental concepts was accomplished within a decade.

Without the vast accumulations of special results, like the data in Darwin's notebooks which suggested the theory of evolution, the problems and methods of abstract algebra might never have emerged. However that may be, interest in the algoristic type of algebra declined rapidly after 1921. Theorems that had been obtained primarily by modes of calculation germane to the particular theory in which the theorems originated, were seen to be instances of underlying structures independent of the particular theory. To cite but one striking example, the Jordan-Hölder theorem and its refinements became more intuitive when analyzed structurally than they had seemed in their traditional settings.

Actually the change might have happened earlier than it did. Hilbert's basis theorem [1] of 1890, in both its statement and its proof, was plainly in an order of ideas different from that of the algebra of its epoch. Yet earlier, in 1877, Dedekind clearly formulated the strategy of abstract algebra, in a statement [2] of the methodology which he had applied in developing his theory of ideals. Having noted that his initial attempt at a general theory of algebraic numbers could, conceivably, be successfully completed, but only (he believed) by overcoming serious obstacles in calculation, Dedekind continued as follows:

"But even if a theory [does not encounter great difficulties when developed algoristically], it seems to me that such a theory, based on calculation, still does not offer the highest degree of perfection. As in the modern theory of functions, it is preferable to seek to extract proofs, not from calculation, but immediately from characteristic fundamental concepts, and to construct the theory in such a manner that, on the contrary, it shall be in a position to predict the results of calculation."

That declaration of intention might well stand as the party manifesto of the modern abstract algebraists. Dedekind himself gave numerous and brilliant examples of his grand strategy. A typical simple specimen, which has passed unaltered into current usage, is his postulational definition of an ideal, and there are many others.

As one of the matters which Dedekind had in mind while distinguishing between mathematics and calculation will appear in the sequel, its nature may be indicated here. It was the theory of composition of forms (homogeneous polynomials) and its outgrowth in that chapter of the classical theory of algebraic number ideals which is devoted to factorable forms of degree n in n indeterminates. Of great historical interest on account of the famous mathematicians who observed some aspect of the whole, but who could not possible have given anything approaching an adequate discussion with the mathematics of their respective eras, the theory of composition still retains its mathematical interest in problems yet unsolved. Its particular significance here, however, is that it was partly responsible for Dedekind's abandonment of calculation in favor of the abstract approach. From Diophantus (c. 250 A. D.) to Brahmegupta (7th century) to Leonardo of Pisa (13th century), to Euler (18th century), to Lagrange (18th century), to Gauss (1777–1855), to Kummer (1810–1893), scattered instances of composition kept foreshadowing something beyond, till Dedekind (1831–1916), rejecting the generalization suggested by Gauss' development for binary quadratic forms and effected by Kummer for cyclotomic fields, abandoned calculation and extracted his proofs "immediately from characteristic fundamental concepts."

To prevent possible misconceptions, it may be observed that few algebraists imagine the current abstract phase to be the climax of algebra. Many things are now much clearer and simpler than they seemed a generation ago, and much has been exposed that was unsuspected then. But the simplicity and clarity frequently refer to theories first worked out by the hard way of calculation from basic discoveries made almost empirically in the course of routine labor. The same kind of drudgery on the higher level of abstractions has been in part responsible for some of the recent accessions; and a single new observation at any time may start another cycle of calculation, discovery, abstraction, calculation. Like the rest of mathematics, algebra needs an occasional transfusion of fresh ideas to keep alive. Nor can algebra go on living on its past forever.

Anyone following the historical evolution of abstract algebra much be impressed by the frequency with which the development was influenced by concepts whose germs, at least, appeared first in the theory of algebraic number fields and their rings of integers. Intensive scrutiny of classical arithmetical ideas, such as integer, unit, prime, divisibility, equivalence relation, domain of integrity, ring, field, module, residue class, ideal, dual group, and many more, suggested abstractions relevant for algebra as a whole.

In the process of abstraction, precise axiomatic technique applied to such arithmetical concepts yielded new ideas, refinements of the concepts from which they may first have been abstracted, and these in turn enriched arithmetic with thitherto unsuspected subtleties. A case in point was the introduction of primary ideals in the theory of commutative rings, which revitalized and amplified a theory whose rudimentary form is as old as Euclid. Another was the outgrowths of Dedekind's uses of mapping and correspondences in a discrete domain, that of algebraic numbers. [3]

Granted that the theory of algebraic numbers has been a suggestive lead in the development of abstract algebra, it may be of interest to see how much of subsequent progress was adumbrated in the work of Gauss, universally acknowledged as the initiator of the theory. To appreciate in some slight degree the novelty and the magnitude of Gauss' achievement in introducing complex (Gaussian) integers into arithmetic, it will be advantageous first to consider briefly the nature of the general problem of which the one that he solved is a very special case.

2. The nature of Gauss' problem

The theory of algebraic numbers is a natural but by no means immediate extension and generalization of rational arithmetic. The primary concern of rational arithmetic is the study of relations between the rational integers.

In any generalization of rational arithmetic, certain fundamentals must be agreed upon before the generalization can significantly be called arithmetic.

In particular, what phenomena of rational arithmetic are to persist in the generalization? The requirement that the generalization, when suitably specialized, shall yield at least part of the classical theory of numbers for rational integers, has been admitted in all generalized arithmetics thus far constructed.

The remaining demands, either explicitly or by tacit assumption, refer to the multiplicative department of arithmetic. The more important of these, in the historical development, appear to have been the distinction between integral and non-integral elements; prime integral elements; unit integral elements; or, concisely, integers, primes, units.

It has not always been immediately evident how the integers, primes, and units were to be selected or defined in some species of algebra for which an arithmetic was desired. The instance of quaternions may be recalled. Generally, the distinction between algebra and arithmetic has been roughly summarized in the dictum that division is only exceptionally impossible in algebra and only exceptionally possible in arithmetic. The quotient of two complex numbers, for example, is a complex number; the quotient of one algebraic integer by another is an algebraic integer only if the second is a divisor of the first. If the elements or generalized "numbers" concerned form a ring with respect to suitably defined operations of "addition" and "multiplication," the integers may be isolated by a system of postulates embodying those characteristics of rational integers which it is desired to preserve. If the addition and multiplication of integral elements do not satisfy all the postulates for the similarly named operations in the ring of rational integers, there is as yet no common agreement on the postulates for integers. Different authors have favored different definitions.

Incidentally, this illustrates the distinction in mathematics between naming a thing and defining it. In the early development of algebraic numbers, the integers were named, but either not defined or defined only inferentially. Explicit definition was not insisted upon until the twentieth century, and then not always. The desirability of a postulational definition of algebraic integers in the first great generalization of rational arithmetic was overlooked, possibly on account of the relative simplicity and familiarity of the material.

With the analysis of divisibility the necessity for precision became apparent. The distinction between primes and indecomposable or irreducible integers was first sharply recognized by Dirichlet [4], about 1843, although probably Gauss had noticed it long before he officially introduced [5] (1831) his complex integers $a+bi$, a, b rational integers, into arithmetic. The distinction being of no significance for rational arithmetic, it proved singularly elusive. Actually it slipped through the grasp of mathematicians as powerful as Lamé [55] and Cauchy [57].

With primes satisfactorily defined, the next capital question was that of units. The definitions of primes and units are interdependent, and this may have concealed the more profitable approach at first. The "natural" definition that a unit is an integer which divides (arithmetically) every integer, may have been easy to guess. But this natural definition is comparatively sterile. Its more productive equivalent in terms of norms, with which Gauss as early as 1808 was acquainted in special cases, can scarcely have been obvious.

There would have been but little point in defining primes and units unless they were to provide a basis for unique factorization. Fortunately for Gauss, in applications of his complex integers to cubic and biquadratic residues [6], his primes satisfied this demand: up to unit factors a Gaussian integer is uniquely the product of a finite number of powers of primes in the integral domain defined by $i^2 + 1 = 0$, and likewise for the domain defined by $\rho^2 + \rho + 1 = 0$. This relative unicity however did not wholly satisfy the exacting Gauss. To appease his arithmetical conscience, Gauss insisted on absolute uniqueness, and attained it for Gaussian integers by a proper classification of associates [7].

In all of his work on algebraic numbers thus far recovered, Gauss relied on the existence of the Euclidean algorithm, for the greatest common divisor of two integers, to give him his unique factorization proofs. The Euclidean algorithm is a sufficient but not necessary condition for unique factorization. Whether Gauss was aware of this distinction seems not to be known.

Progress in the theory of algebraic numbers has gone so far since Gauss took the first bold steps into "a boundless new domain of arithmetic", that the foregoing remarks on integers and the rest may strike a beginner today as platitudes and truisms. Possibly they are; but if so, it should be an easy exercise to construct an acceptable arithmetic for a general skew ring, or to show that there is none.

3. The central difficulty

The anticipations of non-Euclidean geometry and elliptic functions by Gauss are as well established as they are well known. His actual contributions to the early development of algebraic numbers are readily accessible and equally well known. But, so far as seems to have been discovered, Gauss left no such illuminating record of his thoughts on algebraic numbers as he did for elliptic functions and non-Euclidean geometry. Nor, when Kummer [8] in 1845 restored the fundamental theorem of arithmetic (finite and unique factorization into primes) to his cyclotomic fields by the invention of ideal numbers, did Gauss give any sign that he may have been blocked by the central difficulty—the breakdown of the fundamental theorem—in his own researches of a quarter-century earlier.

It would be of interest to know whether Gauss did encounter that apparently insurmountable obstacle—Kummer [9] seems at first (1844) to have despaired of surmounting it—and whether, if he did, he hoped it might be obviated. From what has survived of Gauss' early work on algebraic numbers, it is possible that he recognized the main difficulty but foresaw no possibility of circumventing it. Certainly any way that he may have tried—to judge by what he himself published and the posthumous fragments and correspondence whose publication he could not have foreseen—must have been quite different from Dedekind's ideals. And for reasons that will appear in connection with Galois imaginaries, it is unlikely that Gauss would have recognized Kummer's ideal numbers as legitimate arithmetical concepts, even if they had occurred to him.

Gauss might, conceivably, have started from the higher theory of congruences, which was to have formed the eighth section of the *Disquisitiones Arithmeticæ*. Dedekind [2] at one time believed the end attainable by this approach, but after much thought decided (erroneously) that this was not a feasible way. It was not until the twentieth century that a usable theory [59] in this general direction was constructed. It is not likely that Gauss could have progressed so far, although he does throw out hints for overcoming some of the difficulties he leaves unclarified in his theory of congruences to a double modulus [48]. Still less likely is it that he ever imagined anything like Hensel's theory of p-adic numbers, which provides a theoretical means for finding the discriminant of an algebraic field, but which is difficult to apply. It is scarcely enough in the theory of numbers to prove that a problem can be solved in a finite number of steps, if the time required for taking the necessary steps seriously encroaches on eternity.

That Gauss was aware of at least some of the pitfalls in algebraic number fields as early as 1816 appears—though somewhat cryptically—in a letter (to be quoted later) of that year. Indeed, as will be seen presently, Euler in 1770 had observed one of the apparent anomalies in the behavior of algebraic numbers, in a work with which Gauss was familiar almost from his boyhood. This alone would have sufficed as a signal to proceed with caution.

Some of Gauss' earliest work in arithmetic and algebra can now be re-read in the light of its much later significance. As examples of unconscious foresight, the relevant matters are among the most interesting in the history of nineteenth-century mathematics. These will be noted next.

4. Shapes of things to come

Gauss (1777–1855) began the *Disquisitiones Arithmeticæ* early in 1795. It went to press in April, 1798, and was published in September, 1801. During the printing Gauss made several corrections and additions.

Although the work contains nothing on the theory of algebraic numbers in the current technical sense, it is rich in ideas and theorems that were to prove suggestive to the creators of the theory. The seventh section, on cyclotomy, concerned with the roots of binomial equations though it is, is no exception. The algebraic irrationalities appearing are not recognized as arithmetical beings worth or capable of sharing the austere purity of the rational integers. "The higher arithmetic" for Gauss in 1801 meant a theory of the rational integers and nothing more [10]. Rational fractions were tolerated on occasion as removable blemishes. If Gauss in his private thoughts speculated on the possibility of admitting other "numbers" to the select society of the rational integers, he permitted no hint of what he was thinking to escape into print.

A few items from the many in the *Disquisitiones* and other early works, that were to acquire a broadened meaning when algebraic numbers were recognized as legitimate material for arithmetic, may be recalled as specially significant. There is no question here of anticipations. Gauss never hinted that he had foreseen the potentially wider scope of his early work; nor has any mathematical critic, however partisan in other judgments, asserted that he did.

First in order and second to none in importance, is the proof (Art. 13) of the fundamental theorem of arithmetic—finite and unique decomposition into primes. Until Gauss proved this theorem, it either had been overlooked or taken for granted as obvious. Gauss gave a careful proof, possibly because his mathematical ethics were as rigid as those of Dedekind, who stated [11] the precept that "whatever can be proved in mathematics should not be believed without proof."

The materials for a proof were available to anyone after the late sixteenth century. Euclid [12] had taken the most difficult step, when he proved that if a prime divides neither of two integers, it does not divide their product. This also is obvious in the same deceptive sense that the fundamental theorem is obvious. The snare is in assuming that "prime" and "irreducible" or "indecomposable" necessarily have identical connotations. Gauss saw that the fundamental theorem requires proof and can be proved. His acuity in this detail alone put him in a class above his predecessors and contemporaries in arithmetic. He seems to have been quite irritated (*ibid.* p. 9) by their vagueness in this and other fundamental matters.

Next there is the extraordinarily prolific concept of congruence, with its equally fruitful notation suggesting valid analogies with algebraic equations. This immediately gave arithmetic a new direction. It later (1847) did the like for algebra, when Cauchy [13]"realized" imaginaries by referring their induced real algebra to congruences modulo $i^2 + 1$, a tactic which Kronecker adopted and applied to algebraic number fields in general.

The relation between congruences and norms, though not mentioned in the *Disquisitiones*, possibly because it is almost trivial for rational arithmetic, was exploited by Gauss in connection with complete residue systems for Gaussian integers, specifically in his discussion of biquadratic residues [7]. This led others at the opportune time to the definition of the norm of an ideal. In retrospect, the transition from real to complex residue systems may appear to have been simple, and the means for effecting it natural. It would be interesting to know how much thought it cost Gauss. Whatever the effort, Gauss usually found the straight road to the future.

The ideas of mappings and correspondences, so prominent in modern algebra, entered abstract algebra by way of algebra itself rather than through arithmetic. Certainly homomorphism is nowhere recognized in Gauss' published writings as an independent concept, although instances of it are fully developed in the *Disquisitiones* and elsewhere in the theory of congruences. Dedekind's uses of mapping in his theory [3] of algebraic numbers (1871) were probably inspired by the Galois theory of equations, which he was one of the first to master, rather than by his critical study of the arithmetical works of Gauss, Dirichlet, and Kummer.

If it is mandatory in a sketch like the present to exhume a more or less ancient precedent for every mathematical principle mentioned and now in common use, this may easily be done for homomorphism. Both congruence and homomorphism, as usable abstractions, were commonplaces to the Pythagoreans of the sixth century B. C. Those ingenious numerologists remarked that all numbers, and therefore all things, are contained in the decad. This grand principle of mapping the entire universe on a double handful of positive integers was taken over by Plato and bodily incorporated into his peculiar substitute for science. It ultimately matured in his ideal numbers—not to be mistaken for Kummer's, though there are reasons for suspecting that Kummer had heard of Plato's celestial arithmetic. And if it is desirable to push recognition yet further back, it is sufficient to recall that the Pythagoreans most likely got their decad from the finger-counting of savages.

In another direction, an isolated result in the *Disquisitiones* (Art. 42) is the extremely useful theorem now frequently called the "Gauss Lemma"—not to be confused with Gauss' Lemma for quadratic residues. It is to the effect that if A, B are polynomials in the indeterminate x, each with leading coefficient unity and the remaining coefficients rational numbers but not all integral, the coefficients in the distributed product AB are not all rational integers. An immediate corollary of this is more frequently used. Gauss in his Diary [14] (Item 69) dates the theorem 23 July 1797. The extension to algebraic numbers as coefficients was not far to seek, and Kronecker made powerful use of it in his theory of integral functions.

Another application of the extended Gauss Lemma has become classic in the theory of algebraic number ideals. The basic theorem that, if the ideal α divides the ideal γ, then there is an ideal β such that $\gamma = \alpha\beta$, or, what is equivalent, that for any ideal δ there is an ideal δ' such that $\delta\delta'$ is a principal ideal, cost Dedekind in the late 1860's some effort to prove. In his simplified presentation of parts of Dedekind's theory, Hurwitz [15] obtained (1894) a painless proof based on the extended Lemma.

On a more abstract level, Gauss understood and freely applied the concept of indeterminates. In fact he is usually, but not always by those who have not dipped into his works, regarded as the originator of the concept; certainly nobody preceded him in this. Indeterminates permeate the theory of quadratic forms in the *Disqisitiones*, also the unpublished general theory (1797–8) of congruences [16]. Thus the form $ax^2+2bxy+cy^2$, for example, is symbolized (Art. 153) as (a,b,c); and binary forms are called "functions of two indeterminates x, y." The meaning of the term is carefully explained.

A perhaps more subtle type of use for indeterminates appears in Gauss' second proof (1815) for the so-called fundamental theory of algebra [17]. The critical detail of a "purely analytical" proof occurred to Gauss (Diary, 29 February 1812) in November, 1811, in connection apparently, with his long struggle to complete the proof. Indeterminates are used as such in a proof (Art. 5) of the algebraic independence of the elementary symmetric functions. Here also (Art. 4) is the useful device of ordering the terms of a polynomial in any number of indeterminates lexicographically. Possibly the inner significance of indeterminates was not generally appreciated till Kronecker, in his arithmetical and algebraic philosophy, elevated them almost to the status of a religion.

Binary quadratic forms were to contribute to the development of algebraic numbers in another and possibly more tangible respect. Dirichlet [18] observed (1840) that the Gaussian theory of binary quadratic forms is properly a topic in multiplicative arithmetic. The full significance of this observation was perceived only when Dedekind [3] (1871) established a correspondence between binary quadratic forms of a given discriminant and the ideals of a quadratic field having the same discriminant, the composition of forms corresponding to the multiplication of ideals. The connection between the Gaussian class number, the class numbers for the relevant domains, and unique factorization or the lack of it for the integers concerned, then became apparent, as did also the reason for the existence of theorems giving the number of representations of rational integers in certain quadratic forms as closed formulas in terms of the divisors of the integers represented. The root of the connection was concealed in the Gaussian theory of composition for binary quadratic forms, and the composition of classes, orders, and genera of such forms.

Also in that theory occurs one of the earliest nontrivial arithmetical instances of an Abelian group: composition is the "multiplication" of the group. In his own terminology, Gauss discussed some features of the structure of the group of classes. Naturally he did not communicate his truly profound discoveries in the technical language of groups; for groups had not been officially invented in 1801, however numerous may have been almost trivial occurrences of them in primitive art no less than in elementary mathematics from Euclid to Gauss himself.

A somewhat labored attempt [19] by Poincaré (1885) to extend the Gauss-Dirichlet-Dedekind theory of binary quadratic forms to binary forms of degree higher than the second was only moderately successful, the substitute in terms of ideals for the original rational problem being at least as difficult as the original problem itself. If Gauss (or anyone else) had succeeded in constructing a usable theory of composition for the forms in question, some degree of practical success might have been anticipated. But Gauss never tried, and the problem is still open in all its unnaturalness. Although it may be difficult to characterize an artificial problem, it is generally agreed that Gauss seldom attacked one.

The significance of laws of reciprocity in the early development of algebraic numbers will be noted in another connection. For the moment it is sufficient to recall that the *Disquisitiones* (Arts. 135–45) contains the first complete proof of the Euler-Legendre law of quadratic reciprocity. Gauss records his success in obtaining a proof, after many trials, in his Diary (Item 2), under the date 8 April 1796. The entry for 29 April 1796 announces the generalization of the law rediscovered and published by Jacobi in 1837.

The fundamental theorem of algebra, already mentioned, also received its first acceptable proof (1799) at the hands of Gauss in his doctoral dissertation [20]. Although the theorem in its classical form, as stated and proved by Gauss, is no longer as important in algebra as it was, being replaced in the abstract theory by a purely algebraic equivalent that is almost a truism, the arithmetic of algebraic numbers could hardly have begun without it. Of the four proofs which Gauss devised, the second [21] was to become the most suggestive, especially for Kronecker's development of modular systems. The proof refers the existence of a root to an algebraic identity. From this, and the existence of a real root for an equation of odd degree with real coefficients, the theorem follows, provided certain assumptions of continuity be admitted.

As Gauss himself remarked, he had been anticipated by Euler and de Foncenex in the ingenious algebraic device underlying the proof; so possibly the credit for its suggestiveness should be transferred to them. It is interesting to recall that Clifford [22] (1876) independently outlined a somewhat similar proof. The theorem itself, as Gauss understood it, however proved, is now assigned to analysis rather than to the mathematics of the discrete, and its use in algebra is avoided.

Other items of importance for the development of algebraic numbers are the proof (*Disquisitiones*, Art. 341) of the irreducibility of $(x^n - 1)/(x - 1)$, n prime, and the like for the equation whose roots are the primitive mth roots of unity, m any positive integer. The last was obtained in 1808 but was not published. Its proof presupposes a knowledge—in a special instance—of the Dedekind inversion formula in numerical functions. A similar instance occurs in Gauss' theory (to be noted later) of what are now called Galois fields, generated by the residue systems of double moduli.

The preceding selection of items is by no means exhaustive of those in Gauss' early work which acquired a new significance with the advent of algebraic integers. But they are probably sufficient to indicate that Gauss' thought was in the main current of progress in algebra during the first four decades of the nineteenth century, although he himself may have been unaware of the particular direction progress was to follow.

5. Recognition of algebraic integers

It is no disparagement of Gauss' epochal achievement in introducing algebraic integers into arithmetic, to state the historical fact that there was no shadow of an attempt at an autonomous, general arithmetic of an arbitrary algebraic number field until Dedekind [3], in the late 1860's succeeded with his theory of ideals. Kummer took a gigantic stride beyond Gauss when (1845) he created an arithmetic for the algebraic numbers that appear "naturally" in connection with Fermat's last theorem and higher reciprocity laws. But his ideal numbers were devised specifically for the integers of cyclotomic domains; and although Dedekind [2] gratefully acknowledged his initial indebtedness to Kummer's theory as a source of inspiration, his own general theory is of a radically different nature. Gauss might have imagined (and repudiated) ideal numbers—from his start in higher congruences; ideals, on the contrary, with their solution of a finite arithmetical problem by an appeal to infinite classes of integers, seem to belong to a realm of concepts alien to his mode of thought.

Nevertheless, Gauss took the first and decisive step. To appreciate the revolutionary character of what Gauss did in the early development of algebraic numbers, it is necessary to remember the sparse but significant hints in the work of his predecessors, and to bear in mind that both they and he were interested in irrational algebraic numbers solely for the possibility of applications to rational arithmetic.

One of the earliest applications of irrational algebraic numbers to rational arithmetic was the attempt by Euler (in his *Algebra*, 1770) to prove Fermat's assertion that the only solution of $x^2 + 2 = y^3$ in rational integers is $x = 5$, $y = 3$. Euler assumed that $(x + i\sqrt{2})(x - i\sqrt{2}) = y^3$ implies that each of $x \pm i\sqrt{2}$ is the cube of a number of the form $u + iv\sqrt{2}$, u, v rational integers. The assumption can be justified since the $u + iv\sqrt{2}$ constitute a unique factorization domain. This, of course, was unknown to Euler; and the failure of a similar device for the equation $2x^2 - 5 = y^3$ (which has the solution $x = 4$, $y = 3$) was inexplicable at the time.

Almost simultaneously with Euler, and independently of him, Lagrange [23] conceived "the idea of using irrational and even imaginary factors of formulas [forms] of the second degree to find the conditions which render these [forms] equal to squares or to any powers." He remarks that in 1768 he read an unpublished memoir on the subject before the [Berlin] Academy; a resumé was printed in 1769. Gauss was familiar with this work. The detail of significance here is that the norm of a general algebraic number of degree n is a form of degree n in n indeterminates admitting composition. Incidentally, Lagrange reduced the general number to its canonical form. Elsewhere [24] he showed (1770) that the form $x^2 + ay^2 + bz^2 + abw^2$ repeats under multiplication, thus generalizing Euler's theorem (1748) in which $a = b = 1$. Gauss derived Euler's theorem from multiplication of norms of complex numbers [25].

No trace of a conjecture that irrational algebraic numbers might be subject to some of the laws of rational arithmetic has been detected in the writings of any predecessor of Gauss. What might have suggested the possibility to him? As he never explicitly told anyone, the only recourse is to examine the occurrences of algebraic integers in his work, much of which he did not publish. The relevant material is in those parts of mathematics in which Gauss worked, and to which, during his lifetime, a rapidly developing arithmetic of algebraic numbers was applied.

Three prolific sources of the theory of algebraic integers were attempts to prove Fermat's last theorem ($x^n + y^n = z^n$, $xyz \neq 0$, $n > 2$ is impossible in rational integers x, y, z, n); the laws of $r - ic$ reciprocity, $r > 2$; and the complex multiplication of certain special elliptic functions. It is known that Gauss (1777–1855) proceeded at different times from the first two. The first engaged his attention [26] as early as 1808; the second about 1814, when, according to his own testimony [27], he obtained a proof of the law of biquadratic reciprocity. It is possible though improbable that biquadratic residues induced him to consider (Gaussian) complex integers as early as March, 1797, when he discovered the double periodicity of the lemniscate functions, whose study he had begun in January of the same year [28]. The third main historical source, complex multiplication, is problematical, so far as Gauss is concerned, although Jacobi, as will be seen later, suspected that Gauss had been led to (complex) algebraic integers through his discovery (unknown to Jacobi) of complex multiplication for the lemniscate functions.

The fragments [26] on Fermat's last theorem are inconsiderable in themselves, especially when contrasted with Dirichlet's and Kummer's works, or even with some of Lamé's. in the same direction. But in their intimations of a boundless new province of arithmetic they were of prophetic significance. Probably they were composed not later [29] than 1808. In the second, Gauss gives the material for a proof that Euclid's algorithm for the greatest common divisor of two integers in the natural domain R is applicable in the quadratic domain D defined by an imaginary cube root of unity, and hence that there is unique factorization of integers into primes in D. The unicity is assumed in the first fragment, where the impossibility of Fermat's equation is proved for $n = 3$ in both R and D. A similar proof is outlined for $n = 5$. For $n = 7$ the devices that succeeded for $n = 3, 5$ fail, and Gauss queries "whether it is to be hoped that a proof may be obtainable from the properties of norms ["Determinanten", not "determinants"] and units ["Einheitszahlen"]."

The cases $n = 3$, $n = 5$ of Fermat's last theorem thus inspired the earliest known extensions of the fundamental theorem of arithmetic to domains other than the rational. The relevant fragments have the appearance of memoranda not intended for publication. This may account for the omission of definitions of "integer," "prime," and "unit" in the domains considered. Or it may have seemed to Gauss that the definitions were obvious and their explicit statement superfluous. In any event mathematics in 1808 was less precise than it gradually became after 1900, when Hilbert's insistence on axiomatics began to percolate.

In the first commentary (1825) on biquadratic residues, Gauss states [31] that he began to systematize the theories of cubic and biquadratic residues in 1805. From his posthumously published notes on cubic residues, it appears that he first approached both theories from the side of rational numbers and binary quadratic forms [32]. The forms required were actually norms of complex integers in the relevant quadratic fields. By 1808, Gauss had recognized the (complex) integers in the field generated by a complex cube root of unity, and was thereby enabled to formulate [44] (implicitly) the law of cubic reciprocity in its simplest form [33].

An entry in his Diary for 23 December 1808, records that Gauss began the study of the diophantine equation

$$x^3 + ny^3 + n^2z^3 - 3nxyz = \pm 1$$

on that day. The left member of this equation is the norm of the number $x + \nu y + \nu^2 z$ in the pure cubic field C defined by $\nu^3 = n$, n a rational integer. He proved that n is a cubic residue of each divisor of any rational integer represented in the norm. The similarity between this property of the norm and the like for numbers represented in binary quadratic forms, may account for Gauss' invariable designation of norms as determinants.

Solutions of the diophantine equation furnish units in C. Without indicating the significance—from the standpoint of algebraic integers—of his calculations, Gauss recorded the units for certain values of n, and even in some instances exhibited the fundamental unit [34].

Here is the earliest recorded hint that the "natural" generalization of the theory of binary quadratic forms, to which Gauss devoted so much of his early activity, is not to binary forms of degree higher than the second, but to the composable forms of degree n in n indeterminates arising as norms of algebraic integers of degree n. The fragment on the units in a pure cubic field is a step in the progression from Lagrange to Dirichlet [18], the latter of whom in 1842–46 developed the general theory of algebraic units, incidentally pointing out that his theory includes and generalizes the theory of the Pellian equation. As Gauss had not published his own start in the same direction, Dirichlet was unaware of it. He proceeded directly from the algebraic developments of Euler and Lagrange; and it cannot be said that he was anticipated in any essential manner by Gauss.

From the fragments described and a very few others of a roughly similar nature on norms, it is clear that as early as 1808 Gauss had a practical knowledge of a few of the simpler integral domains with class number unity. The complete absence of notes on the factorization of integers in domains whose class number exceeds unity, may indicate that Gauss had satisfied himself that the Euclidean algorithm is not universally applicable, or, what for him may have been equivalent, that the fundamental theorem of arithmetic fails, in some of the domains he considered [35]. Either by almost miraculous good fortune or by mathematical insight, he did not commit himself in 1808 to plausible but erroneous inferences from an assumed universality of the fundamental theorem, as Lamé [36] and Cauchy [36] were to do in 1847. If the conjecture (by a prominent algebraist who prefers to remain anonymous), that the number of unique factorization domains of algebraic numbers is finite, should be substantiated, it would seem that luck alone was not responsible for Gauss' avoidance of the ingeniously contrived pitfall that deceived his eminent contemporaries [36].

It is not known when, or why, Gauss lost interest in Fermat's last theorem. There is no hint in anything he published himself that he had ever given the theorem a thought. The notes on the cases $n = 3, 5, 7$ have been dated 1808. It is a singular coincidence that in January of that year, Gauss [37] rather needlessly singled out Fermat for special suspicion of having stated results, found inductively, for which he either had no proof or deluded himself into believing he had. The occasion was an abstract of a new proof of the law of quadratic reciprocity. In the *Disquisitiones Arithmeticæ* (Art. 365), Gauss had already paid his disrespects to Fermat as an arithmetician. By stating only the less important half of the truth concerning Fermat's conjecture (1640) regarding the primality of $2^{2^n} + 1$, Gauss succeeded in producing something dangerously close to a whole falsehood. In any event he did Fermat, who valued his reputation for probity more highly than his skill as a mathematician, a serious injustice.

Eight years after Gauss had stopped, baffled, before his special case $n = 7$ of Fermat's last theorem, he disclaimed any serious interest in the theorem, but admitted by implication that he was still attracted by algebraic integers. This appears from his reply [38] to a letter of 7 March 1816 from Olbers, in which Olbers informed him that one of the prize questions proposed for 1818 by the Paris Academy of Sciences was a proof of Fermat's last theorem. Olbers remarked that the question seemed to him to have been specially made "für Sie, lieber Gauss." Gauss answered two weeks later. In view of all the circumstances, his reply can scarcely be considered a model of perspicuity and disingenuousness. The italics are in the original.

"Göttingen, 1816 March 21. I am much obliged to you for your communications concerning the Paris prizes. I confess indeed that the Fermat theorem as an isolated proposition has little interest for me, since a multitude of such propositions, which one can neither prove nor refute, can be easily promulgated. Nevertheless I am induced by this to take up again several old ideas for a *great* extension of the higher arithmetic. Admittedly this theory belongs to those things in which one cannot foresee how far it will succeed in reaching distant, [uncertainly defined] ends. A lucky star must also prevail; and my situation and so many distracting occupations indeed do not permit me so to cling to such meditations as in the fortunate years 1796–1798, when I composed the main items of my *Disquisitiones Arithmeticæ*. However, I am convinced that, should luck do more than I may expect, and prosper me in certain main steps in that theory, then also the Fermat theorem will thereby appear as only one of the least interesting corollaries."

6. Complex multiplication

The last entry in Gauss' Diary (Item 146), dated 1814 Jul. 9, records an unstated connection, discovered by induction, between the theory of biquadratic residues and the lemniscate functions. The putative connection, according to Gauss, is through the total number of solutions, modulo the Gaussian prime $a + bi$, of the congruence $1 \equiv x^2 + y^2 + x^2y^2$, where $2 + 2i$ divides $a - 1 + bi$. Such a connection was explicated and proved from the theory of complex multiplication for the lemniscate functions, by Herglotz in 1921. The proof is quite complicated and long [44]. It is an earlier hint of doubly periodic functions however that is of interest here.

In Article 335 of the *Disquisitiones Arithmeticæ* (1801), introductory to the famous seventh section of the entire work, Gauss states that the principles he is about to explain for the division of the circle may be applied not only to the circular functions, but also to many other transcendental functions [44], for example those which depend on the (lemniscate) integral

$$\int \frac{dx}{\sqrt{1-x^4}},$$

and moreover also to different kinds of congruences. He promises a full account of these functions in a work to be devoted specially to them, and says that congruences will be extensively treated in a sequel to the *Disquisitiones*. Neither project was carried out to the extent of publication.

What Gauss had accomplished but did not publish in the theory of elliptic functions was surpassed and superseded by the independent works of Abel and Jacobi in the 1820's. In his general theory of congruences, at which he was working in 1797–8, Gauss [16] was forestalled by Galois [39] in 1830. Gauss' theory was to have formed the eighth and concluding section of the *Disquisitiones*, but was omitted, partly to save expense in printing. His actual contributions to both theories were recovered from his private papers. They have been minutely analyzed by the editors of his works.

The immediate question here is the possible influence complex multiplication may have had on the early shaping of Gauss' thoughts on algebraic numbers. It has often been remarked that the Gaussian theory of binary quadratic forms might have been specially designed to accommodate the later theory of complex multiplication—a theory which Hilbert considered the most beautiful part of all mathematics. The same theory, specifically the complex multiplication of lemniscate functions, was conjectured by Jacobi [40] in 1839 to have been the true source of the (Gaussian) complex integers $a + bi$, a, b rational integers, which Gauss defined and used in his theory (1831) of biquadratic residues [31].

As Gauss neither affirmed nor denied Jacobi's conjecture, and as he was in vigorous health when it appeared in print, there is no positive means of deciding whether Gauss maintained his silence because the conjecture was baseless or because it was founded on concealed fact. Gauss, incidentally, seems to have had an aversion to Jacobi as a scientific correspondent, ignoring his young emulator's mathematical letters, occasionally to Jacobi's considerable chagrin. It is somewhat remarkable then that Jacobi, who at first was disinclined to cede Gauss any priority for the basic discoveries in elliptic functions, ended by giving him more than has been claimed or justified by the editors of his works.

Whatever the cogency of Jacobi's suggestion, it seems unlikely that Gauss, even as a young man, would have permitted himself the scandal of the "$1 + 4i$ and $1 - 4i$ parts" which Jacobi would foist on him in the passage summarized presently. For at least thirty years, Gauss turned imaginaries over and over in his mind before he would admit that he had finally satisfied himself that they are as legitimate mathematical concepts as negative rational integers. What he might have thought of using imaginary numbers to enumerate the parts of a lemniscate arc, or anything else, must be left, with Jacobi's inexistent "parts," to the imagination. It is clear enough however what Jacobi meant, even if of itself it means nothing. The specious analogy he promoted with so much zeal is at least alluring, and might even make sense to an intelligence higher than the mathematical. Gauss ignored it.

Jacobi begins by recalling that Gauss had introduced numbers of the form $a + bi$ as moduli or divisors in his researches on biquadratic residues, and that he succeeded thereby in stating for two complex primes of the form $a+bi$ a reciprocity law as simple as that for quadratic residues, the so-called gem of the higher arithmetic. But however simple such an introduction of complex numbers may appear now (1839), Jacobi continues, it belongs nevertheless to the profoundest ideas of science. New concepts do not always present themselves in the simplest guise. He then confesses his disbelief that arithmetic disclosed so deeply hidden an idea: it was drawn from the study of elliptic functions, and indeed from that particular kind furnished by the rectification of an arc of a lemniscate.

In the theory of the multiplication and division of the lemniscate, complex numbers of the form $a+bi$, according to Jacobi, play exactly the same role as ordinary numbers. The analogue in the theory of the lemniscate functions for the expression of the circular functions of n times a circular arc, as rational functions of circular functions of the arc, is the rational expression (in terms of the appropriate functions) for $a + bi$ times the arc of a lemniscate. Division of the circumference of a circle into n equal parts can be effected by solving an equation of degree n; division of a lemniscate "into $a + bi$ parts" leads to an equation of degree $a^2 + b^2$—the norm of $a + bi$. Thus if the circle is to be divided into 15 parts, it is first divided into 5 and into 3 parts, from which $(1/15 = (1/3 - 1/5)/2)$ the required division is found; to divide the lemniscate into 17 parts, it is first divided "into $1+4i$ and into $1-4i$ parts which, combined, give the 17 parts." The thought here, if any, seems to be the seductive misconception that

$$\text{“}(1 + 4i)(1 - 4i) = 17\text{”}$$

makes $1+4i$ and $1-4i$ cardinal numbers because their product is the cardinal number 17. From this esoteric revelation it follows, according to Jacobi, that a study of the lemniscate integral must of necessity lead to the introduction of numbers $a+bi$ as moduli. But would Gauss even in a nightmare, have seen himself using imaginary numbers for counting?

Although Gauss may have discovered other values of n beyond the $n = 5$, recorded in his Diary (Item 60, 19 March 1797), for which division of the lemniscate into n equal parts is possible by a Euclidean construction [44], analogously to the division of the circle for $n = 3, 5, 17, \ldots$, Jacobi's ingenious conjecture sounds like a typical example of *post hoc ergo propter hoc*. Gauss, as has been remarked, took no notice of it. But then, there were a good many perfectly sensible speculations regarding his work which he also ignored.

To give Jacobi's conjecture whatever credence it may merit, it is necessary to recall an episode from the early history of elliptic functions. From this, anyone may form his own opinion as to whether Gauss was led to his complex integers by his work on elliptic functions. The introduction of complex numbers as variables in analysis is an entirely different matter; and it is established that Gauss' study of the lemniscate functions made him suspect that the complex variable is necessary for an understanding of algebraic functions and their integrals. But to confuse the two kinds of use which Gauss made of complex numbers, is to deprive him of one of his greatest achievements. Although some will disagree, the recognition of algebraic integers seems to arithmeticians to have required a more penetrating insight than did the recognition of complex variables. In the second, at least partially, Gauss had several predecessors; in the first, none.

After Abel's *Recherches sur les fonctions elliptiques* (§I–§VII), appeared [41] in Crelle's Journal for 1827, Gauss [42] wrote (30 May 1828) to Schumacher that Abel had forestalled him "by at least a third" and, curiously enough, had used partly the same notations as Gauss himself. In a letter of 18 May 1828 to Abel, Crelle [42] reported that when he had begged Gauss for something on elliptic functions (Crelle had heard that Gauss had been occupied with the theory for more than thirty years), Gauss replied substantially as he had to Schumacher. In addition, he stated that "other occupations" prevented him for the moment from putting his researches into shape for publication. Abel, he said, had followed exactly the same path that he (Gauss) had set out on in 1798. To Gauss it was not astonishing then that the major part of Abel's results were the same as his own. And, he concluded, since Abel's development was carried out with "so much sagacity, penetration, and elegance, I believe therefore that I am relieved of the editing of my own researches."

The critical detail for Jacobi's conjecture is what Gauss meant by "at least a third." Abel's first publication (§I–§VII) contains nothing on complex multiplication. It ends with some special results on the lemniscate functions as simple illustrations of the general theory of elliptic functions developed in the preceding sections. So it is possible but improbable that the remaining two thirds of Gauss' anticipations may have included more on complex multiplication than has been found in his papers—all of it very special [44]. This would be conclusively confirmed if it could be proved that Gauss was aware of the contents of the continuation (§VIII–§X) of Abel's memoir, which was received for publication by Crelle on 12 February 1828, and which appeared in his journal for 1828.

In §X, Théorème II, Abel states [43] the conditions that an imaginary value of α shall render the equation

$$\frac{dy}{\sqrt{(1-y^2)(1+\mu y^2)}} = \alpha \frac{dx}{\sqrt{(1-x^2)(1+\mu x^2)}}$$

algebraically integrable: "it is necessary and sufficient that α be of the form $m \pm \sqrt{-1}\cdot\sqrt{n}$, where m and n are rational numbers. In this case the quantity μ is not arbitrary; it must satisfy an equation having an infinity of real and imaginary roots. Each value of μ satisfies the question."

This was the origin of the vast theory of complex multiplication, whose ramifications are not yet all exhaustively explored. Abel illustrates his general theorem by some special cases. No responsible advocate [44] of Gauss' priorities yet has seriously claimed that Gauss preceded Abel in anything approaching this complete generality; nor has anything been recovered from his scientific remains to substantiate such a claim. The closest approximation to a claim is the frequently expressed opinion that, in elliptic functions as in some other departments of mathematics, Gauss probably discovered a great deal more than has yet been found on paper [44]. So for the present at least, Jacobi's conjecture can be ignored as Gauss himself ignored it.

The further history of elliptic functions is irrelevant for Gauss' part in the early development of algebraic numbers. But as he seems to have had a marked partiality for the lemniscate functions and biquadratic residues, some work of Eisenstein's [45] (1846), which disclosed the intimate connection between the two, may be recalled in passing. For sheer ingenuity, if nothing more profound, Eisenstein's deduction of Gauss' law of biquadratic reciprocity from the complex multiplication of the lemniscate functions would be hard to surpass. Except for the case of two real primes, easily handled otherwise, the law is proved for all cases of Gaussian primes. Eisenstein emphasizes that Gauss' partition of a (proper) complex residue system modulo an odd primary complex integer into "quarter-systems" of associates, is basic for his own analysis.

An alternative proof of the biquadratic law, which Eisenstein [46] gave in the preceding year, is even more suggestive, in that the proofs of the laws of quadratic, cubic, and biquadratic reciprocity are unified analytically. The quadratic case, from this point of view, is degenerate, being derived from properties of the circular functions. In the biquadratic case an uneven Gaussian complex prime is the key to the analysis; while for the cubic case, Eisenstein states merely that integers $a + b\xi$, ξ an imaginary cube root of unity, are required. It was noted earlier [33] that Gauss used these same integers in 1808 to prove (implicitly) the law of cubic reciprocity. He made no claim [44] to have anticipated Eisenstein.

It would be interesting to know why Gauss restricted himself to what are now called Gaussian integers in the work which he himself published, when he was familiar with other algebraic integers as well. Possibly he was less bold than Eisenstein, who admitted that "perhaps one may disapprove of the use of circular and elliptic functions in arithmetical reasoning. But it is to be observed that these functions enter, so to say, only *symbolically*, and that it would be possible to get rid of them entirely without destroying the substance and the basis of the proofs." He then indicates how this may be done for the quadratic case, and states that something similar but more complicated exists for the other two, "and one can say... that by no means is the formula for the multiplication of elliptic functions needed. Nevertheless it does not always seem preferable to avoid analytic functions in arithmetical investigations, especially when it is seen *a posteriori* that they do not enter essentially into the demonstrations, and that they serve only to fix ideas and abridge the conclusions." Similar opinions were expressed by Gauss regarding the use of circular functions in cyclotomy.

Gauss gave as his reason for multiplying proofs of the law of quadratic reciprocity, the hope that methods of proof applicable also to biquadratic residues would emerge. He found such methods; but it is unlikely from what has been recovered from his papers that any of them combined algebraic integers and the complex multiplication of elliptic functions. Eisenstein's inspiration seems to have been his own. Gauss thought highly enough of Eisenstein to write an appreciative preface—the impersonal part of which has become classic—to a collection of his mathematical papers [47].

7. Higher congruences

Analogies between algebraic equations and congruences noted in the *Disquisitiones Arithmeticæ* led naturally to the "much higher investigation," of which the theory of congruences is only a part, the posthumously published *Analysis Residuorum*. This was to have formed the eighth section of the *Disquisitiones*. From Gauss' Diary, it is known that this "higher investigation" was undertaken in 1797–8. It is in two parts [48], the first of which is concerned with binomial congruences for a prime modulus. The contents of this part need not be discussed here, except one rather significant detail which seems to have caused Gauss considerable anxiety. The difficulty was of the same genus as one that was to recur in early attempts to construct an unexceptionable theory of decomposition for algebraic integers.

Gauss showed (Art. 251) that once any one of the ef-nomial periods introduced by him into the theory of binomial equations (as in the seventh section of the *Disquisitiones*) is given, all the periods are obtainable by solving a simultaneous system of e linear congruences for a prime modulus. There is no difficulty if the modulus is not a divisor of the determinant of the system. In the contrary case, the analysis is inapplicable, and must be supplemented by other considerations. Because this exceptional case can occur so rarely, Gauss remarks, he does not linger over it—a curious reason for a mathematician like Gauss to give. In the second part, concerning the general theory of congruences, exceptional cases of a somewhat similar nature arise, and Gauss promises (Art. 363) that "it will be shown below how this difficulty may be obviated." But as the work ends abruptly, uncompleted, in the middle of a congruence, the promise is unfulfilled.

It is generally supposed however that Gauss had in mind as the proposed remedy congruences modulo p^a, $a > 1$, p prime, to supplement the case

$a = 1$ actually investigated. An entry in the Diary (Item 68) for 21 July 1797 makes a suggestion to this effect. But that is all, and there is no record of why Gauss thought his suggestion might do what was required, or how it might be put into effect. In fact he seems to have tired of the subject. The attempt to prepare the general theory of congruences for publication was abandoned. Other interests absorbed his attention, and until a complete draft of the promised "higher theory of congruences" is found—an improbable contingency—it will not be known what goal Gauss had in view. The "higher theory" would have been the natural way for him to follow if he had been seeking ultimately to construct an arithmetic for the integers of cyclotomic fields. There is a mere fragment on higher congruences when the modulus is a composite number.

The extension of congruences as they appear in the *Disquisitiones* is the factorization of polynomials in a single indeterminate, with rational integer coefficients, for a prime modulus. This is the theme of the second part of the *Analysis Residuorum* [48]. Prime polynomials modulo a prime are defined, and their number for a given prime is determined by means of a generating function and an unexplained application of inversion. This result is noted in the Diary (Item 75) under 26 August 1797. More significant from the standpoint of general arithmetic is the proof that the Euclidean algorithm applies to these "modular" polynomials. The units in the theory are readily detected, and the fundamental theorem of arithmetic follows for the polynomials.

As already noted [39], Galois (1830) forestalled Gauss in this theory. Gauss never published his outline. It was more in the spirit of Dedekind's [49] exposition (1856) than in that of Galois. With the publication of the outline, it was disclosed that Gauss as early as 1797–8 had foreseen the possibility of using the imaginary roots of congruences invented by Galois.

Gauss at the time was thinking intensively of complex numbers in connection with his first proof (1799) of the fundamental theorem of algebra, and he had invented the notation of congruences partly because it suggested fruitful arithmetical analogies with algebraic equations. But, as appears from his correspondence and his published observations on the "metaphysics "—foundations—of mathematics, he meditated on imaginary numbers for many years before he finally felt that he understood them completely. He was satisfied only when (as he asserted) he reached the same interpretation as Hamilton [50] reached six years later (1837), in which complex numbers are replaced by ordered couples of real numbers. So when, in 1797, another might have been tempted to pursue the analogy between congruences and algebraic equations to its extreme, by endowing congruences with imaginary roots, Gauss refrained. He did so, not for lack of capacity, but most likely because he did not then see how such imaginary roots were to be logically justified.

"It is obvious," he says (Art. 338), "that the congruence $\xi \equiv 0$ has no real roots if ξ has no factors of the first degree; but there is no hindrance, if ξ is resolvable into factors of the second, third, or higher degrees, in ascribing to the congruence *imaginary* roots." And he goes on to say that if he were to permit himself a license similar to that indulged in by recent mathematicians, he would introduce such imaginaries, thereby "incomparably" abridging all of his subsequent investigations. But, he concludes, he prefers to deduce everything "from principles."

Galois was unaware of the predestined iniquity of his imaginaries; for he had been dead all forty-four years when Gauss' condemnation first appeared in print. Dedekind's theory [49], making no use of imaginaries, would have been acceptable to Gauss. The logical difficulties which Gauss might have observed in Galois' imaginaries were removed in 1866 by Serret [51]. There is therefore no longer any licentiousness in indulging in these imaginaries, if desired.

Much later (1920), the serious lacunæ in Gauss' first and fourth proofs of the fundamental theorem of algebra were filled [52]. It is one of the finer ironies of mathematical immortality, that while Galois' imaginaries as now rigorized [53] are unobjectionable in the increasingly popular intuitionist philosophy of mathematics, even the amended proofs for the fundamental theorem are not. However, not all mathematicians are intuitionists.

At last, in 1831, Gauss [7] permitted himself to present his complex integers $a+bi$ to the mathematical public. His meticulous scruples had been overcome; he felt that he now thoroughly understood the imaginaries of algebra; the logic in the crucial proofs had been rigidly secured; and he was confident that history would justify him in his unprecedented boldness. He made no mention of his early adventures in Fermat's last theorem, and dropped no hint that he had proved the law of cubic reciprocity [44] twenty-three years earlier by the use of quadratic integers other than those he now acknowledged. All the credit for the epoch-making inspiration was bestowed on biquadratic residues.

It was indisputable that the introduction of Gaussian integers induced an astonishing simplification in the theory of biquadratic residues. Disparate results with no evident connection now appeared as necessary consequences of the arithmetic of the underlying complex domain. The units and primes for the new integers were defined; the Euclidean algorithm was shown to hold: the new integers were as legitimate elements of arithmetic as the rational integers themselves. A "boundless new domain" of the theory of numbers had been consolidated with a firmness which must have satisfied even its exacting discoverer.

Students of human nature may be interested to speculate why Gauss omitted any mention of his early work in algebraic integers. Perhaps he had forgotten all about it. Or he may have thought it unworthy of him, in comparison with his published work on biquadratic residues, and wished to be remembered for that alone. If so, he is but one more victim of an inquisitive posterity which, never content with the picture a man wished to leave of himself, publishes all the private papers he neglected to destroy. Though the scholarly emoluments from such indelicacy may be colossal, the scientific gain is in general negligible, and the increase in understanding the outraged man nil. Considered objectively, biography is the meanest form of gossip.

8. From the inexistent to the existent

Having formally pronounced the field of algebraic numbers open, Gauss left it. Others might cultivate it, should they find it profitable. Several did, but Gauss never returned to inspect or bless their labors. Perhaps it was as well for his peace of mind that he held himself aloof from the contemporaries of his declining years. If the mere suggestion of imaginary roots of congruences was repugnant to him in 1797, when he was just twenty years old, it is unlikely that he would have participated with gusto in a bacchanal of inexistent ideal numbers half a century later.

The account of Gauss' active part in the theory of algebraic numbers should close, logically and historically, with the memoir of 1831 on biquadratic residues. To link the work of Gauss to the present, however, it is necessary to connect him with Dedekind and Kronecker, especially Dedekind. The connection is through Kummer. Kronecker was Kummer's pupil at the time when Kummer was creating his theory. Dedekind was inspired by Kummer's theory to invent his own. From Dedekind to abstract algebra the way, though long, is straight and clear. It will be sufficient to reach Dedekind, the last personal pupil of Gauss. The pupil owed nothing to his master in the matter concerned: Dedekind struck out on a way of his own, of which there is not a hint in anything attributed to Gauss.

The obvious problem after Gauss' introduction of complex integers $a+bi$ into the study of biquadratic residues, was to construct an arithmetic for cyclotomic fields, in the hope that it would lead to the discovery and proof of higher reciprocity laws. The problem lay disregarded till the 1840's, when Kummer [54] took it up, and through the creation of his ideal numbers (1845), was enabled (1847, 1858) to state—with certain critical exceptions—a general reciprocity law [58].

The 1840's also witnessed a resurgence of Fermat's last theorem. Although the theorem was not disposed of, the abortive attempt (1847) of Lamé [55] to find a general proof early attracted attention to Kummer's fundamentally new work. A profitable debate ensued between some of the leading mathematicians of the time. Lamé's correct proof (1839) for seventh powers [56] had impressed Cauchy, Liouville, and others: Lamé now had to be taken seriously. Although Gauss did not participate in the debate over Lamé's mishap of 1847, he let it be known that in his opinion Lamé was the foremost French mathematician of the day. His own rebuff by the seventh powers which had yielded to Lamé may have caused him to overlook Cauchy, whom he did not acclaim.

Kummer also was an early contestant for Fermatian immortality. About 1843 he submitted to Dirichlet what he had convinced himself was a general proof of Fermat's last theorem. Dirichlet [4] detected the fatal oversight. It was true, as Kummer had stated, that an integer in the relevant cyclotomic field could be resolved into a product of indecomposable integers of the field; but it was false that the resolution was unique, as Kummer had assumed. This initial failure was the beginning of Kummer's greatest success. He set himself to obliterate the defect by creating new elements sufficient to ensure unique factorization.

While Kummer was thus engaged, Lamé was endeavoring to solidify his general proof based on the arithmetic of the field generated by a primitive nth root of unity. Thinking he had succeeded, he submitted his purported proof to the Paris Academy, only to be shown by Liouville, and a little later by Kummer, that he had erred. Acknowledging the error, he persisted, but without success. Cauchy [57] then (1847) tried to span the unbridgeable chasm by a flimsy deduction that the Euclidean algorithm for the greatest common divisor applies to all cyclotomic domains. He finally saw that he had been too optimistic, gave a counterexample from the field defined by $x^{23} = 1$ for one of his deductions, and incontinently dropped the problem. Kummer had the field to himself [58].

Fermat's last theorem had discharged the function for which it was destined. Contrasted with what it helped to instigate, it had at last become the isolated proposition of minor interest which Gauss, perhaps prematurely, in 1816 had declared it to be. Still, until it is settled, it cannot be shrugged off.

A chemical analogy for his ideal numbers, which Kummer himself offered to make them seem less supernatural than his contemporaries found them, may assist the understanding of a later generation. The element fluorine had not been isolated in Kummer's heyday—it was not till 1888; yet chemists inferred not only its existence but also its properties from reactions which, unless such an element existed, were unnatural and inexplicable. There is this cardinal distinction, however, between Kummer's creation and an inferred element: nobody will ever isolate an ideal number, for such a number has no existence. This ghostly unreality of Kummer's ideal numbers was the catalyst which, in Dedekind's mind, transformed inexistence to existence.

Kummer did not exhibit his ideal numbers; it is in fact impossible to do so. But he did define their behavior, and this was sufficient. More precisely, he defined "divisibility by a specific ideal number." Algebraic numbers are "existent" numbers. If an algebraic number (of the kind he considered) satisfies one or more congruences, the existent number in question is said to be divisible by a determinate ideal number corresponding to the property that the algebraic number satisfies the congruences. An existent number is considered as a special case of an ideal number. The "multiplication" of these ideal numbers is defined, and if the products of each of two numbers (ideal or existent) by the same ideal number are both existent, the numbers are said to be equivalent. Such equivalence satisfies the postulates for an equivalence relation as in abstract algebra, so that ideal numbers may be separated by it into classes: a class is the set of all those numbers which are rendered existent by one and the same multiplier. No ideal multiplier is required for existent numbers; the class of all such numbers is called the principal class. To each principal class corresponds an infinity of composable forms of degree n in n indeterminates with existent coefficients. The number of such classes is finite; it plays a fundamental part in the theory and in its applications to Fermat's last theorem and higher reciprocity laws.

Probably enough has been said to suggest the elusiveness of Kummer's ideal numbers and the perspicacity demanded in using them. Even Kummer made mistakes—not mere slips—in handling his own invention, some of which were not detected until well into the twentieth century. The theory also lacked the complete generality desirable in modern mathematics; it referred only to cyclotomic fields, whereas a theory of algebraic numbers should be applicable to any algebraic number field.

Dedekind created a general theory. (Others also produced theories for any algebraic field; but Dedekind's was the one that survived.) His aim was to operate wholly in the realm of "existent" numbers. He had observed that Kummer's theory could be simplified by an application of the theory of higher congruences. Accordingly he started from higher congruences in his first attempt at a general theory. Almost successful, he could not dispose of certain exceptional cases. He therefore abandoned what he considered the way of calculation and proceeded ab initio from general principles. A right approach was found by analyzing the problem into its simplest elements and discarding all inessential details. But to say this is not to explain how he did it, or to account for his triumph. As Poincare observed, it is the man, not the method, that solves a problem.

It is of interest that the way of higher congruences which Dedekind first explored was not hopelessly blocked as he supposed. Technically, the obstacles that stopped him were the extraneous common index divisors. The difficulties were overcome in 1926–8 by Øre [59], who took, instead of congruences modulo a prime p, congruences modulo p^{α}, α a fixed integer $> \delta$ where p^{δ} exactly divides the discriminant of the corresponding equation.

9. New directions

Some measure of a century's progress in algebra and arithmetic, consequent on Gauss' use of complex integers to impart simplicity and coherence to a problem in rational arithmetic, may be inferred from two statements concerning the intimate connections between arithmetic and other parts of mathematics. Only half a century separates them. The first is from the preface which Gauss wrote (1847) for Eisenstein's *Mathematische Abhandlungen*; the second, from Hilbert's preface (1897) to his report on algebraic numbers, *Die Theorie der algebraischen Zahlkörper*.

"The higher arithmetic," Gauss observed, "offers an inexhaustible store of interesting truths, which indeed are not isolated, but stand in a close internal connection, and between which, as science increases, always new and wholly unexpected connections are ascertainable."

With this observation of Gauss' in mind, Hilbert (1862–1943) elucidated the "unexpected connections": "Like Gauss, Jacobi and Lejeune-Dirichlet frequently and emphatically expressed their surprise at the intimate connections between questions concerning numbers and certain algebraic problems, in particular the problem of cyclotomy. Today the intrinsic reason for these connections is completely uncovered. *The theory of algebraic numbers* and the Galois theory of equations have their common roots in the theory of algebraic fields, and this theory of number fields has become the most important part of the modern theory of numbers. The merit for having introduced the first germ of the theory of number fields again belongs to Gauss."

Some further remarks of Hilbert's are of special interest in the present connection: "The theory of number fields is a monument of admirable beauty and incomparable harmony. The most beautiful part of this monument seems to me to be the theory of abelian fields and of relative abelian fields. Kummer and Kronecker revealed it to us, the former by his works on the higher laws of reciprocity, the latter by his investigations on the complex multiplication of elliptic functions. The profound insights given us by the work of these mathematicians in this theory, also show us that in this domain of science an abundance of the most precious treasures still lie hidden, promising a rich reward to the seeker who knows the value of such treasures and who practices the art of discovering them with devotion."

Monuments of "admirable beauty and incomparable harmony" are but seldom erected to the living. Even while Hilbert himself was still scientifically vigorous, both arithmeticians and algebraists began to abandon, at least temporarily, the search for the kind of undiscovered treasures which he described; and his own restless interests changed. A quarter of a century after his eulogy of "the most important part of the modern theory of numbers" appeared, the analytic theory of numbers suddenly began to expand as never before; and the older monuments, cast into the shade, were relatively neglected. Simultaneously, abstract algebra also began to flourish luxuriantly, and presently had overspread the fields of its origin in its rapid growth toward new and unexplored territory. Nobody foresaw either development in 1897 when Hilbert published his report on algebraic numbers. Hilbert's own work in axiomatics at the turn of the century was partly responsible for the swerve away from the classic past; and yet he did not anticipate it.

If an outstanding instigator of radical progress cannot predict the consequences of his own work, it is futile for others to prophesy. Yet it is hard to suppress the feeling that half a century hence, it may happen that some of the treasures prophesied nearly fifty years ago by Hilbert, will be recognizable in the findings of the present generation of abstract algebraists. If much that is now vivid will then have been absorbed and blurred beyond recognition in a newer mathematics, the labor of perfecting it is not necessarily futile. The advance of mathematics has been like the rhythm of an incoming tide: the first wave, with ever slackening speed, reaches its farthest up the sand, hesitates an instant before rushing back to mingle with the following wave, which reaches a little farther than its predecessor, recedes, mingles with its successor, and so on, till the tide turns, and all are swept back to the ocean to await the next tide. In each surge forward there is some remnant of all the tides that went before, though whatever remains may long since have lost its individuality and be no more recognizable for what it was.

References, Notes

W = C. F. Gauss, *Werke*; Crelle = *Journal für die reine und angewandte Mathematik*; Liouville = *Journal des Mathématiques pures et appliquées*.

1. *Mathematische Annalen*, 36, 1890, 473–534.
2. *Bulletin des Sciences mathématiques et astronomiques*, sér. 2, 1, 1877, 92.
3. Supplement XI von Dirichlets *Vorlesungen über Zahlentheorie*, 4. Auflage, (1894), 434–657. (The first version appeared in the second edition, 1871). See also the Anzeige der zweiten ... Zahlentheorie, Göttingische gelehrte Anzeigen, Jhg. 1871, 1481–94. Dedekind began his project of creating a general theory in 1856, the year after the death of Gauss.
4. See Dickson, *History of the theory of numbers*, 2, 1920, 738.
5. Göttingische gelehrte Anzeigen, 1831 April 23; W 2, 181–178.

6. See references 7, 34.
7. Theoria residuorum biquadraticorum, Commentatio secunda, 1831, W 2, 102–148; §§31–76 contains the arithmetic of (Gaussian) complex integers.
8. Crelle, 35, 1847, 319–367.
9. Liouville 12, 1847, 202. Reprint of Kummer's Dissertation, 1844.
10. *Disquisitiones Arithmeticæ*, Praefatio, especially VII–VIII.
11. Was sind und was sollen die Zahlen, 1888, 1.
12. *Elements*, Book VII, Prop. 32.
13. *Exercises d'analyse*, etc., 4, 1847, 87–110.
14. Facsimile and transliteration with comments in W 10, Abt. 1, 483–572.
15. *Göttingen Nachrichten*, 1894, 291–8.
16. W, 2, 212–240, posthumous.
17. Reference 21.
18. G. Lejeune Dirichlet's *Werke*, 1, 619–23, 633–44.
19. *Bull. Soc. Math. de France*, 12, 1885, 162–94.
20. W, 3, 1.
21. W, 3, 105.
22. *Mathematical Papers of W. K. C.*, 1882, 20.
23. *Œuvres* 2, 527–32; 7, 170–9.
24. *Œuvres*, 3, 189—201, especially 201, ($-B$, $-C$ for a, b).
25. W 3, 383–4.
26. W 2, 387–91.
27. W 2, 516; letter of 30 May 1828 to Dirichlet.
28. Diary, Item 51.
29. According to Bachmann, in W 10, Abt. 2, 60, the fragments belong to the last third of the year 1808. This he infers from Schering's note, W 3, 398. See reference 34.
30. Reference 26, 391.
31. W 2, 67. The first commentary was transmitted to the Royal Society of Göttingen in April 1825, and was printed in 1828; the second, transmitted in April 1831, was printed in 1832.
32. W 8, 5–19.
33. The law was assembled (1900) by Fricke [32] in the course of his duties as an editor, from the *membra disjecta* strewn through Gauss' fragmentary notes. See reference 44.
34. W 8, 21–24.
35. See references 26, 32–4.
36. See references 55, 57.
37. W 2, 152.
38. W 10, erste Abt., 75–6.
39. Liouville, 11, 1846, 398–407.
40. Crelle, 19, 1839, 314.
41. Crelle, 2, 1827, 101–181.
42. Niels Henrik Abel, Memorial, etc., 1902, Correspondence, Lettre XXXI (65). (Slightly different in W 3).
43. Crelle, 3, 1828, 181. Ibid., 195, for Jacobi's partial statement, received for publication, 2 April 1828—seven weeks later than Abel's complete statement.
44. G. Herglotz, Zur letzten Eintragung im Gausschen Tagebuch. *Berichte ... der Sächsischen Akademie ... zu Leipzig*, 73, 1921, 271–6.

Gauss' P, Q, p, q are Weierstrass' σ, σ_3, σ_1, σ_2. In the special case of the lemniscate functions, Gauss proceeds (W 3, 411–12) from $Pi\rho = iP\phi$, $Qi\phi = Q\phi$, $pi\phi = p\phi$, $qi\phi = q\phi$ to the calculation of $P(a+bi)\phi, \ldots, q(a+bi)\phi$ for $(a,b) = (1,1), (2,1), (3,1)$ that of the P, Q functions for $(a,b) = (1,2), (1,3), (1,4)$, and that of the P function for $(a,b) = (1,n)$, n a positive integer. The calculations are straightforward from the addition theorem for the lemniscate functions. If Gauss knew the general addition theorem for elliptic functions, he did not record it in any of his papers yet examined. If he did not know the general addition theorem, it was impossible for him to get anything approaching Abel's result by the methods he used. Even for the lemniscate functions no general conclusion is stated.

In the matter of the division of the lemniscate, Abel's letter of December, 1826, to Holmboe announces the complete solution: "I have written a big memoir on elliptic functions, which contains certain rather curious things, and which cannot fail, I flatter myself, to fix the attention of the literary world. Among other things it treats the division of the lemniscate. Ah! how magnificent it is! I have found that by means of compass and straightedge the arc of the lemniscate is divisible into $2^n + 1$ equal parts, whenever $2^n + 1$ is prime. The division depends on an equation of degree $(2^n+1)^2 - 1$. but I have found its complete solution by means of square roots. That at the same time has enabled me to penetrate the mystery which has enveloped the theory of M. Gauss on the division of the circumference of the circle. I see, clear as day, how that comes about."

Regarding speculations as to what Gauss may have known, Schlesinger (W 10, Abt. 2, Abhand. 2, 9), emphasizes the subjective element in conjectural restorations for which there is no documentary evidence. Whether there is any of this element in the claims (such as Jacobi's) made in Gauss' behalf respecting complex multiplication, is for the individual commentators to settle according to their own judgments. In a similar connection, it has been objected by some that there is a great deal more of Fricke than there is of Gauss in the "restored" proof [33] of the law of cubic reciprocity; and there are several further instances, for example some concerning modular functions.

45. Crelle, 30, 1846, 191–4.

46. Crelle, 39, 1845, 179–183.

47. *Mathematische Abhandlungen*, u.s.w., von Dr. G. Eisenstein, u.s.w., mit einer Vorrede von Prof. Dr. Gauss, u.s.w., Berlin, 1847.

48. W 2, 199–211; 212–240.

49. Crelle, 54, 1857, 1-26. Dedekind reduced the theory to its arithmetical residuum, avoiding Galois imaginaries, also the appeal to the fundamental theorem of algebra, as by Schönemann in 1844 (Crelle, 31, 1846, 269–325.)

50. *Trans. Roy. Irish Acad.*, 17, 293. There is no hint in Gauss' works (or in his published correspondence) of the current interpretation of negative integers and rational numbers through number couples of positive integers. This seems to have originated in Kronecker's theory of modular systems.

51. Reproduced in his *Cours d'algèbre supérieure*, ed. 4, 2, 1879, 122–189.

52. By Ostrowski; reprinted as a supplement, 18 pp., to W 10, Abt. 2.

53. And their theory extended, as in Dickson, Linear groups with an exposition of the Galois field theory, Leipzig, 1901, 1–71.

54. Reference 9, and Crelle, 30, 35, 40, 44; Liouville, 16, 1851, 377–498, especially Arts. VI, VIII.

55. Liouville, 12, 1847, 137–171, 172–184.

56. Liouville, 5, 1840, 195–211.

57. The relevant papers are reprinted in his *Œuvres*, 10.

58. *Berichte*, Berlin Akademie der Wissenschaften, 1847, 132–9.

59. Papers in *Mathematische Annalen*, 96, 1926, 315–52; 97, 1927, 569–598; 99, 1928, 84–117. See also the exposition in Fricke, *Lehrbuch der Algebra*, 3, 1928.

Part III

The 1950s

The Harmony of the World

Morris Kline

Mathematics Magazine
27 (1953/54), 127–39

Editor's Note: Morris Kline was for many years Professor of Mathematics at New York University's Courant Institute of the Mathematical Sciences. He had received a PhD from NYU in 1930. Though initially his work was in topology, he moved into applied mathematics following a stay at the Institute for Advanced Study in Princeton and for some years he directed the Courant Institute's Division of Electromagnetic Research.

In 1953 he published an influential and popular book on the role of mathematics in society, *Mathematics in Western Culture*. This was followed in subsequent years by a series of books in this general area: *Mathematics and the Physical World* (1959); a history, *Mathematical Thought from Ancient to Modern Times* (1972); some textbooks, *Mathematics for Liberal Arts* (1967); *Mathematics: a Cultural Approach* (1962); *Calculus, an Intuitive and Physical Approach* (1967); *Mathematics: The Loss of Certainty* (1980); and a book on mathematics education, *Why Johnny Can't Add: The Failure of the New Math* (1973). The last was provocative and stirred up a great deal of controversy among mathematicians and mathematics educators. Another book was aimed at the university establishment, *Why the Professor Can't Teach: Mathematics and the Dilemma of University Education* (1977).

In this early paper we have a preview of the 1953 book on Western culture. Oddly enough, though Kline's own mathematical interests were largely in mathematical physics, in this book he makes a case for mathematics as a cultural component that should be in everyone's background, not just for students in the physical sciences and engineering but also those interested in intellectual history and the evolution of Western thought. In this particular article he concentrates on astronomy and cosmology.

From Harmony, from heavenly harmony,
This universal frame began:

Dryden

Among the many contributions of mathematics to modern civilization the most valuable are not those which serve the physicist and the engineer but rather those which have fashioned our culture and our intellectual climate. Not enough people are aware of the latter contributions, not even of the role mathematics played in the greatest revolution in the history of human thought—the establishment of the heliocentric theory of planetary motions. It is a fact of history that mathematics forged this theory and was the sole argument for it at the time that it was advanced. No more impressive illustration of the enormous influence mathematics has had on modern culture can be found than that which it exerted through its contributions to the heliocentric doctrine.

The first publication on this theory was Copernicus's *On the Revolutions of the Heavenly Spheres* (1543). The title page of this work gave advance notice of the important role mathematics was to play for on this page appeared the legend originally inscribed on the entrance to Plato's academy: "Let no one ignorant of geometry enter here." With this publication the Renaissance gave to the world one of its finest fruits.

Perhaps the enterprising merchants of the Italian towns got more than they bargained for when they aided the revival of Greek culture. They sought merely to promote a freer atmosphere; they reaped a whirlwind. Instead of continuing to dwell and prosper on firm ground, the *terra firma* of an immovable earth, they found themselves clinging precariously to a rapidly spinning globe which was speeding about the sun at an inconceivable rate. It was probably sorry recompense to these merchants that the very same theory which shook the earth free also freed the mind of man.

The fertile soil for these new blossoms of the mind was the reviving Italian universities. There Nicolaus Copernicus became imbued with the Greek conviction that nature is a harmonious medley of mathematical laws and there too he learned of the suggestion of the ancient Greek, Aristarchus, that the behavior of the planets might be describable by regarding them as moving about a stationary sun and by introducing daily rotation of the earth. In Copernicus's mind these two ideas coalesced. Harmony in the universe demanded a heliocentric theory and he became willing to move heaven and earth in order to establish it.

Copernicus was born in Poland. After studying mathematics and science at the University of Cracow, he decided to go to Bologna, where learning was more widespread. There he studied astronomy under the influential teacher Novara, a foremost Pythagorean. It was at Bologna, also, that he came into contact with Greek culture and Greek astronomy.

In the year 1500 Copernicus was appointed canon of the Cathedral of Frauenberg in East Prussia but got leave to continue his studies in Italy. For the next few years, he studied medicine and canon law, securing a doctor's degree in each field, and continued his thinking on astronomical problems. In 1512 he finally assumed the duties of his position at Frauenberg, but spent much time during the remaining thirty-one years of his life in a little tower on the wall of the cathedral closely observing the planets with naked eye and making untold measurements with crude home-made instruments. The rest of his spare time he devoted to improving his new theory on the motions of heavenly bodies.

After years of mathematical reflection and observation, Copernicus finally circulated a manuscript describing his theory and his work on it. The reigning pope, Clement VII, approved of the work and requested publication. But Copernicus hesitated. The tenure of office of the Renaissance popes was rather brief and a liberal pope might readily be succeeded by a reactionary one.

Ten years later Copernicus's friend Rheticus persuaded him to allow the publication, which Rheticus then undertook. A Lutheran pastor, Osiander, took it upon himself to add in press an apologetic preface claiming for the work merely the status of a hypothesis which facilitated astronomical calculations. This preface, Osiander believed, would preclude suppression of the book and permit it to make its way in a hostile world.

While lying paralyzed from an apopleptic stroke, Copernicus received a copy of his book. It is unlikely that he was able to read it for he never recovered. He died shortly afterwards, in the year 1543, having contributed to the world in overflowing measure for the seventy years allotted to him.

At the time that Copernicus delved into astronomy the science was practically in the state in which Ptolemy had left it. However, it had become increasingly difficult to include under the Ptolemaic heavens the knowledge and observations of earth and sky accumulated, largely by the Arabians, during the succeeding centuries. By Copernicus's time it was necessary to invoke a total of seventy-seven mathematical circles in order to account for the motion of the sun, moon, and five planets under the geocentric scheme created by Hipparchus and Ptolemy. No wonder that Copernicus grasped at the possibilities in the Greek idea of planetary motion about a stationary sun.

Copernicus took over some other ideas from the Greeks. Because the latter believed that the natural motion of bodies was circular, he, too, used the circle as the basic curve on which to build his explanation of the motions of the heavenly bodies. For a mystic reason similar to that held by the Greeks he also retained the notion that the speed of any heavenly object must be constant. A change in speed, he reasoned, could be caused only by a change in motive power, and since God, the cause of the motion, was constant, the effect could not be otherwise.

Then Copernicus proceeded to do what no Greek had ever attempted: he carried out the mathematical analysis required by the heliocentric hypothesis. His basic idea was still that used by Hipparchus and Ptolemy, that is, epicycles. In accordance with this scheme the motion of a planet was mathematically described by having it move along a circle with constant speed while the center of the circle also moved with constant speed on another circle. The center of this latter circle was either the Sun or did itself move on a circle around the Sun. However, merely by using the Sun where Hipparchus and Ptolemy had used the Earth, Copernicus found that he was able to reduce the total number of circles involved from seventy-seven to thirty-one. Later, to secure better accord with observations, he refined this idea somewhat by putting the sun near, but not quite at, the center of some of these aggregations of circles.

When Copernicus surveyed the extraordinary mathematical simplification which the heliocentric hypothesis afforded, his satisfaction and enthusiasm were unbounded. He had found a simpler mathematical account of the motions of the heavens and, hence, one which must therefore be preferred, for Copernicus, like all scientists of the Renaissance, was convinced that, "Nature is pleased with simplicity, and affects not the pomp of superfluous causes". God had so designed the universe. Copernicus could pride himself, too, that he had dared to think through what others, including Archimedes, had rejected as absurd.

Copernicus did not finish the job he set out to do. Though the hypothesis of a stationary sun considerably simplified astronomical theory and calculations, the epicyclic paths of the planets did not quite fit observations and Copernicus's few attempts to patch up his theory, always on the basis of circular motions, did not succeed.

It remained for the German, Johann Kepler, some fifty years later, to complete and extend the work of Copernicus. Like most youths of those days who showed some interest in learning he was headed for the ministry. While studying at the University of Tübingen he obtained private lessons in Copernican theory from a teacher with whom he had become friendly. The simplicity of this theory impressed Kepler very much. Perhaps this leaning awakened suspicions in the superiors of the Lutheran Church, for they questioned Kepler's devoutness, cut short his ministerial career, and assigned him to the professorship of Mathematics and Morals at the University of Gratz. This position called for a knowledge of astrology, and so he set out to master the rules of that "art". By way of practice he checked its predictions with his own fortunes.

As an extracurricular activity he applied mathematics to matrimony. While at Gratz he had married a wealthy heiress. When this wife died he listed the young ladies eligible for the vacancy, rated each on a series of qualities, and averaged the grades. Women being notoriously less rational than nature, the highest ranking prospect refused to obey the dictates of mathematics and declined the honor of being Mrs. Kepler. Only by substituting a smaller numerical value was he able to balance the equation of matrimony.

Kepler's interest in astronomy continued and he left Gratz to become an assistant to that most famous observer, Tycho Brahe. On Brahe's death Kepler succeeded him as astronomer, part of his duties being once again of an astrological nature for he was required to cast horoscopes for worthies at the court of his employer, Rudolph II. He reconciled himself to this work with the philosophical view that nature provided astronomers with astrology just as she had provided all animals with a means of existence. He was wont to refer to astrology as the daughter of astronomy who nursed her own mother.

During the years he spent as astronomer to the Emperor Rudolph II, Kepler did his most serious work. It is extremely interesting that neither he nor Copernicus ever succeeded in ridding himself of the scholasticism from which their age was emerging, Kepler, in particular, mingled science and mathematics with theology and mysticism in his approach to astronomy, just as he combined wonderful imaginative power with meticulous care and extraordinary patience.

Moved by the beauty and harmonious relationships of the Copernican system he decided to devote himself to the search for whatever additional geometrical harmonies the data supplied by Tycho Brahe's observations might suggest, and, beyond that, to find the mathematical relationships binding all the phenomena of nature to each other.

However, his predilection for fitting the universe into a preconceived mathematical pattern led him to spend years in following up false trails. In the preface to his *Mystery of the Cosmos* we find him writing:

> I undertake to prove that God, in creating the universe and regulating the order of the cosmos, had in view the five regular bodies of geometry as known since the days of Pythagoras and Plato, and that he has fixed according to those dimensions, the number of heavens, their proportions, and the relations of their movements.

And so he postulated that the radii of the orbits of the planets were the radii of spheres related to the five regular solids in the following way. The largest radius was that of the orbit of Saturn. In a sphere of this radius he supposed a cube to be inscribed. In this cube a sphere was inscribed whose radius was that of the orbit of Jupiter. In this sphere he supposed a tetrahedron to be inscribed and in this another sphere whose radius was that of the orbit of Mars, and so on through the five regular solids. The scheme called for six spheres, just enough for the number of planets then known. The beauty and neatness of the scheme overwhelmed him so completely that he insisted for some time on the existence of just six planets because there were only five regular solids to determine the distances between them.

While publication of this "scientific" hypothesis brought fame to Kepler and makes fascinating reading even today, the deductions from it were, unfortunately, not in accord with observations. He reluctantly abandoned the idea, but not before he had made extraordinary efforts to apply it in modified form.

If the attempt to use the five regular solids to ferret out nature's secrets did not succeed, Kepler was eminently successful in later efforts to find harmonious mathematical relations. His most famous and important results are known today as Kepler's three laws of planetary motion. These laws became so famous and so valuable to science that he earned for

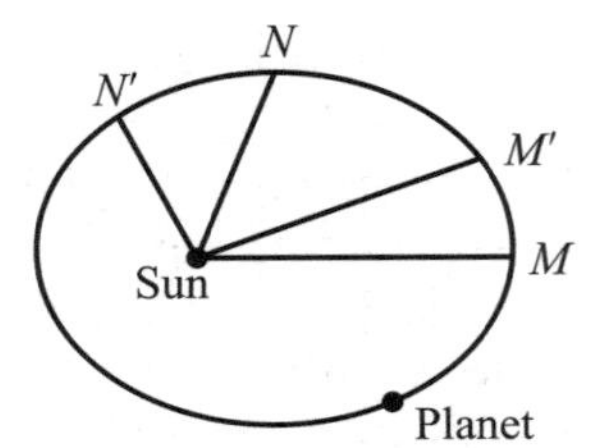

Figure 1.

himself the title of "legislator of the sky."

The first of these laws says that the path of each planet is not a circle but an ellipse with the sun at one focus of the ellipse. Hence one ellipse replaced the several circular motions superimposed on each other which even Copernicus's epicyclic theory had required to describe the motion of a planet. The simplicity gained thereby convinced Kepler that he must abandon attempts to use circular motions.

Kepler's second law is best understood with the aid of a diagram, (Figure 1). Copernicus, we saw, believed that each planet moves on its immediate circle at a constant speed, that is, that it covers equal distances in equal intervals of time. Kepler at first held firmly to the doctrine that each planet moves along its ellipse at a constant speed, but his observations finally compelled him to abandon this cherished belief. His joy was great when he discovered that he could replace it by an equally pleasing law, for thereby his conviction was reaffirmed that nature was mathematically designed.

If MM' and NN' (Figure 1) are distances traversed by a planet in equal intervals of time, then, according to the principle of constant speed, MM' and NN' would have to be equal distances. However, according to Kepler's second law, MM' and NN' are generally not equal but the areas OMM' and ONN' are equal. Thus Kepler replaced equal distances by equal areas and the mathematical rationality of the universe remained unshaken. To wrest such a secret from the heavens was indeed a triumph, for the relationship described is by no means as easily discernible as it may appear to be here on paper. These two laws Kepler published in a book entitled *On the Motions of the Planet Mars.*

Kepler's third law is as famous as his first two. It says that the square of the time of revolution of any planet is equal to the cube of its average distance from the sun (provided the time of the earth's revolution and the earth's distance from the sun are the units of time and distance.)

It is clear that mathematical concepts and mathematical laws are the essence of the new theory. But what is even more pertinent is that the mathematical excellence of the new theory recommended it to Copernicus and Kepler despite many weighty arguments against it. Indeed had Copernicus or Kepler been less the mathematician and more the scientist, or had they been blind religionists, or even what the world calls "sensible" men, they could never have stood their ground. As scientist neither could answer Ptolemy's logical objections to a moving earth. Why, for example, if the earth rotates from west to east does not an object thrown up into the air fall back to the west of its original position? If, as all scientists since Greek times believed, the motion of an object is proportional to its mass, why doesn't the earth leave behind it objects of lesser weight? Why doesn't the earth's rotation cause objects on it to fly off into space just as an object whirled at the end of a string tends to fly off into space? These objections remained unanswered even at the time of Kepler's death.

No less a personage than Francis Bacon, the father of empirical science, summed up in 1622 the scientific arguments against Copernicanism:

> In the system of Copernicus there are found many and great inconveniences: for both the loading of the earth with a triple motion is very incommodius and the separation of the sun from the company of the planets with which it has so many passions in common is likewise a difficulty and the introduction of so much immobility into nature by representing the sun and the stars are immovable ... all these are the speculations of one, who cares not what fictions he introduces into nature, provided his calculations answer.

While the clarity of Bacon's arguments could be surpassed, the opposition of a man of his reputation and ability could not be lightly brushed aside. Bacon's conservatism was due, incidentally, to his persistent inability to appreciate the importance of exact measurement in spite of his insistence on observation.

Were Copernicus and Kepler more "sensible," "practical," men they would never have defied their senses. We do not feel either the rotation or the revolution of the earth despite the fact that Copernican theory has us rotating through about 3/10 of a mile per second and revolving around the sun at the rate of about 18 miles per second. On the other hand we do see the motion of the sun. To the famous astronomical observer, Tycho Brahe, these arguments were conclusive proof that the earth must be stationary. In the words of Henry More, "sense pleads for Ptolemee."

Were Copernicus and Kepler religionists, they would not have been willing even to investigate the possibilities of a heliocentric hypothesis. Medieval theology, buttressed by the Ptolemaic system, held that man was at the center of the universe and the chief object of God's attention. By putting the sun at the center of the universe, the heliocentric theory denied this comforting dogma of the Church. It made man appear to be one of a possible host of wanderers on many planets drifting through a cold sky and around a burning ball. He was an insignificant speck of dust on a whirling globe instead of chief actor on the central stage. Unlikely was it that he was born to live gloriously and to attain paradise upon his death, or that he was the object of God's ministrations. Banished from the sky was the seat of God, the destination of saints and of a Deity ascended from the earth, and the paradise to which the good people could aspire. In short, the undermining of the Ptolemaic order of the universe removed a cornerstone of the Catholic edifice and threatened to topple the whole structure.

Copernicus's willingness to battle entrenched religious thinking is well evidenced by a passage in a letter to Pope Paul III:

> If perhaps there are babblers [he says] who, although completely ignorant of mathematics, nevertheless take it upon themselves to pass judgment on mathematical questions and, improperly distorting some passages of the Scriptures to their purpose, dare to find fault with my system and censure it, I disregard them even to the extent of despising their judgment as uninformed.

Religion, physical science, and common sense bowed to mathematics at the behest of Copernicus and Kepler. So did even astronomy. The hypothesis of a moving earth calls for motion of the stars relative to the earth. But observations of the sixteenth century failed to detect this relative motion. Now no scientific hypothesis which is inconsistent with even one fact is really tenable. Nevertheless Copernicus and Kepler held to their heliocentric view. These moonstruck lovers of mathematics were designing a beautiful theory. If the theory didn't fit all the facts, too bad for the facts. Copernicus, though deliberately vague on the question of the motion of the earth relative to the stars, disposed of the problem by stating that the stars were at an infinite distance. Apparently not too satisfied himself with this statement he assigned the problem to the philosophers. The true explanation, namely, that the stars were very far from the earth, so far as to render their relative motion undetectable, was not acceptable to the Renaissance "Greeks" who still believed in a closed and limited universe. The true distances involved were utterly beyond any figure which they would have thought reasonable. Actually the problem of accounting for the motion of the stars was not solved until 1838 when the mathematician Bessel finally measured the parallax of the nearest star and found it to be 0.76".

In view of all these arguments and forces working against the new theory why did Copernicus and Kepler advocate it? To answer this question we must examine their philosophical position.

It is necessary to dispose, first of all, of a commonly held belief that practical problems motivate and determine the course of mathematical and scientific thinking. Knowing that the great explorations of the fifteenth and sixteenth centuries demanded a better astronomy one is tempted to ascribe the motivation for their work to the need for more reliable geographical information and improved techniques in navigation. But Copernicus and Kepler were not at all concerned with the pressing practical problems of their age. What these men do owe to their times was the opportunity to come into contact with Greek thought, an opportunity furnished by the revival of learning in Italy. Copernicus, we saw, studied there and Kepler benefitted by Copernicus's work. Also both men owe to their times an atmosphere certainly more favorable to the acceptance of new ideas than the one which prevailed two centuries earlier. The geographical explorations, the Protestant Revolution, and so many other exciting movements were challenging conservatism and complacency, that one new theory did not have to bear the brunt of the natural opposition to change.

Actually Copernicus and Kepler investigated their most revolutionary theory to satisfy certain philosophical and religious interests. Having become convinced of the Pythagorean doctrine that the universe is a systematic, harmonious structure whose essence is mathematical law, they set about discovering this essence. Copernicus's published works give unmistakable, if indirect, indications of his reasons for devoting himself to astronomy. He values his theory of planetary motion not because it improves navigational procedures but because it reveals the true harmony, symmetry, and design in the divine workshop. It is wonderful and overpowering evidence of God's presence. Writing of his achievement, which was thirty years in the making, Copernicus expresses his gratification:

> We find, therefore, under this orderly arrangement, a wonderful symmetry in the universe, and a definite relation of harmony—in the motion and magnitude of the orbs, of a kind that it is not possible to obtain in any other way.

He does mention in the preface to the work already mentioned that he was asked by the Lateran Council to help in reforming the calendar which had become deranged over a period of many centuries. Though he writes that he kept this problem in mind it is quite apparent that it never dominated his thinking.

Kepler, too, makes clear his dearest interests. His published work, the fruit of his labors, attests to the sincerity of his search for harmony and law in the creations of the divine power. In the preface to his *Mystery of the Cosmos* he says:

> Happy the man who devotes himself to the study of the heavens: he learns to set less value on what the world admires the most: the works of God are for him above all else, and their study will furnish him with the purest of enjoyments.

A major treatise entitled *The Harmony of the World*, which Kepler published in 1619, actually expounds a system of heavenly harmonies, a new "music of the spheres," which makes use of the varying velocities of the six planets. These harmonies are enjoyed by the sun which Kepler endowed with a soul specifically for this purpose. Lest it be supposed that this treatise was just a lapse into poetic mysticism, one should note that it also announced his celebrated third law of motion.

The work of Copernicus and Kepler is the work of men searching the universe for the harmony which their commingled religious and scientific beliefs assured them must exist, and exist in aesthetically satisfying mathematical form.

Having discovered new mathematical laws of the universe Copernicus and Kepler were irresistibly attracted by their simplicity. To these men, who were convinced that an omnipotent being designing a mathematical universe would necessarily prefer simplicity, this feature of the new theory attested to its truth. Indeed only a mathematician assured that the universe was rationally and simply ordered would have had the mental fortitude to set at naught the prevailing philosophical, religious, and scientific beliefs and the perseverance to work out the mathematics of so revolutionary an astronomy. Only one possessed of unshakeable convictions as to the importance of mathematics in the design of the universe would have dared to uphold the new theory against the powerful opposition it was sure to encounter. As a matter of history Copernicus did address himself to mathematicians because he expected that only these would understand him, and in this respect he was not disappointed.

Granted that it was the superior mathematics of the new theory which inspired Copernicus and Kepler, and later Galileo, to repudiate religious convictions, scientific arguments, common sense, and well entrenched habits of thought, how did the theory help to shape modern times?

First of all, Copernican theory has done more to determine the content of modern science than is generally recognized. The most powerful and most useful single law of science is Newton's law of gravitation. Without undertaking here the discussion reserved for a more appropriate place we can say that the best experimental evidence for this law, the evidence which established it, depends entirely upon the heliocentric theory.

Second, Copernican theory is responsible for a new trend in science and human thought, barely perceived at the time but all important today. Since one's eyes do not see, nor one's body feel, the rotation and revolution of the earth, the new theory rejected the evidence of the senses and based itself on reason alone. Copernicus and Kepler thereby set the precedent which guides modern science, namely, that reason and mathematics are more important in understanding and interpreting the universe than the evidence of the senses. Vast portions of electrical and atomic theory and the whole theory of relativity would never have been conceived had scientists not come to accept the reliance upon reason which Copernican theory first exemplified. In this very significant sense Copernicus and Kepler begin the Age of Reason, while they also fulfilled one of the cardinal functions of scientists and mathematicians, namely, to provide a rational comprehension of natural phenomena.

By deflating the stock of *homo sapiens* Copernican theory reopened questions which the guardians of Western civilization were answering dogmatically upon the basis of Christian theology. Once there was only one answer; now there are ten or twenty to such basic questions as: *Why does man desire to live and for what purpose? Why should he be moral and principled? Why seek to preserve the race?* It is one thing for man to answer such questions in the belief that he is the child and ward of a generous, powerful, and provident God. It is another to answer them knowing that he is a speck of dust in a cyclone.

Copernican theory flung such questions in the faces of all thinking men and women, and thinking beings could not reject the challenge. Their struggles to recover mental equilibrium which was even further upset by the mathematical and scientific work following Copernicus and Kepler provide the key to the history of thought of the last few centuries.

Much evidence can be found in literature since Kepler's times of the agitation aroused by the new disturbing thoughts. The metaphysical John Donne, though trained in and content with the encyclopedic and systematic scholasticism is compelled to acknowledge the undesirable complexity to which Ptolemaic theory had led:

> We think the heavens enjoy their spherical
> Their round proportion, embracing all;
> But yet their various and perplexed course,
> Observed in divers ages, doth enforce
> Men to find out so many eccentric parts,
> Such diverse downright lines, such overthwarts,
> As disproportion that pure form.

Though the argument for Copernicanism is clear to him he can only deplore that fact that the sun and stars no longer run in circles around the earth.

Milton, also, ponders the challenge to Ptolemaic theory but manages to retain it nevertheless. Unable to meet the new mathematics on its own ground he turns instead to rebuking its creators. Man should admire, not question, the works of God.

> From Man or Angel the great Architect
> Did wisely to conceal, and not divulge
> His secrets to be scann'd by them who ought
> Rather admire; ...

Yet even Milton was unconsciously moved to accept a more mysterious and a vaster space than the compact, thoroughly defined space of Dante, for example.

The gentle remonstrations of the milder poets, Ben Jonson's satire, Bacon's scientific arguments as well as personal jealousy, the ridicule of professors, the mathematical arguments of the brilliant Cardan, the resentment of astrologers who feared for their livelihood, Montaigne's scepticism, complete rejection from Shakespeare, and condescending mention from John Milton earned for Copernicus a reputation as a new *Duns Scotus*, the learned crazy one. In 1597 Galileo wrote to Kepler describing Copernicus as one "who though he has obtained immortal fame among the few, is nevertheless, ridiculed and hissed by the many, who are fools."

Nevertheless the opinion of the few prevailed. The cultural revolution gained momentum: people were compelled to think, to challenge existing dogmas, and to re-examine long accepted beliefs. And from criticism and re-examination of long established doctrines emerged many of the philosophical, religious, and ethical principles now accepted in Western civilization.

By far the greatest value of the heliocentric theory to modern times is the contribution it made to the battle for freedom of thought and expression. Because man is conservative, a creature of habit, and convinced of his own importance, the new theory was decidedly unwelcome. The vested interests of well entrenched scholars and religious leaders caused them to oppose it. The most momentous battle in history, the battle for the freedom of the human mind, was joined on the issue of the right to advocate heliocentrism.

The self-appointed representatives of God entered the fray with vicious attacks on Copernicanism. Martin Luther called Copernicus an "upstart astrologer" and a "fool who wishes to reverse the entire science of astronomy". Calvin thundered: "who will venture to place the authority of Copernicus above that of the Holy Spirit?" Do not Scriptures say that Joshua commanded the sun and not the earth to stand still? That the sun runs from one end of the heavens to the other? That the foundations of the earth are fixed and cannot be moved? The Inquisition condemned the theory as "that false Pythagorean doctrine utterly contrary to the Holy Scriptures," and in 1616 the Index banned all publications dealing with Copernicanism. Indeed if the fury and high office of the opposition is a good indication of the importance of an idea no more valuable one was ever advanced.

So shackled did the spirit of inquiry become in that age that when Galileo discovered the four satellites of Jupiter with his small telescope, some religionists refused to look through his instrument to see those bodies for themselves. And many who did tempt the devil by looking refused to believe their own eyes. It was this bigoted attitude that made it dangerous to advocate the new theory. One risked the fate of Giordano Bruno who was put to death by the Inquisition "as mercifully as possible and without the shedding of blood," the horrible formula for burning a prisoner at the stake.

Despite the earlier ecclesiastic prohibition of works on Copernicanism, Pope Urban VIII did give Galileo permission to publish a book on the subject, for the Pope believed that there was no danger that any one would ever prove the new theory necessarily true. Accordingly, in 1632 Galileo published his *Dialogue on the Two Chief Systems of the World*, in which he compared the Ptolemaic and Copernican doctrines. To please the Church and so pass the censors he incorporated a preface to the effect that the heliocentric idea was only a product of the imagination. Unfortunately Galileo wrote too well and the Pope began to fear that the argument for Copernicanism, like a bomb wrapped in silver foil, could still do a great deal of damage to the faith. The Church aroused itself once more to do battle against a heresy "more scandalous, more detestable, and more pernicious to Christianity than any contained in the books of Calvin, of Luther, and of all other heretics put together." Galileo was again called by the Roman Inquisition and compelled on the threat of torture to declare: "The falsity of the Copernican system cannot be doubted, especially by us Catholics.... "

Burning faggots, the wheel, the rack, the gallows, and other ingenious refinements of torture were more conducive to orthodoxy than to scientific progress. When he heard of Galileo's persecution, Descartes, who was a nervous and timid individual, refrained from advocating the new theory and actually destroyed one of his own works on it.

However the heliocentric theory became a powerful weapon with which to fight the suppression of free thought. The truth (at least to the seventeenth and eighteenth centuries) of the new theory and its incomparable simplicity won more and more adherents as people gradually realized that the teachings of religious leaders could be fallible. It soon became impossible for these leaders to impose their authority over all Europe and the way was prepared for freer thought in all spheres.

The import of this battle and its favorable outcome should not be lost to us. No modern dictatorship has done more to suppress free thought than did the religious and secular powers of sixteenth and seventeenth century Europe. No modern dictatorship has perpetrated more fiendish torture, terror, and murder to impose its will on its subjects. Those who still enjoy and those who have lost the freedoms so recently acquired in Western civilization cannot fail to appreciate how much was at stake in the battle to advance the heliocentric theory and how much we owe to the men of gigantic intellect and stout heart who carried the fight.

Fortunately for us the very fires which consumed the martyrs to free inquiry dispelled the darkness of the Middle Ages. The fight to establish the heliocentric theory broke the stranglehold which ecclesiasticism held on the minds of men. The mathematical argument proved more compelling than the theological one and the battle for the freedom to think, speak and write was finally won. The scientific Declaration of Independence is a collection of mathematical theorems.

What Mathematics Has Meant to Me

E. T. Bell

Mathematics Magazine
24 (1950/51), 161

Editors' Note: This was the first of what was intended to be a series of such "testimonials" in the *Magazine*. It seems there was only one additional contribution to the series, What Mathematics Means to Me, by Lewis Bayard Robinson (25 (1951–52), 115). Bell was best known for his *Men of Mathematics*, but he was a prolific author of many other books and a significant research mathematician in number theory, algebra, and combinatorics. He wrote these paragraphs late in his career as Professor of Mathematics at Caltech. Bell died in 1960.

The Editor has asked for about 400 words on "what mathematics has meant to me." Notice the "me"—not someone else. This will account for all the "I" and "me" in what follows, for which I apologise. I am as embarrassed as if I had inadvertently stood up in church to tell the congregation how and why I had been saved. You may be even more embarrassed in witnessing my testimony.

My interest in mathematics began with two school prizes, one in Greek, the other for physical laboratory, both richly bound in full calf. The Greek Prize was Clerk Maxwell's classic on electricity and magnetism, the other, Homer's *Odyssey*. My cousin got the prize for Greek, I got the other. He read mine, I tried, and failed, to read his. The integral signs were particularly baffling to one who had not gone beyond the binomial theorem for a positive integral exponent. The calculus was not a school subject at the time, so my mother paid for private lessons from a man—the late E. M. Langley—who was the best teacher I ever had. From him I learned what dy/dx and $\int y\,dx$ mean. The rest was comparatively easy, and I found myself in possession of a key that unlocks a hundred doors. Although I have never done anything in mathematical physics, I have been able to read some of the great classics which, without the calculus, would have been incomprehensible. This has been one thing that has made life interesting. How some philosophers of science and others have the audacity to write on relativity and the quantum theory without a reading knowledge of the calculus is the wonder of the ages.

Another thing I got from mathematics has meant more to me than I can say. No man who has not a decently skeptical mind can claim to be civilized. Euclid taught me that without assumptions there is no proof. Therefore, in any argument, examine the assumptions. Then, in the alleged proof, be alert for inexplicit assumptions. Euclid's notorious oversights drove this lesson home. Thanks to him, I am (I hope!) immune to all propaganda, including that of mathematics itself. Mathematical 'truth' is no 'truer' than any other, and Pilate's question[1] is still meaningless. There are no absolutes, even in mathematics.

[1] The reference is to Pontius Pilate's question, "What is truth?," a response to Jesus' statement "Every one that is of the truth heareth my voice" (John 18:37). (Ed.)

Mathematics and Mathematicians from Abel to Zermelo

Einar Hille

Mathematics Magazine
26 (1952/53), 127–46

Editors' Note: Here one of the foremost twentieth century researchers in analysis, Einar Hille of Yale University, tries his hand at a historical overview of largely nineteenth century mathematics. Much of the material will be known to mathematical history buffs, but Hille has some interesting observations, insights, and stories that make this a worthwhile piece.

Though born to Swedish parents in New York in 1894, he returned with his mother to Sweden when he was two and received his education there, earning his PhD from the University of Stockholm in 1918. After appointments at Harvard and Princeton, he went to Yale where he remained until retirement. He married the sister of the eminent Norwegian number theorist, Øystein Øre, who was also a professor of mathematics at Yale and a biographer of Abel.

Hille was described as being "almost unique among mathematicians in applying functional analysis to investigate classical problems, rather than simply considering abstract situations for their own sake." (Reference [9] at the MacTutor website.) Hille himself said in the preface to his book, *Methods in Classical and Functional Analysis* (Addison-Wesley,1972), "If the book has a thesis, it is that a functional analyst is an analyst, first and foremost, and not a degenerate species of a topologist. His problems come from analysis and his results should throw light on analysis ... It seemed to me that I could do some useful work in giving the student a historical perspective and in showing how the multitude of abstract concepts have arisen and are present in Euclidean spaces."

From the historical point of view, mathematics may be considered as a series of events in space-time associated with definite persons or, if you prefer, as a type of finite geometry where the coordinates measure, for instance, person, place, time, quality, and field of research. We should be able to get some idea of the structure of this geometry by considering suitable chosen cross sections. Two such cross sections will be used below.

In the first study the time coordinate is held approximately constant at $t = 1852$ A.D. and we shall list the prominent mathematicians of a century ago together with brief accounts of their work and background. In the second study, the field of research is held approximately constant and the time coordinate is restricted. Here we shall follow the development of analysis, in particular, that of complex function theory, over a period of a century, roughly from 1820 to the time of the first world war.

I. Who Was Who in Mathematics in 1852?

1. France

We shall try to reconstruct a Who's Who in Mathematics around 1850. We have to keep in mind that there were much fewer mathematicians in those days, fewer periodicals, little personal contact, no mathematical societies, no scientific meetings or congresses. Mathematics flourished in France, Germany, and Great Britain; outside of these countries research mathematicians were few and far between. There was interchange of ideas, however, mathematicians did write to each other, young students traveled abroad to study and so forth.

In this survey it is reasonable to start with France where there is a splendid mathematical tradition going back to the seventeenth century. The Revolution, having abolished the old Académie des Sciences, founded in 1666, and having liquidated most of its members, found it necessary for military reasons to revive the Academy as the Institut National in 1795 and to found the École Normale Supérieure and the École Polytechnique. The latter became the cradle of French mathematics, a role that it kept for about a century until this function was taken over by the École Normale. Organized and led by the geometer Gaspard Monge (1746–1818), the École Polytechnique gave an intense two year course concentrating on mathematics, descriptive geometry and drawing. Entrance was severely restricted, only 150 students could be accepted each year on the basis of searching examinations. Galois failed twice in this examination, during the first century of its existence. J. Hadamard held the highest score for admission with 1875 points out of a possible 2000. Only the best brains could secure admission, only hard work could carry the student through the intense training and the grueling final examinations. Most French mathematicians of the nineteenth century belonged to this school in one capacity or another as student, instructor, professor or examiner.

The time of the revolution and the first empire saw the flourishing of mathematics in France. J. L. Lagrange (1736–1813) spent the last twenty years of his wandering life in France, men like J. Fourier (1768–1830), P. S. Laplace (1749–1827), A. M. Legendre (1752–1833), and S. D. Poisson (1781–1840) brought French mathematics to new heights. A reaction is noticeable around 1830 and lasted for almost fifty years. The old heroes were dying off and the replacements are perhaps a shade less impressive.

The big name in Paris in 1850 was Augustin-Louis Cauchy (1789–1857). As an ardent royalist Cauchy had given up his professorship at the École Polytechnique after the July revolution in 1830 and had gone in exile with the Bourbons; they had trouble with loyalty oaths in those days too. He returned to Paris in 1838 and taught at a Jesuit college. In 1848 the February revolution opened the way to a professorship at the Sorbonne without any oath and Napoleon III did not interfere with him when the constitution was changed again in 1852. Cauchy was above all an analyst, the first rigorous analyst, the founder of function theory, of the theory of ordinary and partial differential equations, who also made important contributions to elasticity and optics as well as to algebra. We shall return to the work of Cauchy in Part II of this paper.

At the Collège de France in Paris we would have found Joseph Liouville (1809–1882), the man who started the mathematical theory of boundary value problems for linear second order differential equations, who produced the first integral equation and the first resolvent, but also the founder of the theory of transcendental numbers. He was the editor of the *Journal de Mathématiques Pures et Appliquées*, started in 1835 and still know as the *Journal de Liouville*. In connection with Liouville we often think of Jacques-Charles-François Sturm (1803–1855); he was Poisson's successor as professor of mechanics at the Sorbonne in 1850. Lesser lights, but famous in their days, were August-Albert Briot (1817–1882) and Jean-Claude Bouquet (1819–1885), great friends who collaborated in the theory of differential equations and elliptic functions. A young man by the name of Charles Hermite (1822–1901) was struggling for recognition, he still had to wait until 1869 for a professorship. Those mentioned so far were all analysts, but there were also geometers around: Jean-Victor Poncelet (1788–1867), the father of projective geometer, Charles Dupin (1784–1873) and Michel Chasles (1793–1880) were active. The algebraic geometer Chasles, who created the enumerative methods of geometry, devoted almost twenty-five years of his life to banking in his home town Chartres; in 1846 the Sorbonne created a professorship in higher geometry for him. The private fortune that he had amassed before returning to mathematics led him into collecting autographs and he was swindled into buying numerous forgeries. His most famous acquisition was a letter supposedly written by Mary Magdalene from Marseille to Saint Peter in Rome! Even Chasles ultimately had to admit that he had been swindled. Among French mathematicians we find few algebraists, the analysts attended to the theory of equations, and group theory was in its infancy. The publication of the work of Évariste Galois (1811–1832) did not take place until 1846.

A characteristic feature of French mathematics, then as well as now, is the strong centralization. Paris, that is, the Sorbonne, Collège de France, École Polytechnique, École Normale Supérieure, and the many technical schools, has most of the desirable positions while the provincial universities vary in importance, but cannot hold their own against Paris. There were numerous facilities for publication in France. Joseph-Diaz Gergonne (1771–1859), who competed with Poncelet in discovering the principle of duality, published the first mathematical journal in the world, the *Annales des mathématiques pures et appliquées*, in Nîmes during 1810–31 where he was then professor at the local lycée. Liouville started his journal in 1835. Notes could be published quickly in the *Comptes Rendus de l'Académie des Sciences* at Paris which has appeared weekly since August 1835. The restriction to short notes (originally four pages, now two) was due to Cauchy flooding the Academy with his publications. The École Polytechnique and later also the École Normale had their own journals.

2. Germany

Turning now to Germany we find a totally different picture. German mathematics is unimportant during the eighteenth century and is essentially a one-man-show during the first quarter of the nineteenth. From then on German mathematics is in a steady upward march until the time of the first world war and even during the empire this movement is not centralized as in France but is attached to a number of local centers. The Germany of 1850 was a geographical and cultural unit but not a political one. It was split into a large number of autonomous kingdoms, grand-duchies, duchies, principalities, free cities and what have you, and this kaleidoscopic picture reflected itself in a large number of independent and thriving universities. Apparently the many political boundaries interfered very little with the students and the professors who moved freely around. Prussia had the largest number of universities of all German lands: Berlin was the largest and was outstanding in mathematics but there were also important schools in Bonn and Königsberg. In the kingdom of Hanover we find the Georgia Augusta University at Göttingen. Incidentally, it was founded two hundred years ago by the then elector of Hanover who as King George II of Great Britain and Ireland advanced the cause of learning in the then colonies by the founding of the College of New Jersey (Princeton), King's College (Columbia), and Queen's College (Rutgers). Leipzig was the intellectual center of Saxony, in Bavaria Erlangen, in Baden Freiburg and Heidelberg, and in the Thuringian maze of principalities Jena kept the torch burning.

The eighteenth century was the time when the Swiss Leonard Euler and the Frenchman Lagrange, born in Italy, lived and worked in Berlin and filled the memoirs of the Prussian Academy, but of native German mathematicians there were very few. The nineteenth century starts out with Carl Friedrich Gauss (1777–1855), a giant in mathematics, astronomy, geodesy and magnetism, who got his doctor's degree in Helmstadt in 1799 and from 1807 on spent his life in Göttingen as director of the astronomical observatory. His publications in number theory, algebra, and differential geometry were basic, but many later discoveries in function theory and geometry had been anticipated by Gauss who published only sparingly.

Though Gauss had few direct pupils, he had started something in Germany which was of lasting importance for mathematics. This development gathered momentum in the eighteen twenties and led to a flourishing of German mathematics which lasted until the days of Hitler. For the next phase of the development we have to turn to Berlin and Königsberg. An outward sign of the changing times was the founding in Berlin in 1826 of the *Journal für die reine und angewandte Mathematik* by August Leopold Crelle (1780–1855), usually known as *Crelle's Journal* (1855–1880 *Borchardt's Journal* after the then editor). Crelle started out with a scoop: five memoirs by the young Norwegian mathematician Niels Henrik Abel (1802–1829) but there were also papers by Carl Gustav Jacob Jacobi (1804–1851) and by Jacob Steiner (1796–1863). In volume 3 appear the names of Peter Gustav Lejeune Dirichlet (1805–1859), August Ferdinand Möbius (1790–1868), and Julius Plücker (1801–1868). *Crelle's Journal* soon became known as the *Journal für die reine und angewandte Mathematik*, much to the embarrassment of its editor who was much more prominent as an engineer than as a mathematician. In addition to Crelle, the German mathematicians had at their disposal a large number of academy publications. The *Mathematische Annalen* did not start until 1868.

Jacobi got his degree in Berlin but came to Königsberg in 1826 where he did his fundamental work on elliptic functions in competition with Abel (see further in Part II). He became ordinary professor in 1831, but retired owing to ill health in 1842, whereupon he moved back to Berlin where he got a research chair without teaching duties. Jacobi is the founder of the Königsberg school to which belonged men like L. O. Hesse (1811–1874) and R. F. A. Clebsch (1833–1872), followed in later years by A. Hurwitz (1859–1919), D. Hilbert (1862–1943) and H. Minkowski (1864–1909). In more recent years men like K. Knopp (1882–1957)[1] and G. Szegő (1895–1985) held professorships there. The last mathematician of note to get his training in Königsberg seems to have been Th. Kaluza now in Göttingen. Perhaps there will also be a Kaliningrad mathematical school. Who knows?

Dirichlet was born in the Rhineland as the son of French emigrants. He spent the years 1822–27 as a teacher in a private family in Paris where he came under the influence of the great French analysts of the period. He also read and reread Gauss's *Disquisitiones Arithmeticæ*. Both facts are strongly reflected in his mathematical life. After a short time in Breslau he came to Berlin in 1829 as Privatdozent and became ordinary professor ten years later, the first of the great masters of the Berlin school. It is typical for German conditions, however, that he accepted a call

[1]For those who have died since the publication of this article in 1952, we have taken the liberty of entering the year of death. (Ed.)

to Göttingen to become Gauss's successor in 1855. He died four years later. Dirichlet worked in number theory, in particular analytic number theory (Dirichlet's series commemorates this fact), functions of a real variable, Fourier series, potential theory, etc. Dirichlet was an excellent expositor and exercised a strong influence on the younger men who came to Berlin such as Leopold Kronecker (1823–1891), F. G. Eisenstein (1823–1852), Richard Dedekind (1831–1916), and Bernhard Riemann (1826–1866). At Berlin we also find the brilliant Swiss born geometer Jacob Steiner who, however, never got beyond the extraordinary professorship to which he was appointed in 1834. Kronecker, a man of wealth, came to Berlin in 1855 and became attached to the University in 1861. In 1856 Ernst Eduard Kummer (1810–1893) was called to Berlin as the successor of Dirichlet. He was the father of ideal theory and also a geometer of note. The same year Karl Weierstrass (1815–1897) was called to Berlin: he had twelve hours a week at the Gewerbeakademie (a college for trade and commerce) and an extraordinariat at the university. The ordinary professorship did not come until 1864. The reader is referred to Part II for more details about Weierstrass.

Let us now turn to Göttingen and the year 1850. Gauss is still active. A young man by the name of Riemann has just returned from a three year stay in Berlin and is now working on a dissertation which gave him the doctor's degree in 1851 and ultimate fame and glory. At this stage Riemann is scarcely influenced by Gauss, but he worked with Wilhelm Weber, a famous physicist and one of the discoverers of the electric telegraph. From Dirichlet and Weber he picked up an interest in mathematical physics which influenced both his research and his lectures, later published in book form as *Differentialgleichungen der Physik*, first edited by Hattendorf, from the fourth edition by Heinrich Weber, and the seventh edition of 1925–26 by Philipp Frank and Richard v. Mises. Riemann qualified for the *venia legendi* in 1854, that is, for the right to give lectures as a Privatdozent. This required writing another dissertation (Habilitationsschrift) and giving a lecture (Habilitationsvortrag) before the faculty on one out of three preassigned topics. The former was his paper on trigonometric series, the latter dealt with the basic hypotheses of geometry. Riemann became ordinary professor in 1859. With Gauss, Dirichlet and Riemann as fathers of the Göttingen mathematical school a star was born which has never set.

There are few names to be added from the other German universities, Möbius (with the strip!) was professor of astronomy in Leipzig. He was a pupil of Gauss who turned him into an astronomer, but most of his production was in mathematics: number theory, combinatorics, and the barycentric calculus, his *magnum opus* of 1827, while the non-orientable manifolds was work done in his old age appearing in 1858. In Bonn we find another famous geometer, Julius Plücker, professor of physics and mathematics from 1838 until his death in 1868. As a physicist he was an experimental one and did pioneering work in spectral analysis which took up most of his time. But his contributions to geometry such as the Plücker coordinates in line geometry, a subject created by him, and the Plücker formulas in algebraic geometry, are fundamental. Plücker had several famous pupils, the physicist Hittorf was outstanding in spectral analysis, the mathematician Felix Klein (1849–1925) published his collected works. Finally we have to mention Christian v. Staudt (1798–1867), professor in Erlangen since 1835 until his death in 1867. Among his main achievements are his making projective geometry independent of metric considerations and the introduction of imaginary elements in geometry. His ideas ripened slowly, his *Geometrie der Lage* appeared in 1847, followed by three further *Beiträge* in 1856, 1857 and 1860.

3. Great Britain and Ireland

Due to the priority fights between Newton's pupils and those of Leibnitz, mathematics in the British isles had enjoyed a splendid but unhealthy isolation for over a century, the after-effects of which were not overcome until after 1900. One of the consequences of this was an almost complete lack of feeling for analysis per se. What there was of analysis went into mathematical physics. As an example we could take George Green (1793–1841) who applied analysis to electricity and magnetism. Drink carried him off to an early death, but his name lives (Green's theorem, Green's function, etc.). George Gabriel Stokes (1819–1903), another Cambridge don, made fundamental contributions to optics and hydrodynamics. There is a theorem of Stokes in the Calculus, there is also a phenomenon of Stokes in the theory of differential equations. To the same tradition belongs the work of the famous physicists James Clerk Maxwell (1831–1879) and William Thomson, Lord Kelvin (1824–1907). Sir William Rowan Hamilton (1805–1865) was professor of astronomy at Trinity College in Dublin, but is more famous for his work in optics and mechanics (the Hamiltonian equations) and as creator of vector analysis and the theory of quaternions.

But there is also another direction in British mathematics which breaks through around 1850, represented by the three names Arthur Cayley (1821–1895), George Salmon (1819–1904), and James Joseph Sylvester (1814–1897). They were personal friends. Cayley and Sylvester read for the bar together in London. Cayley and Salmon had a joint paper on the twenty-seven lines on the cubic surface in 1849 and Cayley helped Salmon to revise his Higher Plane Curves. Salmon stayed all his life at Trinity, he was a fellow from 1840 on, became professor of divinity in 1866, and provost in 1888. He is best known for his books on geometry and algebra which were much read and translated into French and German, but he also wrote books on theology.

Cayley was one of the most prolific mathematicians of all time: starting production in 1841 he produced a total of 887 papers. And this intense activity was kept up while he was reading for the bar and during the fourteen years he practiced law—he was a conveyancer—as a matter of fact, most of his basic ideas stem from this period. In 1863 he retired from practice and accepted the Sadlerian professorship of pure mathematics in Cambridge which he held until his death. Cayley's nine memoirs on quantics (= forms) are famous, he started the theory of matrices and, together with Sylvester, the theory of invariants. The very name of invariant is due to Sylvester who also introduced covariants, contragradient, discriminant, etc. Silvester used to refer to himself as the new Adam since he had named so many things. Sylvester held a multitude of positions, including a brief stay at the University of Virginia in the early forties. He had been an actuary, he was called to the bar in 1850, from 1855 to 1870 he taught at the Military Academy at Woolwich, 1876–1883 he spent at The Johns Hopkins University where he started the *American Journal of Mathematics* as well as the high traditions of Johns Hopkins in mathematics. At seventy he accepted a call to Oxford as successor of the number theory man H. J. S. Smith (1826–1883) as Savilian professor of geometry.

If to this list we add the logicians Augustus de Morgan (1806–1871) and George Boole (1815–1864) we have a fairly good picture of British mathematics at the middle of the last century. The *Cambridge Journal of Mathematics* started in 1837, after four volumes it became the *Cambridge and Dublin Mathematical Journal*, later (in 1855) the *Quarterly Journal of Pure and Applied Mathematics*. The London Mathematical Society, the first of its kind, was organized in 1865.

4. Other countries

The compiler of our Who's Who would have found a single research mathematician in this country, namely Benjamin Peirce (1809–1880), professor at Harvard since 1833, whose work on linear associative algebra was really fundamental. In the seventies things are beginning to stir elsewhere also; we have already referred to Sylvester's stay at John Hopkins. In 1871 Josiah Willard Gibbs (1839–1903) became professor of mathematical physics at Yale; his main work on thermodynamics, equilibria and the phase rule date from 1873–78. His work on vector analysis was of some importance and one or two of his students made mathematical history. The real awakening did not come until the eighteen nineties, however.

Back in Europe again, let us look briefly at the situation in Italy where the mathematical tradition goes back to the year 1200. The beginning of the nineteenth century is meager, however. After the death of Paolo Ruffini (1765–1822), there is not much mathematics in Italy, but the awakening in mathematics came with the general national awakening and shortly before the unification. As father of modern Italian mathematics it is customary to acknowledge Francesco Brioschi (1824–1897), Enrico Betti (1823–1892), Felice Casorati (1835–1890) and Luigi Cremona (1830–1903). Brioschi became professor of applied mathematics in Pavia in 1852; in 1858 he made a journey to France and Germany together with Betti and Casorati and it is from this journey that the revival is dated. Brioschi organized the technological institute of Milan from 1862 on, Cremona played a similar part in Rome, and Betti became the director of the Scuola Normale Superiore of Pisa. To these names should be added that of Eugenio Beltrami (1835–1900). Most of the Italian mathematicians turned towards algebraic geometry.

Euler spent over twenty years in Russia but, beyond filling the memoirs of the Saint Petersburg Academy (for which purpose he had been imported), there is little trace of his activities. Around 1850 we would have found at least three prominent mathematicians in Russia: Victor Jacovlevich Bouniakovski (1804–1889), Pafnuti Livovich Chebicheff (1821–1894), both in St. Petersburg (= Petrograd

= Leningrad), and Nicolay Ivanovich Lobatchevsky (1793–1856) in Kazan. Bouniakovsky's inequality was the customary name in Russia for what we know as Schwarz's inequality (usually misspelt Schwartz in this country). Chebicheff started the investigation of extremal problems in function theory and problem of best approximation. He also worked with prime numbers. Lobatchevsky had been pondering over the parallel postulate since 1815; around 1826 he arrived at the alternate postulate of two parallels and his investigations were published in seven memoirs 1829–1856.

Turning to Hungary we find the other discoverer of non-euclidean geometry Janos (= John or Johann) Bolyai (1802–1860) whose discovery was published in an appendix to a mathematical memoir of his father's, Farkas (= Wolfgang) Bolyai (1775–1856), in 1832–1835. It is well known that Gauss had been thinking along similar lines; it is perhaps more than a coincidence that both Wolfgang Bolyai and Lobatchevsky's teacher Bartels in Kazan were personal friends of Gauss.

There is little activity elsewhere in Europe. Switzerland had overexerted itself during the eighteenth century with all the Bernoullis and with Euler and was recovering from the strain. Scandinavia had produced Abel and new talent was in the making both in Norway and in Sweden, but there is nothing to report until after 1860 to 1870.

II. The development of analysis, primarily complex function theory, until the time of the first world war

5. Cauchy

The outstanding mathematical creation of the nineteenth century is hard to single out and the choice will largely depend upon individual preference: projective geometry, invariant theory, group theory, theory of sets, complex function theory, each of these will have its spokesmen, and the list is not exhausted. My vote is cast for complex function theory, the development of which I propose to describe on the following pages.

We have already mentioned Augustin-Louis Cauchy (1789–1857) in Part I of this paper. Cauchy laid the foundations of complex function theory in his *Cours d'Analyse* of 1821 where he developed the theory of the elementary functions in the complex plane. This was followed by a small memoir on complex integration in 1825, but the general integral theorem is considerably later, about 1840. That holomorphic functions can be expanded in power series he proved in 1831, while in exile in Turin. In 1843 P. A. Laurent (1813–1854) proved the expansion theorem for the annulus and in 1850 V. A. Puiseux (1820–1883) obtained the expansion in fractional powers valid at algebraic branch points.

6. Abel and Jacobi

Parallel with this general development and at the time overshadowing it, there was a special trend dealing with elliptic and Abelian functions. To get the background here we have to go back to the latter half of the eighteenth century. Mathematicians like Euler, Lagrange and Legendre were then struggling with integrals of algebraic functions, in particular with integrals of rational functions of x and y where $y^2 = P(x)$, a polynomial of degree n without multiple roots. If $n = 1$ or 2, the integration could be carried out in finite form, the result being a rational function of the values of x and y in the upper limit plus a sum of multiples of logarithms of such functions. Such an expression will be called logarithmico-rational in the following. But if n was 3 or 4, all they could do was to reduce the integral to logarithmico-rational functions plus one or more normal forms which defied further reduction. An interesting observation made by Euler became basic for the later development: the sum of three or more integrals of the same rational function of x and y with upper limits $x_1, \ldots, x_k$, is a logarithmico-rational function of the upper limits provided x_k is a particular rational function of $x_1, \ldots, x_{k-1}, y_1, \ldots, y_{k-1}$.

Order was made in this chaos by Niels Henrik Abel (1802–1829), the son of a poor Norwegian minister, himself desperately poor, shy, depressed, and ailing in health. Mathematically self-taught, he made quite a stir at the young University of Christiania (= Oslo) by his investigations on the algebraic resolution of algebraic equations. To start with he thought that he had proved that every algebraic equation could be solved by radicals, but he discovered the error and was able to use the method for a proof that the general equation of degree greater than four cannot be solved by radicals. On the basis of this discovery he was awarded a traveling fellowship which took him to Berlin in 1825 and to Paris the next year. In Berlin Crelle befriended him and got him as a collaborator for his new *Journal für Mathematik* in the first volume of which there are six papers by Abel, among others the paper on the non-solvability by radicals. Paris was a disappointment to him, though of some importance to his scientific development. He presented a big manuscript to the Académie des Sciences containing what was later to be called Abel's theorem. It was not printed until 1841 and then only after diplomatic representations had been made.

Abel's theorem is the direct generalization of the theorem of Euler mentioned above. Suppose that we are dealing with integrals of a fixed rational function $R(x, y)$ of x and y where y is the root of a polynomial equation $P(x, y) = 0$ so that y is an algebraic function of x. Again there exist sums of integrals which add up to a logarithmico-rational function of the values of x and y in the upper limits, provided certain relations hold between these limits. But it is no longer necessarily true that one limit is a rational function of the others. Instead there is a characteristic number p of these limits which have to be algebraic functions of the other limits. This number p was determined by Abel in a few cases. It was later called the genus of the curve $P(x, y) = 0$ or of the corresponding Riemann surface. It depends only on P and not on R.

Abel did not stop at this point, however. In Paris he had become acquainted with Cauchy's work and the advantages to be gained by studying functions in the complex plane. Such considerations were not new to him, he had probably just finished his very detailed study of the binomial series for complex values of variable and exponent. He now tackled the inversion problem for the elliptic integral of the first kind, that is,

$$u = E(x) = \sum \int_0^x [P(t)]^{-1/2}\, dt,$$

where $P(t)$ is a polynomial of degree four. Here Euler's formula would give $E(x_1) + E(x_2) = E(x_3)$, where x_3 is an explicitly known rational function of x_1, x_2, y_1, y_2. Abel may have kept in mind the corresponding situation when $P(t) = 1 - t^2$. The corresponding integral, $A(x)$ say, defines arcsin x and here Euler's formula gives

$$A(x_1) + A(x_2) = A(x_1 y_2 + x_2 y_1).$$

But in this case a great simplification is obtained by introducing the inverse function $x = \sin u$ which is single-valued and Euler's formula simply becomes the addition theorem for the sine function. There was a chance that similar considerations would work for $u = E(x)$. Perhaps x is a single-valued function of u having an addition theorem to be read off from Euler's formula. This turned out to be the case; Abel met with complete success and the results appeared in a long memoir in volumes 2 and 3 of Crelle followed by a number of other papers some of which appeared after his death. Abel proved that inversion of the elliptic integral of the first kind led to a single-valued meromorphic double-periodic function and for this function he obtained expansions in partial fractions and as the quotient of two infinite double products. For the latter he determined representations as simple products of trigonometric functions as well as expansions in trigonometric series.

Abel met with little recognition while he was living; the University of Berlin made a professorship for him, but the call arrived a few days after his death. The Paris Academy gave him its Grand Prix also after his death.

As it so often happens, it takes a long time for an idea to ripen, but then it comes to several people at the same time. Abel's competitor was Carl Gustav Jacob Jacobi (1804–1851) whose first publication in this field also appeared in 1827, followed by an organized theory in 1829. Jacobi arrived at the same results as Abel, but he has more time to delve into the theory and his interests were also slightly different from those of Abel. Jacobi based his theory on what he called the theta functions which were essentially the trigonometric series of Abel. But in this case as well as in the rest of the theory, it was Jacobi's definitions and notation which came to be accepted.

The elliptic case having been settled, it was natural to proceed to the hyper-elliptic one: $\int R(x, y)dx$ where $y^2 = P(x)$ and the degree of $P(x)$ is at least five. Now in the elliptic case, $n = 4$, Abel and Jacobi worked with the integral $\int dx/y$ which is bounded as soon as n exceeds one. But if $n = 5$ or 6 there is one more integral of the first kind, namely $\int x\, dx/y$, so they should perhaps be considered simultaneously. Actually each integral alone is hopeless for the various determinations of the integral for a fixed upper limit are everywhere dense in the complex plane so the existence of a single-valued inverse is out of the question. Guided by Abel's theorem, Jacobi finally came to a solution in 1834. Considering the equations

$$\int_0^{x_1} [P(t)]^{-1/2}\, dt + \int_0^{x_2} [P(t)]^{-1/2}\, dt = u_1,$$

$$\int_0^{x_1} t[P(t)]^{-1/2}\, dt + \int_0^{x_2} t[P(t)]^{-1/2}\, dt = u_2,$$

Jacobi succeeded in showing that the symmetric functions $x_1 + x_2$ and x_1x_2 of the upper limits are single-valued functions of the two variables u_1 and u_2 with a system of four periods. Jacobi surmised that such functions could be expressed as quotients of theta functions in the two variables u_1 and u_2. This was proved by one of Jacobi's pupils Johann Georg Rosenhain (1816–1887) in a prize memoir of 1846 published by the French Academy in 1851.

7. Weierstrass

Karl Weierstrass (1815–1897) was born in Ostenfelde not far from Münster in Westphalia. He had made the acquaintance of some of Steiner's work already as a high school student, but his father wanted him to enter the civil service so he was sent off to the University of Bonn to study law 1834–38. Here he became a typical Corps student, enjoying the beer and the singing and unexcelled at the Mensur (= formal duelling). Of law he learnt none, the mathematics lecturers were unattractive, but he started to read Jacobi's theory of elliptic functions of 1829 by himself and decided to go into mathematics. Having heard that Christoffer Gudermann (1798–1852) lectured on Jacobi's theory at the Akademie (later university) of Münster, he finally succeeded in overcoming his father's resistance and moved to Münster in 1839 where he qualified for a teacher's certificate in 1841. Weierstrass became a high school teacher, first a year on probation in Münster, 1842–1848 at the catholic Progymnasium in Deutsch-Crone (then in West Prussia, Polish since 1919), 1848–1855 at the Collegium Hoseanum, a catholic seminary in Braunsberg (then in East Prussia, Polish since 1945); both Weierstrass and Gudermann were catholics. In 1854 Königsberg gave him a doctor's degree honoris causa. He was called to Berlin in 1856, became a member of the Berlin Academy the following year and ordinary professor at the university in 1864. In Berlin he rose to the highest fame, for about twenty-five years he was an inspiring teacher and the final authority in all questions mathematical. He was succeeded in 1892 by his pupil Hermann Amandus Schwarz (1843–1921).

Weierstrass learned two things from Gudermann: Jacobi's theory of elliptic functions and to work with power series, and he learned both lessons very well. At his own request Gudermann had assigned Weierstrass a real research problem as topic for the essay that was required for the teacher's certificate. This problem was to represent the elliptic functions as quotients of power series in the variable u. This he did and he returned to the problem repeatedly in later days. Ultimately it led him to the theory of the sigma functions which served as the point of departure for his theory of doubly-periodic functions. The fifteen years during which Weierstrass was a school teacher, he spent very well. He read Abel's papers: "Lesen Sie Abel!" was his standing advice to his students in later years. He set himself the task of solving the inversion problem for general Abelian integrals and to put the whole investigation on a firm basis.

Now Weierstrass was a methodical, painstaking man, and a logician. The brilliant flashes of intuition, so characteristic of Abel, Jacobi and also Riemann, but rarely visited him. He distrusted intuition and tried to put mathematical reasoning on a firm basis. He was the first to criticize the naive notion of a number and to give a formal introduction of the real number system. This was never published, however, a publication by H. Kossak in 1872, claiming to contain the theory of Weierstrass, was disowned by the latter in strong terms. Incidentally, 1872 is important in the history of the real number system for in this year appeared also the papers of Georg Cantor (1845–1918), E. Heine (1821–1881), Charles Méray (1835–1911), and Richard Dedekind (1831–1916) on the same subject. Both Cantor, of set theory fame, and Heine were pupils of Weierstrass; Méray was the apostle of arithmetization and power series in France and was independent of Weierstrass.

Once the real numbers were in order, Weierstrass could proceed to build up a theory of analytic functions based upon power series and the process of analytic continuation, for one as well as for several variables. These theories were published at least in part in the memoirs of the Berlin Academy in 1876–1881 and re-issued in book form together with other material as *Abhandlungen aus der Functionenlehre* in 1886. These memoirs contain a number of results, now-a-days much better known than the main body of Weierstrass's work. Thus we find here the first example of a continuous, nowhere differentiable function, the first example of a power series with the circle of convergence as natural boundary, further Weierstrass's approximation theorem for ordinary and trigonometric polynomials. There is also an example of what Weierstrass called an analytic expression, representing different analytic functions in different parts of the plane. The factorization theorem for entire functions occurs in the first memoir (of 1876) forming part of the book. Weierstrass had discovered Laurent series in 1841, the following year he studied the existence of solutions of algebraic differential equations and in this connection he formulated the principle of analytic continuation.

All this material—the number system, function theory, and differential equations—was grist of the mill that was to grind out the solution of the inversion problem for Abelian integrals. A preliminary communication appeared in the Braunsberg school program of 1849, further indications were given in volume 47 of *Crelle's Journal* in 1853 (the reason for the honorary degree in 1854) and in a long memoir in 1856, Crelle volume 52. In 1902, long after his death, appeared a reconstruction of the whole theory, based on lecture notes, as volume IV of his *Gesammelte Werke*, a tome of 624 pages.

Weierstrass started the theory of entire and meromorphic functions and the serious study of power series per se. The theory of approximation goes back to him and his work led to a new era in the calculus of variations and in the theory of minimal surfaces. It was not all analysis: Weierstrass even gave a purely geometrical proof of the fundamental theorem of projective geometry. His work was finished and polished, there is nothing to correct and, within the limits set by the author, it is complete. There are no loose ends and few ideas are floating around.

8. Riemann

Bernhard Riemann (1826–1866) is a contrast to Weierstrass in every respect. Personally he was shy and clumsy and his health was poor; he suffered from vertigo and died young from tuberculosis. Weierstrass on the other hand was a jolly soul, quite robust, and lived to a high age. Their approach to mathematics is fundamentally different: Weierstrass had the local point of view, Riemann the global one. Weierstrass had one tool which he had perfected and used with great skill; he built slowly and with extreme care. Riemann took what tools he needed without worrying too much about them as long as the tool appeared appropriate for the purpose. His function theory is largely based upon physical and geometrical considerations. Thus the real and the imaginary parts of an analytic function are functions of the logarithmic potential and his long association with Wilhelm Weber had given him almost a physicist's point of view, the physical situation showed the existence of the functions he needed and that was enough. His geometrical intuition was highly developed and he may be regarded as one of the fathers of topology, *analysis situs* as he called it.

There is also quite a difference between the character of their work. Weierstrass finished what he started. Riemann on the other hand, started much more than he ever could finish; he has provided steady work for mathematicians over a century. We are still digesting and developing his ideas and trying to prove his conjectures. Just try to make a census of the concepts and problems to which the name of Riemann is attached.

You start with the Riemann integral and sums, concepts which are not quite outmoded yet. There is a Riemannian theory of trigonometric series which has led to "Riemannian" theories of other orthogonal series. We have Riemannian or geometric function theory, the Cauchy-Riemann equations, Riemann surfaces, and Riemann matrices in his theory of Abelian integrals. There is a Riemann conformal mapping problem, solved by W. F. Osgood, E. H. Taylor, P. Koebe, and C. Carathéodory around 1912–1913 for the case of simply connected domains. There is also a Riemann problem of constructing a function holomorphic in a domain bounded by a smooth curve when the real or the imaginary part of the boundary values is given or are related by a linear equation. The first general solution was found by David Hilbert (1862–1943) in 1904 as one of the early applications of the theory of integral equations. A more general Riemann problem was solved by G. D. Birkhoff (1884–1944) in 1913. Both Hilbert and Birkhoff applied their results to another Riemann problem: that of finding a linear differential equation with preassigned regular singular points and given monodromy group. Birkhoff also extended this to irregular singular points and to difference and q-difference equations. The original Riemann problem was formulated for a second order linear differential equation with three regular singular points and led to the hypergeometric equation (Riemann's fourth paper of 1857). The Riemann zeta function (considered by Euler for real variables) appeared in his seventh paper in 1859; there were originally six unproved statements or conjectures in this paper, all but the main one were proved by 1905 through the work of J. Hadamard and H. von Mangoldt, but the famous Riemann hypothesis that all complex zeros have real part $1/2$ is still unproved. G. H. Hardy (1877–1947) in 1915 proved the existence of infinitely many zeros on this line. The prime problem in other rings leads to other zeta functions and other "Riemann hypotheses" mostly unproved. Riemann's eighth published paper on the propagation of waves in air contains the solution of a boundary problem for a wave equation involving a "function of Riemann". Finally, his habilitation lecture of 1854 on the basic hypotheses of geometry, published in 1867, gave rise to the Riemann variant of non-euclidean geometry where there are no parallels, to various analyses of the space problem, to Riemannian geometry, metric, and curvature tensor.

Winston Churchill's famous epigram "Never ... was so much owed by so many to so few!" also accurately describes the relation between present day mathematicians and the papers of Riemann.

9. Hermite

It is time to turn and take a look at the development in France. Here the leading figure during the third quarter of the nineteenth century is Charles Hermite (1822–1901). His most spectacular achievements were the solution of the general quintic by elliptic functions in 1858 and the proof that e is transcendental in 1873. He was one of the pioneers of invariant theory and quadratic forms in France and contributed to the theory of elliptic and modular functions. Hermite's name is connected with several important concepts: he would have recognized what we call a polynomial of Hermite and an Hermitian form, but an Hermitian operator would probably have puzzled him.

Hermite's influence in the mathematical world was great. This was only partly due to family relations (his wife was a sister of J. Bertrand (1822–1900) and Émile Picard was his son-in-law); above all his fairmindedness, urbanity and benevolent interest in younger mathematicians accounted for his power. He used his international relations and his influence in France for the good of mathematics and mathematicians. A couple of examples will illustrate this. Hermite and Weierstrass encouraged the Swedish mathematician Gösta Mittag-Leffler (1846–1927) to start the first international mathematical journal *Acta Mathematica* in 1882 to provide a forum where French and German mathematicians could meet and thus improve their relations which were strained after the Franco-Prussian war of 1870–71. Incidentally, Mittag-Leffler had come to Hermite to work with him in 1873, but the latter sent him off to Weierstrass with the statement that the latter was the master of all mathematicians. Another famous case is the correspondence which started in 1882 between Hermite and a young Dutch astronomer Jan Thomas Stieltjes (1856–1894) which lasted until the latter's death. Urged by Hermite to come to Paris, Stieltjes got his doctor's degree there in 1885 and a professorship was found for him in Toulouse. Stieltjes' most famous investigation, a memoir on continued fractions and the associated moment problem appeared partly after his premature death. Except for the Stieltjes integral, one of the tools he had to create for his problem, the contents of this memoir remained a sealed book for well nigh twenty-five years.

In passing let us mention some more of Hermite's contemporaries. The name of the differential geometer Ossian Bonnet (1819–1892) survives in the formulas of Gauss-Bonnet. Edmond Laguerre (1834–1886) did beautiful work in theory of entire functions, theory of equations, projective geometry where he introduced the angle as the logarithm of a cross-ratio (Laguerre polynomials were known to Abel). Émile-Léonard Mathieu (1835–1890) will be remembered for his functions, solutions of a certain differential equation with periodic coefficients, and for his group theoretic investigations. Camille Jordan (1838–1922) is famous for the Jordan curve theorem, even if he did not prove it, for the theory of functions of bounded variation, and the Jordan-Hölder theorem in group theory. Jordan's *Cours d'Analyse* and his *Traité des substitutions* were both classics and strongly influenced mathematicians of his day. The eminent differential geometer Gaston Darboux (1842–1917) also made important contributions to analysis (functions of large numbers).

10. Poincaré

The crop of great analysts with which France recovered the leadership in analysis started to ripen in the late seventies with Henri Poincaré (1854–1912), Paul Appell (1855–1930), Émile Picard (1856–1941), Édouard Goursat (1858–1936), later followed by Paul Painlevé (1863–1933), Jacques Hadamard (1865-1963), Émile Borel (1871–1956), and Maurice Fréchet (1878–1973).

Poincaré was encyclopedic in his interests, a genius full of ideas and led by strong intuition, he created first class mathematics with the greatest of ease, and produced around 500 papers from 1878 until his premature death. He worked in the different branches of the theory of ordinary differential equations, real and complex, infinite determinants, continuous and discontinuous groups and noneuclidean geometry. With Felix Klein (1849–1925) as chief competitor, he created the theory of automorphic functions during the early eighties. He was also one of the creators of modern topology (analysis situs in his terminology) to which he devoted a series of six memoirs in 1895–1904. He had been led to such questions through his work of 1880 on the shape of the integral curves of first order differential equations. He lectured on all fields of mathematical physics as well as celestial mechanics to which he contributed copiously. He wrote several books on philosophy, especially on the theory of knowledge. There is much in Poincaré's work which is sketchy and fragmentary; once he had seen through the difficulties, he had a royal disdain for pesky details.

If Sylvester was a second Adam, Poincaré tried hard to be a third, but he did not belong to the ant-school. His specialty was to name problems, functions, and concepts after famous mathematicians who were remotely connected with the idea. Thus he introduced a Neumann problem in potential theory which had never been considered by Carl Neumann (1832–1925). In one of his papers on analysis situs he introduced the Betti numbers in honor of Enrico Betti, with what other justification is unknown to me. The first class of automorphic functions which he considered was connected with the inversion problem for the quotient of two linearly independent solutions of linear second order differential equations of the Fuchsian class so he called them "fonctions fuchsiennes". When Felix Klein protested that Lasarus Fuchs (1833–1902) had never considered such functions while he, Klein, had, Poincaré promptly called the next class that he encountered "fonctions kleinéennes," because, as some wit observed, Klein had never considered them.

It is necessary to mention briefly the definition of automorphic functions. Given a group S of linear fractional substitutions on the variable z such that the set of transforms of z under S is nowhere dense in the plane. It is required to find an analytic function $f(z)$, invariant under S, that is, such that $f(Sz) = f(z)$ for every substitution of the group. If S is finite there are rational automorphic functions, if S is a group of translations, periodic or doubly-periodic functions will do, but for the general situation the construction problem is rather difficult. As observed above, these functions have interesting connections with the theory of linear second order differential equations. They also arise in the problem of mapping the interior of a polygon bounded by circular arcs on the interior of a half-plane (or circle). They solve a number of uniformization problems. In particular, if $C : f(z, w) = 0$ is an algebraic curve, then it is possible to find automorphic functions $G(t)$, $H(t)$ such that $w = G(t)$, $z = H(t)$ is a parametric representation of C by means of single-valued functions meromorphic in their domain of existence.

11. Picard

Let us now pass over to Émile Picard who made his first striking discovery in 1879, published in detail in the *Annales de l'École Normale Supérieure*, Series 2, volume 9 (1880). Here Picard showed that an entire function can omit at most one finite value without reducing to a constant, and if there exist two values each of which is taken on only a finite number of times, then the function is a polynomial. If the function is meromorphic instead, infinity being an admissible value, at most two values can be omitted without the function reducing to a constant. In the same paper he extended the conclusions to the roots of the equations $F(z) = a$, $F(z) = b$ in the neighborhood of an essential singular point. The proof is based on properties of the modular function and its inverse. A few years later (1883, 1887), Picard proved that if $f(z, w) = 0$ is an algebraic curve of genus greater than one, then it is not possible to satisfy the equation by two single-valued functions $w = G(t)$, $z = H(t)$ having isolated essential singular points, in particular not by functions meromorphic in the finite plane. Picard also studied discontinuous groups in space, algebraic integrals of several variables, partial differential equations, integral equations, the extension of the Galois theory to differential equations, and various questions of mathematical physics.

Picard's theorem exercised a tremendous influence on the development of analysis which is still lasting. For a good many years attempts were made to obtain a simpler proof without the use of the modular function. This was achieved by Émile Borel in 1896 and led to very surprising consequences, in particular, to a theorem of Edmund Landau (1877–1938) of 1904 according to which there is a circle whose radius depends only on the first two coefficients of a given power series such that the function defined by the series either takes on the value zero or the value one or has a singular point inside or on the circle. This was considered a great triumph of the elementary methods, but the following year Constantin Carathéodory (1873–1950) showed that the correct value of the radius was expressible in terms of the modular function. The whole question of exceptional values was put on a much more general basis through the investigations of the brothers Frithiof and Rolf Nevanlinna (born 1894 and 1895 respectively). These investigations, having started in 1922, lie outside the scope of this paper, however.

The nineties and the following decade saw a number of investigations devoted to the theory of entire functions, relations between the rate of growth of the maximum modulus, the frequency of the zeros, and the decrease of the coefficients. A multitude of names could be mentioned, we restrict ourselves to those of Borel, Hadamard, Ernst Lindelöf (1870–1946), and Edvard Phragmén (1863–1937). The latter are chiefly famous for an extension of the maximum principle proved in 1908 and based on previous work by Phragmén alone in 1904. It is one of the most powerful tools available in function theory. G. Mittag-Leffler had started a very successful school of mathematicians at the new University of Stockholm in 1882; Phragmén belonged to his early pupils and also held a professorship there before going into life insurance where he held a number of influential positions and was instrumental in establishing the close relations between research mathematics and the life insurance companies which is characteristic for Scandinavian conditions. Lindelöf as a professor at the University of Helsingfors founded a flourishing mathematical school in Finland. The Nevanlinnas mentioned are among his most distinguished pupils.

12. Hilbert

The turn of the century saw two basic discoveries in analysis which were to dominate the future development, one was the Lebesgue integral presented in his dissertation of 1902 and the first of a number of integrals, each more abstract then the previous one. Lebesgue's own work on Fourier series having demonstrated the usefulness of the new notion, it gradually penetrated all branches of real analysis. A contributing factor of the first order of magnitude was the work by F. Riesz [(1880–1956), Riesz Frigyes in Hungarian, Frederick, Frédéric or Friedrich Riesz according to the language of the paper] on functions integrable together with their squares or pth powers (1907, 1910). Through this work on Lebesgue spaces, on functionals of continuous functions, and his book on linear equations in infinitely many unknowns (1913), F. Riesz influenced the whole development of what became known as functional analysis.

The second discovery was the theory of integral equations, one of the forerunners of functional analysis and the theory of operators. Integral equations with variable upper limits occur in the works of Abel and of Liouville, but the first general theory is due to Vito Volterra (1860–1940) starting in 1896. Volterra also started the theory of functionals. Integral equations with a constant interval of integration were the creation of Mittag-Leffler's most famous pupil Ivar Fredholm (1866–1927) by a preliminary communication in 1900 and a finished theory in 1903. Fredholm was led by the analogy with systems of linear equations, but instead of carrying out the limiting process he boldly wrote down the resulting infinite determinants and verified that they satisfied. It is no accident that Fredholm used infinite determinants for another of Mittag-Leffler's pupils Helge von Koch (1870–1924) had developed this theory to a perfection.

The further development was taken over by David Hilbert (1862–1943). Hilbert, one of the giants of our science, was born in Königsberg in East Prussia where he got his doctor's degree in 1885. His early work was devoted to the theory of forms and invariants. In 1892 Hilbert turned to algebraic number theory, culminating in his enormous report on algebraic fields for the Deutsche Mathematiker Verein in 1897. Hilbert had a knack of changing his interests in mathematics suddenly and completely: in 1899 appeared his *Grundlagen der Geometrie* which led to a revival of the postulational method, first in geometry, later in all other fields. The same year he turned away from geometry and was preparing an "Ehrenrettung" of Dirichlet's principle in the calculus of variations. These papers belong to the period 1899–1906. Hilbert had moved from Königsberg to Göttingen in 1895 where Felix Klein represented the Riemannian tradition in function theory. After Weierstrass's very justified criticism of Dirichet's principle, the Riemann school lacked a basis for their existence theorems and it was essential that somebody put the theory back on a firm basis. This Hilbert succeeded in doing, much to the relief of his colleagues.

An accidental seminary lecture (by E. Holmgren of Uppsala) on Fredholm's work directed Hilbert's attention to the field of integral equations in 1901 and he now turned most of his energy into new channels. He created the theory of integral equations with a symmetric kernel, the prototype of all theories of symmetric operators, and showed how it tied in on one hand with a theory of quadratic forms in infinitely many unknowns and on the other with the theory of complete orthogonal systems and the question of expansions in terms of such systems. This period lasted until about 1912, during the latter part of which Hilbert used the new tools for an attack on the problems of mathematical physics.

To this period belongs also Hilbert's solution of Waring's problem on the representation of integers as sums of nth powers in 1909. Hilbert gave a mere finiteness theorem and the actual determination of limits for the necessary number of powers required was left for the powerful analytical methods invented in 1916 by G. H. Hardy and J. E. Littlewood (1885–1977).

Another of the great events at the beginning of the century was the formulation of the axiom of choice by Ernst Zermelo (1871–1953) in 1904 and his proof of the well-ordering theorem based on this axiom. It would take us too far to follow the vicissitudes of this axiom and the closely related conflict between formalists and intuitionists in mathematical logic which occupied much of Hilbert's attention during the last twenty years of his life.

This paper is based upon two talks presented to the Mathematics Colloquium of Yale University in May 1952. It is hoped that the contents might prove to be of interest to a wider public. The material is culled from F. Cajori's and G. Loria's histories of mathematics and F. Klein, *Vorlesungen über die Entwicklung der Mathematik im 19. Jahrhundert.*

Inequalities[1]

Richard Bellman

Mathematics Magazine
28 (1954/55), 21–46

Editors' Note: Richard Ernest Bellman was born in 1920. He attended Brooklyn College as an undergraduate and went on to graduate studies at Wisconsin at Madison for a master's and Princeton University, where he received his PhD in 1947, under the direction of Solomon Lefschetz. He spent some time at Stanford University and the University of Southern California, but his career was mainly at the Rand Corporation.

Though he originally worked in analytic number theory he moved into applied mathematics, specifically linear and nonlinear differential equations, stochastic processes and dynamic programming. He became one of the foremost applied mathematicians in the U.S., described in the National Academy's Memorial Tributes as "a towering figure among the contributors to modern control theory and systems analysis...." He wrote many books, among them *Introduction to Matrix Analysis* (1960); *Adaptive Control Processes* (1961); *Applied Dynamic Programming* (with Stuart Dreyfus) (1962); *Differential-Difference Equations* (with Kenneth L. Cooke) (1963); *Methods of Nonlinear Analysis* (1970); and *Partial Differential Equations: New Methods for Their Treatment and Solution* (1985). In the New Mathematical Library of the MAA we find *An Introduction to Inequalities* (with Edwin Beckenbach) (1961). Shortly before his death, at the early age of 63, he wrote his autobiography, *The Eye of the Hurricane* (World Scientific, 1984).

Much honored during his lifetime, he won the Norbert Wiener Prize of the AMS and SIAM; the Dickson Prize from Carnegie-Mellon University; the John von Neumann Award of the Institute of Management Sciences and the Operations Research Society of America; and the Medal of Honor for his invention of dynamic programming, from the Institute of Electrical and Electronics Engineering.

In this article Bellman uses a "backwards" induction to prove the arithmetic-geometric mean inequality, a little known technique (though it goes all the way back to Cauchy) but which in recent years has been used by computer scientists. In *Introduction to Algorithms/A Creative Approach* (Addison-Wesley, 1989, pp. 244–46), Udi Manber uses what he calls reversed induction to prove that certain very dense graphs have a Hamiltonian cycle. The induction is on the number of edges in the graph, with the base case corresponding to the family of complete graphs (where finding a Hamiltonian cycle is trivial). Manber's proof actually yields an algorithm for finding a Hamiltonian cycle. He points out that reversed induction is tailor-made for designing algorithms, since "it is almost always easy to go from n to $n-1$, namely, to solve the problem for small inputs" by, for example, introducing "dummy" inputs that do not affect the outcome.

It has been said that mathematics is the science of tautology, which is to say that mathematicians spend their time proving that equal quantities are equal. This statement is wrong on two counts: In the first place, mathematics is not a science, it is an art; in the second place, it is fundamentally the study of inequalities rather than equalities.

I would like today to discuss a number of the basic inequalities of analysis, presenting first an algebraic proof of the inequality between the arithmetic and geometric means, and then a most elegant geometric technique due to Young. In passing we will observe how the theory of inequalities may be used to supplant the calculus in many common types of maximization and minimization problems. Finally, I shall show how Young's inequality leads naturally to Hölder's inequality, and Hölder's inequality to Minkowski's.

Since it has become unfashionable in educational circles to pose problems in the spirit of a spelling bee, but rather to motivate the student by relating the problem to our everyday pursuits, we shall consider the following question which is perhaps typical of the way in which the theory of inequalities can enter into our ordinary pursuits.

A football player of some renown, having gone into stocks and bonds and made a substantial score there also, stipulated in his will that his coffin be enclosed in a giant football. His executors, of an economical turn of mind, were confronted with the problem of finding the dimensions of the smallest football which would meet the terms of the will.

Indulging in the usual mathematical license, we may consider the football to be an ellipsoid and reduce the problem to that of finding the coffin of

[1]Delivered before a Teachers Conference at UCLA.

maximum volume which will fit into a given ellipsoid.

Taking the ellipsoid to have the equation

$$\frac{x^2}{a^2} + \frac{y^2}{b^2} + \frac{z^2}{c^2} = 1,$$

we see the analytical equivalent of the practical problem above is that of finding the maximum of $v = 8xyz$ subject to the above constraint on x, y and z.

Let us first observe that we can simplify by observing that v and $v^2/a^2b^2c^2$ are maximized simultaneously. Hence replacing x^2/a^2 by u, y^2/b^2 by v, z^2/w^2 by w, the problem reduces to maximizing uvw subject to $u + v + w = 1$, $u, v, w \geq 0$. It is intuitive now that the symmetric point $u = v = w = 1/3$ should play a distinguished role, as either a minimum or a maximum. Since it is clearly not a minimum it follows that it furnishes the desired maximum.

2. An algebraic approach

In order to prove this, in a purely algebraic fashion without the aid of calculus, we shall derive a general inequality connecting the arithmetic mean of n positive quantities, $(a_1 + a_2 + \cdots + a_n)/n$, and the geometric mean, $\sqrt[n]{a_1 a_2 \cdots a_n}$, namely

$$\sqrt[n]{a_1 a_2 \cdots a_n} \leq \frac{a_1 + a_2 + \cdots + a_n}{n}$$

with equality occurring only if

$$a_1 = a_2 = \cdots = a_n.$$

There are literally hundreds of proofs of this basic inequality, many of which are actually quite different. The proof I will present is perhaps not the simplest, but it is one of the most ingenious. It is perhaps the only application of a particular form of mathematical induction and I think that it will be interesting for that reason.

The proof begins in a very simple manner. The most basic inequality, which I must confess is a tautology, is that a nonnegative number is greater than or equal to zero. The simplest nonnegative number, and one which is invariably nonnegative, is a square of a number. Taking, for our own purposes (and this is where ingenuity enters) the number $a - b$ and squaring it, we have the inequality

$$(a - b)^2 \geq 0$$

Multiplying out and transposing, we have the well-known inequality,

$$\frac{a^2 + b^2}{2} \geq ab \tag{1}$$

together with the important addition that equality can occur if and only if $a = b$. Setting $a^2 = a_1$, $b^2 = b_1$, we have the well-known result that the arithmetic mean of 2 positive quantities is greater than or equal to their geometric mean.

Let us now replace a_1 by $(a_1 + a_2)/2$ and b_1 by $(a_3 + a_4)/2$ obtaining

$$\frac{a_1 + a_2 + a_3 + a_4}{4} \geq \sqrt{\left(\frac{a_1 + a_2}{2}\right)\left(\frac{a_3 + a_4}{2}\right)}$$

and apply the separate inequalities

$$\frac{a_1 + a_2}{2} \geq \sqrt{a_1 a_2}, \quad \frac{a_3 + a_4}{2} \geq \sqrt{a_3 a_4}$$

obtaining

$$\frac{a_1 + a_2 + a_3 + a_4}{4} \geq \sqrt[4]{a_1 a_2 a_3 a_4}.$$

Retracing our steps we see that we still have the important fact that equality can occur only if $a_1 = a_2 = a_3 = a_4$.

Continuing in this way we obtain, for n a power of two, the inequality

$$\frac{a_1 + a_2 + \cdots + a_n}{n} \geq \sqrt[n]{a_1 a_2 \cdots a_n} \tag{2}$$

with equality occurring only if all the variables are equal.

We still cannot apply this to our problem since 3 is not a power of two. We want to show that the same inequality holds for $n = 3$. Once more we require some ingenuity. Let us take the case $n = 4$, and set

$$a_1 = a_1, \quad a_2 = a_2,$$

$$a_3 = a_3, \quad a_4 = \frac{a_1 + a_2 + a_3}{3}.$$

The resulting inequality is

$$\left(\frac{a_1 + a_2 + a_3}{3}\right) \geq \sqrt[4]{a_1 a_2 a_3 \left(\frac{a_1 + a_2 + a_3}{3}\right)}$$

which simplifies to

$$\frac{a_1 + a_2 + a_3}{3} \geq \sqrt[3]{a_1 a_2 a_3}$$

the desired inequality for three. Retracing our steps we see that equality can occur only if we have $a_1 = a_2 = a_3$.

This technique is perfectly general and yields the inequality for $n-1$ whenever it has been established for n. Since we have established it for the integers 2, 4, 8, etc., we see that induction yields it for all n. Observe, however, that this is a backward induction rather than the usual forward induction.

Turning to our original problem we see that

$$\frac{1}{3} = \frac{u+v+w}{3} \geq \sqrt[3]{uvw}$$

unless $u = v = w = 1/3$. In terms of the original variables, x, y, and z this yields

$$x = \frac{a}{\sqrt{3}}, \quad y = \frac{b}{\sqrt{3}}, \quad z = \frac{c}{\sqrt{3}}$$

as the solution of our maximization problem.

Returning to (2), and taking $n = 10$, we obtain an interesting inequality by grouping the variables as follows

$$a_1 = a_2 = b_1$$
$$a_3 = a_4 = a_5 = b_2$$
$$a_6 = a_7 = a_8 = a_9 = a_{10} = b_3$$

The result is

$$\frac{2b_1}{10} + \frac{3b_3}{10} + \frac{5b_2}{10} \geq b_1^{2/10} b_2^{3/10} b_3^{5/10}$$

which is a particular case of the general inequality

$$\left(\frac{n_1 b_1 + n_2 b_2 + \cdots + n_k b_k}{n_1 + n_2 + \cdots + n_k}\right)^{(n_1+n_2+\cdots+n_k)} \geq b_1^{n_1} b_2^{n_2} \cdots b_k^{n_k} \quad (3)$$

where the $n_1, n_2, \ldots, n_k$ are positive integers, $b_1, b_2, \ldots, b_k$ are positive, and equality occurs only if $b_1 = b_2 = \cdots = b_k$.

The limiting form of (3), namely

$$a_1 b_1 + a_2 b_2 + \cdots + a_k b_k \geq b_1^{a_1} b_2^{a_2} \cdots b_k^{a_k}$$

where $a_i \geq 0$, $a_1 + a_2 + \cdots + a_k = 1$ is also valid, but is, of course, no longer a purely algebraic theorem.

I leave it to the reader to use the above inequality to determine the maximum of xyz subject to $x^2 + y^3 + z^5 = 1$, $x, y, z \geq 0$.

3. A Geometric Approach

Let us now turn to an alternate approach to the theory of inequalities. Let $f(x)$ be a monotone-increasing function with $f(0) = 0$ and consider the diagram below. The area under OAP is given by $\int_0^a f(x)\,dx$,

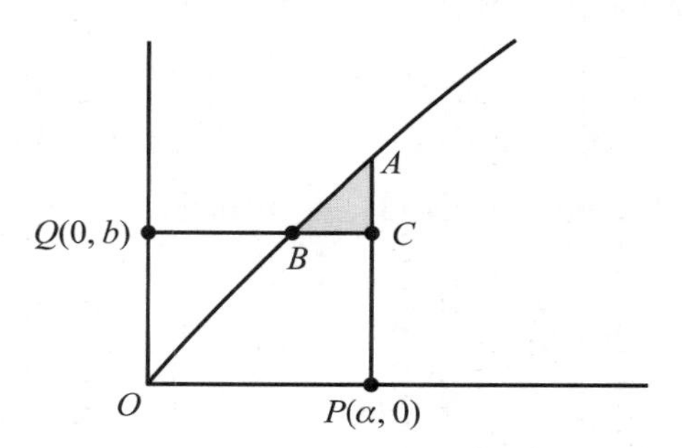

while that under OQB is given by $\int_0^b f^{-1}(x)\,dx$ (where $f^{-1}(x)$ denotes the inverse function) clearly the sum of these two areas is greater than or equal to that of the rectangle $OQCP$, with equality occurring if and only if $b = f(a)$. Writing this statement in analytical terms we obtain the inequality of Young,

$$\int_0^a f(x)\,dx + \int_0^b f^{-1}(x)\,dx \geq ab,$$

for $a, b \geq 0$.

Taking $f(x) = x$, we obtain (1) above. Taking $f(x) = x^{p-1}$, where $p > 1$, we obtain

$$\frac{a^p}{p} + \frac{b^{p'}}{p'} \geq ab \quad (4)$$

where $p' = \frac{p}{p-1}$ (note that $\frac{1}{p} + \frac{1}{p'} = 1$), If $f(x) = \log(1+x)$, we obtain, after some simplification

$$(1+a)\big(\log(1+a)\big) - (1+a) + e^b - b \geq ab, \quad a, b \geq 0 \quad (5)$$

an inequality which is important in the theory of Fourier series.

4. Hölder's Inequality

Let us now show how (5) yields one of the most useful inequalities of analysis, the classical inequality of Hölder,

$$(a_1^p + a_2^p + \cdots + a_n^p)^{1/p} (b_1^{p'} + b_2^{p'} + \cdots + b_n^{p'})^{1/p'} \geq (a_1 b_1 + \cdots + a_n b_n) \quad (6)$$

for $a_i, b_i \geq 0$, $p > 1$.

In (4) set successively

$$a = a_i/(a_1^p + a_2^p + \cdots + a_n^p)^{1/p}$$
$$b = b_i/(b_1^{p'} + b_2^{p'} + \cdots + b_n^{p'})^{1/p'}$$

and add. On the right side we obtain

$$\frac{a_1b_1 + \cdots + a_nb_n}{(a_1^p + a_2^p + \cdots + a_n^p)^{1/p}(b_1^{p'} + b_2^{p'} + \cdots + b_n^{p'})^{1/p'}}$$

while on the left-hand side we obtain

$$\frac{1}{p}\left(\frac{a_1^p + a_2^p + \cdots + a_n^p}{a_1^p + a_2^p + \cdots + a_n^p}\right) + \frac{1}{p'}\left(\frac{b_1^{p'} + b_2^{p'} + \cdots + b_n^{p'}}{b_1^{p'} + b_2^{p'} + \cdots + b_n^{p'}}\right) = \frac{1}{p} + \frac{1}{p'} = 1$$

This yields Hölder's inequality and shows, using the condition of equality in (4) that equality holds in (6) if and only if $b_i = a_i^{p-1}$ for $i = 1, 2, \ldots, n$.

An alternate form of Hölder's inequality which we shall use below to derive Minkowski's inequality is

$$\operatorname*{Max}_B(a_1b_1 + \cdots + a_nb_n) = (a_1^p + a_2^p + \cdots + a_n^p)^{1/p} \tag{7}$$

where B represents the domain: $b_i \geq 0$, $b_1^{p'} + b_2^{p'} + \cdots + b_n^{p'} = 1$, and $q_i \geq 0$. This follows from the observation that the right-hand side is an upper bound according to Hölder's inequality and is attained for $b_i = a_i^{p-1}$.

5. Minkowski's Inequality

Using (7) we have, for $x_k, y_k \geq 0$,

$$\begin{aligned}
\left[(x_1 + y_1)^p + (x_2 + y_2)^p + \cdots + (x_n + y_n)^p\right]^{1/p} &= \operatorname*{Max}_B\left[(x_1 + y_1)b_1 + (x_2 + y_2)b_2 + \cdots + (x_n + y_n)b_n\right] \\
&= \operatorname*{Max}_B\left[(x_1b_1 + x_2b_2 + \cdots + x_nb_n) + (y_1b_1 + y_2b_2 + \cdots + y_nb_n)\right] \\
&= \operatorname*{Max}_B(x_1b_1 + x_2b_2 + \cdots + x_nb_n) + \operatorname*{Max}_B(y_1b_1 + y_2b_2 + \cdots + y_nb_n) \\
&= (x_1^p + x_2^p + \cdots + x_n^p)^{1/p} + (y_1^p + y_2^p + \cdots + y_n^p)^{1/p}
\end{aligned}$$

We have thus established the classical inequality of Minkowski, which is for $p = p' = 2$, the famous "triangle inequality" of Euclid which states that the sum of two sides of a triangle is greater than the third. Put another way, a straight line is the shortest distance between two points.

6. In Conclusion

In presenting these two approaches to the theory of inequalities, I have neglected perhaps the most powerful approach, that based upon the concept of a convex function. This, however, deserves its own presentation.

For those interested in learning more about inequalities, I refer to that fascinating book by Hardy, Littlewood and Pólya entitled quite simply *Inequalities*.

A Number System with an Irrational Base

George Bergman

Mathematics Magazine
31 (1957/58), 98–110, 282

Editors' Note: This article was written by the author when he was a 12-year old student at Junior High School 246 in Brooklyn, New York. Here he explores using the golden mean, which he calls τ, more commonly now called φ, as the base of a number system. Bergman later went on to get a PhD at Harvard under the direction of John Tate and has had a distinguished career as a mathematician at the University of California, Berkeley. As a result of this article, however, he was interviewed by Mike Wallace for the *New York Post* in 1958, before Wallace's name became synonymous with CBS's *60 Minutes*. When Wallace asked this "tall, gawky, talkative" boy genius, by that time 14 and with an IQ of 205, "George, when you wake up in the morning, what's the first thing you think about? Mathematics?". Young George answered, "Oh, don't be silly. I think about breakfast."

Bergman's work here led to an exercise in Donald Knuth's *The Art of Computer Programming*, vol. 1, 2nd ed., 1973, p. 85, #35, that investigates representations of 1 in the φ number system. As noted by Arthur Benjamin in his comments on this article in the *Mathematics Magazine* index online, http://www.math.hmc.edu/journalsearch/, one of the basic results concerning this curious system is that every positive integer has a finite expansion.

The publication of the paper prompted Bergman's mother to write the following letter to the Editor of the *Magazine*:

"Dear Mr. James:

"The paper presented is the work of my twelve year old boy who took more than a year to gather courage to submit it for editorial scrutiny.

"You may be interested to know that when my son first received a subscription to *Mathematics Magazine* about two and a half years ago, he was aghast to note that he couldn't understand a single thing in it. With each successive issue, however, his understanding unfolded (he is a self-taught mathematician) until now he awaits each issue with eagerness, and recently was able to submit his solution to one of the Proposals published in your last issue. He considers his subscription to *Mathematics Magazine* one of the finest presents he ever received!

Sincerely yours,
Sylvia Bergman"

The reader is probably familiar with the binary system and the decimal system and probably understands the basis for any others of that type, such as the trinary or duodecimal. However, I have developed a system that is based, not on an integer, or even a rational number, but on the irrational number τ (tau), otherwise known as the "golden section," approximately 1.618033989 in value, and equal to $(1+\sqrt{5})/2$.

In order to understand this system, one must comprehend two peculiarities of the number τ. They are based on tau's distinctive property[1] that

$$\tau^n = \tau^{n-1} + \tau^{n-2}.$$

(a) Take any approximation (A_1) of τ. Taking the reciprocal, we get a number (a_1) that is proportionately the same distance from $1/\tau$ as A_1 was from τ, but arithmetically nearer. Adding 1,[2] we get a number (A_2) that is proportionately nearer τ than a_1 was to $1/\tau$ but arithmetically just as near. Since a_1 is arithmetically nearer than A_1, A_2 is nearer in both respects to τ than A_1. Repeating the process of taking the reciprocal and adding 1, we approach τ. Now, taking 1 as A_1, and expressing our approximations of τ (i.e., A_1, A_2, A_3, etc.) as fractions, we get

$$\frac{1}{1}\ \frac{2}{1}\ \frac{3}{2}\ \frac{5}{3}\ \frac{8}{5}\ \frac{13}{8}\ \cdots.$$

Taking either the numerators or the denominators, we get what is known as the Fibonacci Series, *each term of which is formed by adding the two previous terms*[3]; for

$$\frac{f_{n+2}}{f_{n+1}} = 1 + \frac{1}{\left(\frac{f_{n+1}}{f_n}\right)} = 1 + \frac{f_n}{f_{n+1}} = \frac{f_{n+1}+f_n}{f_{n+1}}$$

[1] Also true of $-1/\tau$; there are other numbers which have similar properties, e.g., there is a number S between 1 and 2 for which $S^3 = S^2 + S + 1$. Ed.

[2] Because $\tau^{-1} + \tau^0 = \tau$.

[3] This is the basic property defining the Fibonacci Series.

so

$$f_{n+2} = f_{n+1} + f_n.$$

(We designate the nth term of the Fibonacci Series by f_n, setting $f_1 = 1$, $f_2 = 1$. This practice shall be used through the article.)

(b) Any integral power of τ can be expressed in the form $\tau^n = A\tau + B$, where A and B are integers and, in fact, numbers in the Fibonacci Series. The explanation of this startling fact is really rather simple: Since

$$\tau^1 = 1\tau + 0 \quad \text{and} \quad \tau^2 = 1\tau + 1,$$

and since

$$\tau^3 = \tau^2 + \tau^1;\ \tau^3 = (1\tau + 0) + (1\tau + 1) = 2\tau + 1.$$

In the same way

$$\tau^4 = \tau^2 + \tau^3 = (1\tau + 1) + (2\tau + 1) = 3\tau + 2.$$

In general

$$\begin{aligned}\tau^n &= \tau^{n-1} + \tau^{n-2}\\ &= (f_{n-1}\tau + f_{n-2}) + f_{n-2}\tau + f_{n-3})\\ &= (f_{n-1} + f_{n-2})\tau + (f_{n-2} + f_{n-3})\\ &= f_n\tau + f_{n-1}.^4 \qquad \text{(see Note 2.)}\end{aligned}$$

Can this be applied to negative powers of τ? We don't know any Fibonacci numbers before 1, but it is easy to see how we can find them: Taking 1 and 1 as our first two, we can see that the term before must be 0, since that is the only number which, when added to 1 gives 1. In the same way, the number before that must be 1 since 1 is the only number that, when added to 0 gives 1; and the next term must be -1, since no other number gives 0 when added to 1. Continuing this process, we get $0, 1, -1, 2, -3, 5, -8, 13, -21, \ldots$. Obviously, this is alternately $+1$ and -1 times the corresponding Fibonacci numbers. But can this be proved to be true in all cases? It can by induction. The rule we want to prove, expressed as an equation, is:

$$f_{-y} = (-1)^{y+1} f_y.$$

Let us assume it true for $y = 1, 2, \ldots, n$. Now by the basic property of the Fibonacci Series:

$$\begin{aligned}f_{-n} + f_{-n-1} &= f_{-n+1}\\ f_{-n-1} &= f_{-n+1} - f_{-n}\\ f_{-(n+1)} &= (-1)^n f_{n-1} - (-1)^{n+1} f_n\\ f_{-(n+1)} &= (-1)^{n+2}(f_{n-1} + f_n)\\ f_{-(n+1)} &= (-1)^{n+2} f_{n+1}.\end{aligned}$$

The inductive proof is completed by the examples already cited.

Applying this to powers of τ, we make a list of them from τ^{-5} to τ^5:

$$\begin{aligned}&\tau^{-5} = 5\tau - 8 && \tau^0 = 0\tau + 1\\ &\tau^{-4} = -3\tau + 5 && \tau^1 = 1\tau + 0\\ &\tau^{-3} = 2\tau - 3 && \tau^2 = 1\tau + 1\\ &\tau^{-2} = -1\tau + 2 && \tau^3 = 2\tau + 1\\ &\tau^{-1} = 1\tau - 1 && \tau^4 = 3\tau + 2\\ & && \tau^5 = 5\tau + 3\end{aligned}$$

Now, at last, we shall get back to our concept of a system based on τ. Like the binary system, it can have only two symbols: 1 and 0. But, unlike the binary system, it has the rule (2): $100 = 011$[5] (place the decimal point anywhere—it's a general rule). But how do we find the numbers? We know that 1 is τ^0 or 1.0. Next, looking at the table of powers of τ, one notices that

$$\tau^1 = 1\tau + 0 \quad \text{and} \quad \tau^{-2} = -1\tau + 2.$$

Adding them together, one gets

$$\tau^1 + \tau^{-2} = (1\tau + 0) + (-1\tau + 2) = 2.$$

Therefore, $2 = 10.01$ (in this system). Of course, because of rule (2), this can also be expressed as 1.11,[6] 10.0011, 10.001011, 1.101011, etc., but 10.01 is what I call the simplest form (that form in which there are no two 1s in succession, and which, therefore, cannot be acted upon by the reverse of rule (2), called simplification ($11 = 100$). To convert a number to its simplest form, repeatedly simplify the leftmost pair of consecutive 1s.

To continue with our "translation" of numbers into this system, we notice (after a careful examination of the table) that $\tau^2 = 1\tau + 1$ and $\tau^{-2} = -1\tau + 2$, and adding them together $\tau^2 + \tau^{-2} = 3$, and so 3 is 100.01 in this system. What about 4? Well, since $\tau^2 + \tau^{-2} = 3$, $\tau^2 + \tau^{-2} + \tau^0(101.01)$ must equal 4,

[4]There is also a more complex proof which involves multiplying the expressions like $2\tau + 1$ by τ, giving $2\tau^2 + \tau$, and expanding τ^2into $\tau + 1$.

[5]This is a restatement of $\tau^n = \tau^{n-1} + \tau^{n-2}$.

[6]By changing the 1 in the τ column to 1.1.

since $\tau^0 = 1$. Can this method of adding 1 be used for other numbers? The answer is "yes"; just convert the number into the form in which there is a zero in the units column and place a 1 in it. If the method of conversion is not obvious, use this method:

a. Change to the simplest form.
b. If there is no 1 in the units column, you are finished. If there is, look in the τ^{-2} column (there can't be any in the τ^{-1} column because it is in its simplest form and there is a 1 in the column next to it.); if there is a 0 there, expand[7] the 1 into the τ^{-1} and τ^{-2} columns (that's all); if there is 1, look in the τ^{-4} column; if there is a zero there, expand the 1 in the τ^{-2} column into the τ^{-3} and τ^{-4} columns and the 1 in the units column into the τ^{-1} and τ^{-2} columns. If there is a 1, look in the τ^{-6} column; if there is a zero, expand the 1 in the τ^{-4} column into the τ^{-5} and τ^{-6} columns, the 1 in the τ^{-2} column into the τ^{-3} and τ^{-4} columns, and the 1 in the units column into the τ^{-1} and τ^{-2} columns; if on the other hand there is a 1, look in the τ^{-8} column, etc. If it is the endless fraction 1.01010101..., change it to 10.000000....

We can now construct a table of integers in this system. Here are those from 0 to 14 (in their simplest forms)

0 – 0	8 – 10001.0001
1 – 1	9 – 10010.0101
2 – 10.01	10 – 10100.0101
3 – 100.01	11 – 10101.0101
4 – 101.01	12 – 100000.101001
5 – 1000.1001	13 – 100010.001001
6 – 1010.0001	14 – 100100.110110
7 – 10000.0001	

Examples of the basic processes

1. Change 100101.111001 (equals 16) to the simplest form. The first pair (farthest left) is in the units and τ^{-1} columns, so we simplify it into the τ^1 column, giving 100110.011001. This time the pair farthest to the left is the one we have just created with our new 1 in the τ^1 column (the result of our simplification), added to the 1 already in the τ^2 column. This we simplify into a 1 in the τ^3 column, which gives us 101000.011001. Finally, we change the last remaining pair (in the τ^{-2} and τ^{-3} columns) into a 1 in the τ^{-1} column, arriving at our final answer: 101000.100001.

2. Change 101.01 (4) to a form with a zero in the units column. (i.e., a form to which 1 can be added). One can see that it is already in its simplest form. However, there is a 1 in the units column, and we must remove it. The first thing we do is look in the τ^{-2} column; since there is a 1 there, we look in the τ^{-4} column. This is empty, and so we expand the 1 in the τ^{-2} column into the τ^{-3} and τ^{-4} columns, getting 101.0011. Now that the τ^{-2} column is empty, we can expand the unit into a pair in the τ^{-2} and τ^{-1} columns, getting 100.1111, which has a zero in the units column. We can now add 1 to it:

$$100.1111 + 1 = 101.1111 = 110.0111$$
$$= 1000.0111 = 1000.1001 \quad \text{(five)}.$$

The Arithmetic Operations

The arithmetical operations, although they are basically the same as in any other system, are, in practice, quite different because of the peculiarities of this system. As our first step in all of them, we eliminate zeros, which would only hinder us, and show the place values of 1s by actual placement in columns. For instance, four (101.01) would be |1| |1‖ |1|, the heavy line representing the "decimal point." The necessity for this step results from the fact that, though in the systems to which we are accustomed, the steps of addition are simple enough to be performed mentally, this is not so in the tau system, nor is each column nondependent on the one to the left of it. It is thus necessary to have lined columns in which to carry out the work.

Now for the actual processes, we shall start with addition. The example

$$\begin{array}{r} 10010.0101 \\ +\ 1010.0001 \end{array} \qquad \begin{pmatrix} 9 \\ +6 \end{pmatrix}$$

would be represented by

1			1	‖		1	1	
	1		1	‖			1	

In this set-up it can be seen that we have a pair, consisting of a 1 in the τ^3 column and one in the τ^4 column. This we simplify into a 1 in the τ^5 column.

1	~~1~~			1	‖		1		1	
		~~1~~		1	‖				1	

[7] If you come across words (like "expand") used in an unfamiliar way, look for them in the list of definitions at the end of this article. I have put there all words which I have had to invent or alter for use in this system, so as not to break up the text by explaining them.

Now, however, we have no obvious way to continue. We are left with two 1s in the same column. We can neither add them together to give 2 (as we would in the decimal system), nor is there any simple "carrying" operation. We must, therefore, change this to a form not having two 1s in the same column. We will start by expanding one of the 1s in the τ^1 column:

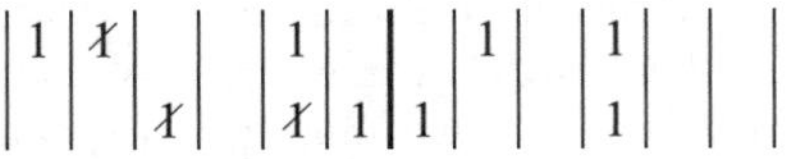

1	~~1~~			1			1		1		
		~~1~~		~~1~~	1	1			1		

Now we can simplify the pair we have just created in the τ^1 and units columns:

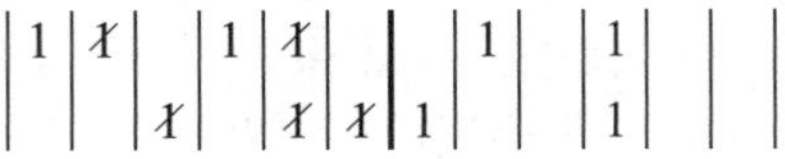

1	~~1~~		1	~~1~~			1		1		
		~~1~~		~~1~~	~~1~~	1			1		

and the one in the τ^{-1} and τ^{-2} columns:

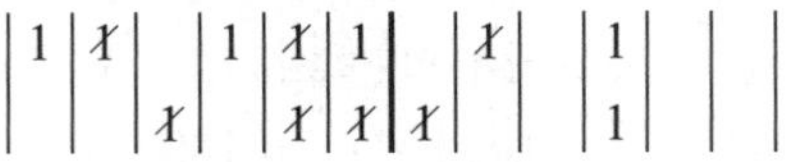

1	~~1~~		1	~~1~~	1		~~1~~		1		
		~~1~~		~~1~~	~~1~~	~~1~~			1		

We shall now use the same type procedure for the 1s in the τ^{-4} column. We expand one of the 1s there:

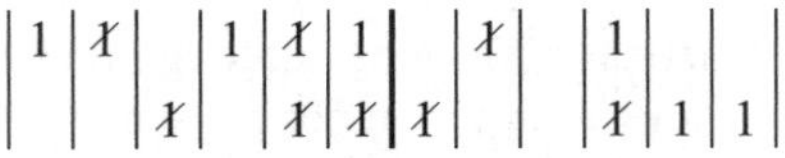

1	~~1~~		1	~~1~~	1		~~1~~		1		
		~~1~~		~~1~~	~~1~~	~~1~~			~~1~~	1	1

and simplify in the τ^{-4} and τ^{-5} columns:

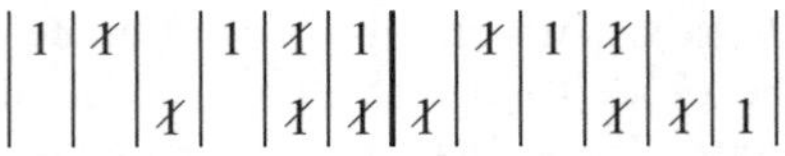

1	~~1~~		1	~~1~~	1		~~1~~	1	~~1~~		
		~~1~~		~~1~~	~~1~~	~~1~~			~~1~~	~~1~~	1

and express our answer in ordinary form, writing 1s in columns with an un-crossed-out 1 and 0s in the columns where all have been crossed out:

100101.001001 (15)

For general rules as to procedure, I believe that these will do in most cases. (These general rules and the ones for the other processes are not the types of rules that, if disobeyed, would give the wrong answer, but merely guides to the quickest way to get the right one):

a) Expand only when that is the only way to remove a 1 from the same column as another, regardless of whether this will result in the same situation in another column, but only if no more simplification can be done.

b) Simplify whenever possible, and, if there are two or more pairs, always simplify the one farthest to the left first. Because of this rule, simplification should never result in two 1s in the same column, i.e., |1|1|1| should be simplified into |1|~~1~~|~~1~~|1| rather than

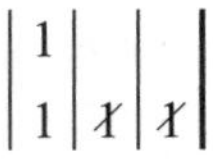

1		
1	~~1~~	~~1~~

Subtraction

Subtraction is the next process I shall describe. As in addition, we set up the numbers in columns, but here we shall assign negative values to the 1s from the subtrahend. For instance, to find (11 – 6) we set up

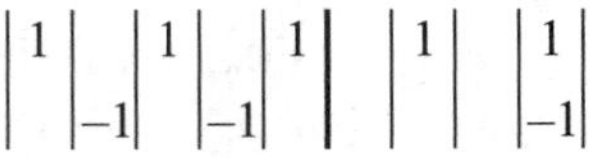

1		1		1		1		1
	–1		–1					–1

We now "cancel" the 1 and the –1 in the τ^{-4} column, giving

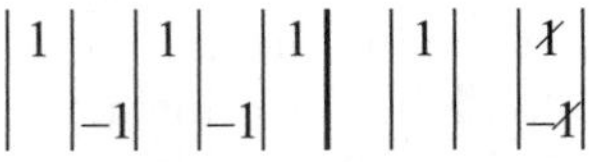

1		1		1		1		~~1~~
	–1		–1					~~–1~~

Next, we expand the 1 in the τ^2 column, getting

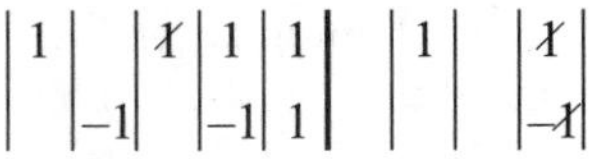

1		~~1~~	1	1		1		~~1~~
	–1		–1	1				~~–1~~

Again, we cancel, this time in the τ^1 column,

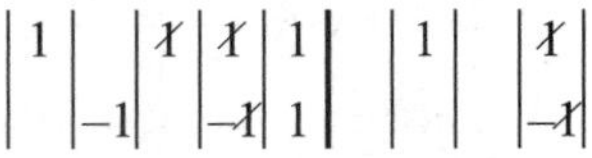

1		~~1~~	~~1~~	1		1		~~1~~
	–1		~~–1~~	1				~~–1~~

And again we expand, this time the 1 in the τ^4 column, getting

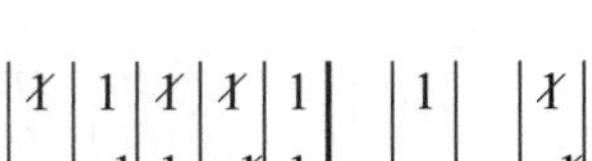

~~1~~	1	~~1~~	~~1~~	1		1		~~1~~
	–1	1	~~–1~~	1				~~–1~~

For the third time we cancel, (in the τ^3 column) giving

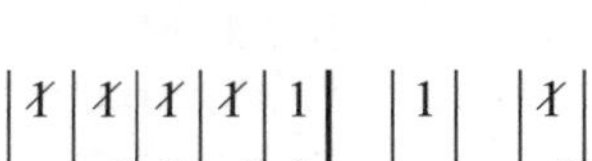

~~1~~	~~1~~	~~1~~	~~1~~	1		1		~~1~~
	~~–1~~	1	~~–1~~	1				~~–1~~

This we treat just as we would if we were adding and arrived at this stage; by expanding one of the 1s in the units column we get

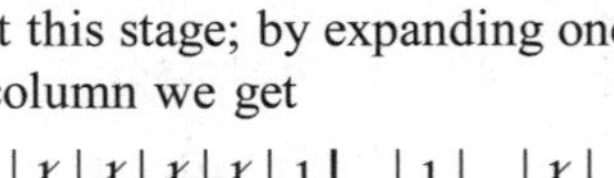

~~1~~	~~1~~	~~1~~	~~1~~	1		1		~~1~~
	~~–1~~	1	~~–1~~	~~1~~	1	1		~~–1~~

Next we simplify the pair in the units and τ^{-1}columns getting

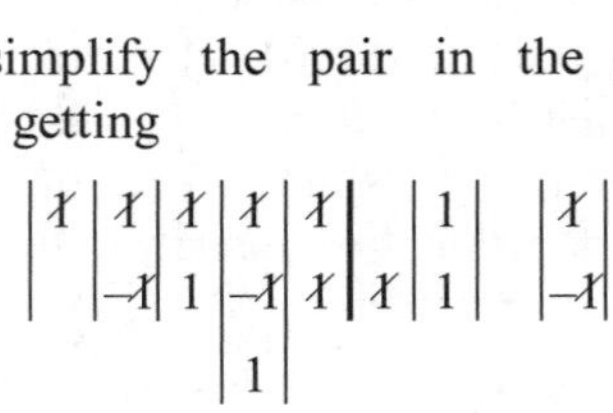

~~1~~	~~1~~	~~1~~	~~1~~	~~1~~		1		~~1~~
	~~–1~~	1	~~–1~~	~~1~~	~~1~~	1		~~–1~~
			1					

and the pair we thereby form in the $\tau^1$and τ^2 columns getting

~~1~~	~~1~~	~~1~~	~~1~~	~~1~~		1		~~1~~
	~~−1~~	~~1~~	~~−1~~	~~1~~	~~1~~	1		~~−1~~
	1		~~1~~					

Finally, we expand one of the 1s in the τ^{-2}column, getting

~~1~~	~~1~~	~~1~~	~~1~~	~~1~~		1	1	~~1~~
	~~−1~~	~~1~~	~~−1~~	~~1~~	~~1~~	~~1~~		~~−1~~
	1		~~1~~					1

and simplify the resulting pair (in the τ^{-2} and τ^{-3} columns) getting as our final answer

~~1~~	~~1~~	~~1~~	~~1~~	~~1~~	1	~~1~~	~~1~~	~~1~~
	~~−1~~	~~1~~	~~−1~~	~~1~~	~~1~~	~~1~~		~~−1~~
	1		~~1~~					1

or 1000.1001 (5).

For subtraction it is harder to formulate a general rule, but I think it would suffice to say: Cancel whenever possible and simplify or expand whenever that would permit cancellation (also remember not to confuse a 1 with a −1 (|1|1|−1| ≠ |1|~~1~~|~~−1~~|) and not to make the mistake of "expanding" a −1 into two +1s.) After all −1s have been removed by cancellation, proceed as you would with an addition example.

Multiplication

Multiplication involves nothing new. We simply place the partial products as we do in the decimal system, and add. For instance:

$$\begin{array}{r} 101.01 \\ \times\ 100.01 \end{array} \qquad (3 \times 4)$$

Setting up the partial products, we get
$$\begin{array}{r} 10101 \\ 10101\ \ \end{array}$$
or

				1		1		1		
1		1		1						

and (from now on it is simple addition) expanding one of the 1s in the units column:

				1		1		1		
1		1		~~1~~	1	1				

We now simplify the pair in the units and τ^{-1} columns

			1	~~1~~		1		1		
1		1		~~1~~	~~1~~	1				

and the pair we thus produce:

			~~1~~	~~1~~		1		1		
1	1	~~1~~		~~1~~	~~1~~	1				

and again the pair this simplification produces, getting:

				~~1~~	~~1~~		1		1		
1	~~1~~	~~1~~	~~1~~		~~1~~	~~1~~	1				

Next we expand one of the 1s in the τ^{-2}column, and one of the 1s in the τ^{-4} column:

				~~1~~	~~1~~		~~1~~		~~1~~	1	1
1	~~1~~	~~1~~	~~1~~		~~1~~	~~1~~	1	1	1		

Finally, we simplify the pair in the τ^{-2} and τ^{-3} columns, and then the one in the τ^{-4} and τ^{-5} columns, giving our final answer:

				~~1~~	~~1~~	1	~~1~~	1	~~1~~	~~1~~	1
1	~~1~~	~~1~~	~~1~~		~~1~~	~~1~~	~~1~~	~~1~~	~~1~~		
1	0	0	0	0	0.	1	0	1	0	0	1

or 100000.101001

Division

Division is quite different in this system, and is, in fact, rather odd. The only things it has in common with ordinary division are the basic principles behind it, the way the example looks, and the movement of the "decimal point" to eliminate any figures to the right of it in the divisor. It is best explained by an example: 12 divided by 2, or

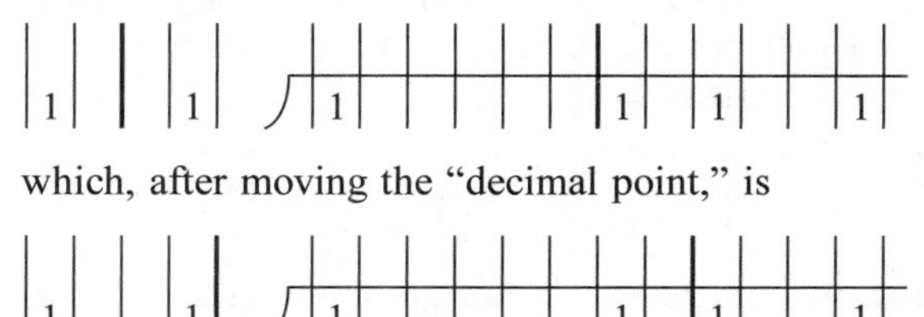

which, after moving the "decimal point," is

Now, since we are dividing by |1| | |1|, if there are anywhere two 1s with two spaces between them (the spaces can be empty or full), they can be crossed out and a 1 placed in the quotient above the rightmost of the two. This crossing out in no way signifies that the 1s should not be there, but is merely equivalent to, in long division (decimal) the subtraction of the product of the number placed in the quotient and the divisor from the dividend. Since the number placed in the quotient can only be 1 (placed in any column, of course), we merely subtract the dividend (placed in that same column), i.e., cross it out. It is obvious that once the whole dividend has been crossed out, the group of 1s in the quotient, after being changed to the simplest form, will be the complete quotient. Getting back to our original problem, we see that we do have just such a set of 1s in the τ^{-1} and τ^{-4} column, and so we cross it off and place a 1 in the quotient, getting

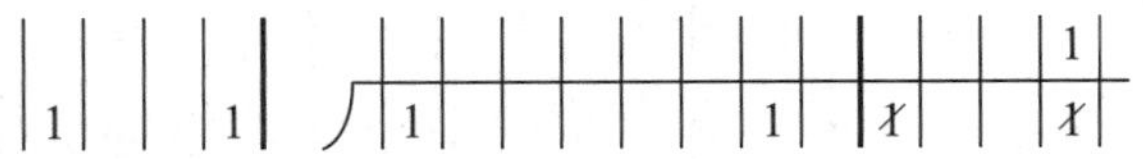

But now, you may say, there are no more pairs of 1s spaced in that way; what shall we do? The answer is our old pair of friends, expansion and simplification. Since they do not change the value of a number, if either of those processes yields a set of 1s spaced correctly, that set can be crossed off and a 1 placed in the quotient just as though that set were part of the original number. Since in our problem it is so far impossible to simplify, we shall expand. Expanding the 1 in the τ^7 column, we get

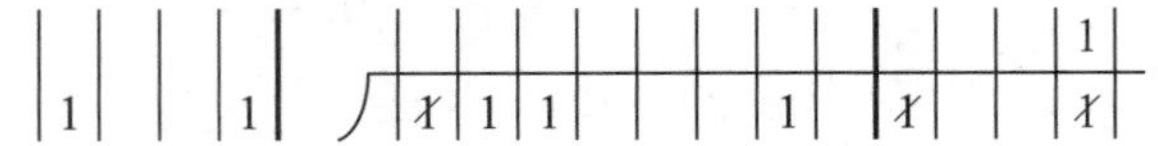

No such set yet. However, when we expand the 1 we've just placed in the τ^5 column, giving

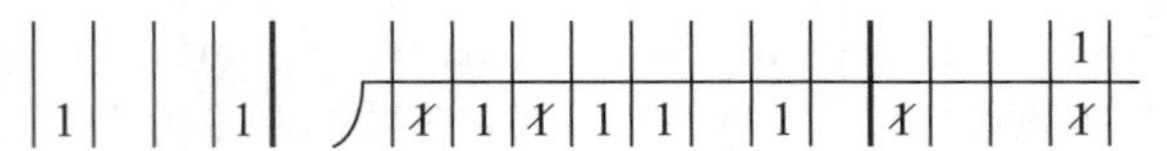

we have not one but two such sets (remember that all we need is two 1s with that certain separation, regardless of intervening and crossed out 1s), one made of the 1s in the τ^1 and τ^4 columns, and the other of the 1's in the τ^3 and τ^6 columns. Crossing them both out and placing the 1s in the correct places in the quotient, we get

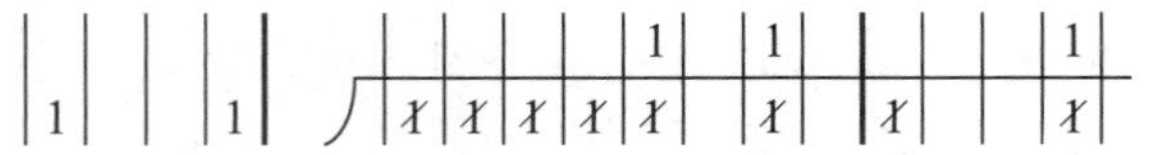

and since there are no more 1s in the dividend, our number in the quotient is the complete quotient, and so 1010.0001, or 6, is our answer. This time the general rule is: Always take that course of action that will place your next 1 (i.e., a 1 in the quotient—set in the dividend) farthest to the left. By a course of action, I mean a series of expansions and simplifications and the exchange of a set for a 1 in the quotient that follows; or simply that exchange, if the set is already there. (I did not obey this rule in my demonstration so that I could show the process in a simpler way.) This is so that the answer be in its simplest form.

By the way, the processes of addition, subtraction, and often multiplication, can be performed together by writing the addends, the subtrahends, and the partial products in one set of columns; for instance: $2 \times 3 + 4 + 3 - 5$:

$$\begin{array}{l} 10.01 \\ \times\ 100.01 + 101.01 + 100.01 - 1000.1001 \end{array}$$

and working it out:

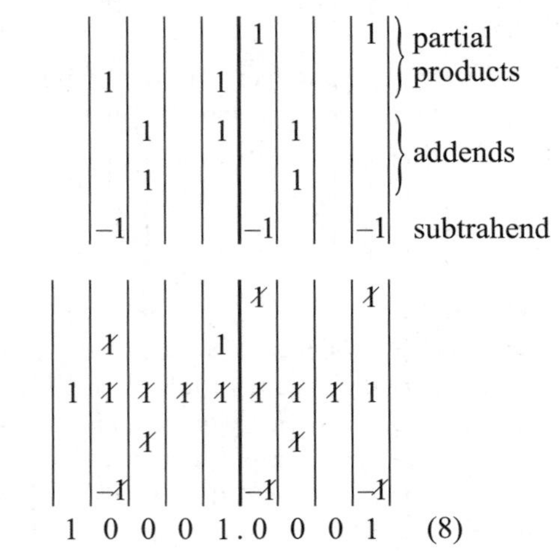

Now that we know these four processes, we have a much better way of finding a number in this system than merely repeatedly adding 1s until we reach it. For instance, to find thirty-seven, we can multiply 6×6 and add 1; to check the arithmetic, we multiply 7×5 and add 2:

$$\begin{array}{lcl} 1010.0001 & \text{(to check)} & 1000.1001 \\ \times\ 1010.0001 + 1 & & \times\ 10000.0001 + 10.01 \end{array}$$

What's more, we are not only able to find integers, but, since we can divide, we ought to be able to find fractions also. Let us try. First, we shall attempt to find 1/2. We begin by setting up our division:

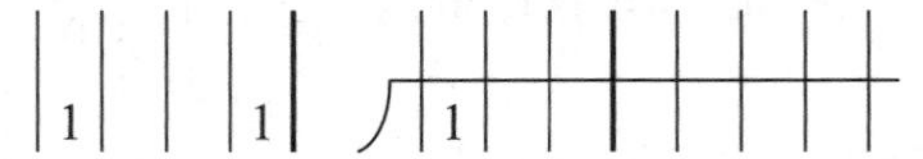

Our first set, as can be seen, will have its leftmost 1 in the τ^1 column. We therefore expand the 1 in the τ^2 column:

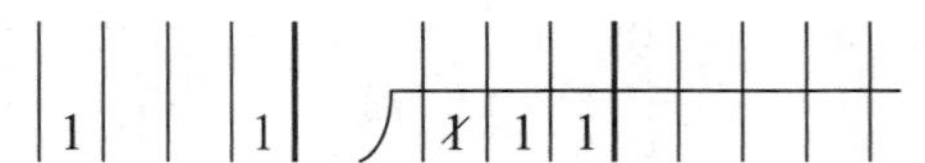

The other one we need to complete the set is a 1 in the τ^{-2} column; this we get by expanding the 1 in the units column:

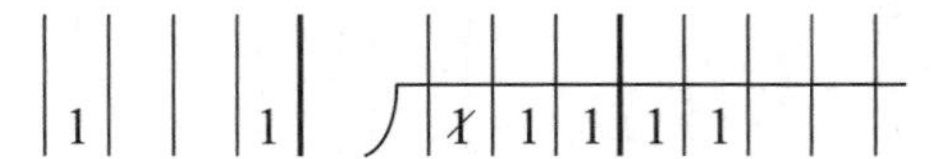

$$\frac{1}{2} = .\underline{010}010010010\ldots$$

$$\frac{1}{3} = .\underline{00101000}0010100000101000\ldots$$

$$\frac{1}{4} = .\underline{001000}001000001000001000\ldots$$

$$\frac{1}{5} = .\underline{0001001010100100100}0001001010100100100\ldots$$

$$\frac{1}{10} = .\underline{00001000010001010000101000101010100010010100000100100010000}00001\ldots$$

Table 1. List of fractions

After we exchange our set for a 1 in the quotient (τ^{-2} column), we notice that our remainder is 1 (in the τ^{-1} column). Since 1 is the number we started with, the next figure in the quotient and the next remainder should be the same as these. However, the question is, where in the quotient shall we place it? Since our first 1 was in the τ^2 column (because we moved the decimal point) and our remainder is three places to the right of it, in the τ^{-1} column, our next 1 in the quotient should be three places to the right of the first 1 there. Since the next remainder will bear the same relationship to the first remainder as the first did to our original 1, the following 1 in the quotient will be three places to the right of our second 1. Since this can be carried on indefinitely, it appears that $1/2$ expressed in the Tau System is .01001001001... (any "doubting Thomases" may carry it out a few places to see).

Before we go on to other fractions, it would be wise to mention something about 1 in the Tau System. As you can easily see, $1 = .11 = .1011 = .101011 = .10101011$ etc. It is, therefore, equal to the endless "fraction" .10101010... (just as in the decimal system $1 = .9999999\ldots$). If we can now take this fraction and expand the leftmost 1, and then expand the 1 in the τ^{-3} column, so as to prevent the occurrence of two 1s in the same column, and then expand the 1 in the τ^{-5} column so that there are not two 1s in that column, etc., we will get .01111111.... If, on the other hand, we start by expanding 1s in other columns, we get: .100111111..., .10100111111..., .101010011111..., etc. Therefore, if you multiply .01001001001... (1/2) by 10.01 (2) and get .1001111111..., this does not mean that $2 \times (1/2) =$ a fraction, but merely shows a different way of representing 1.

To get back to fractions, we can make a list of them just as we did of integers before (see the table above) Of course, finding these fractions is immeasurably harder than finding $1/2$, and with $1/10$ I had to work it out five or ten times before I got the correct answer, as there is much room for error.

By the way, no fraction can be terminating in this system, since that would mean that it could be expressed as the sum of a group of integral powers of tau. Since all the powers of tau can be expressed as the sum of an integer and an integral multiple of tau, if the integral multiples "cancel" (e.g., $\tau^{-1} + \tau^{-3} + \tau^{-4} = 1\tau - 1 + 2\tau - 3 - 3\tau + 5 = 1$) the result will be an integer, and if they don't (e.g., $2\tau - 3 - 3\tau + 5 = -\tau + 2$), it will naturally be irrational. However, when we have an endless series, this paradox is detoured by admitting the fact that Lim $A\tau + B$ with A and B always integral can be a rational fraction if A and $B \to \infty$.

The Tau System has a good many other interesting and unusual characteristics, and investigation by the readers of some, such as the frequency, occurrence, and nature of numbers with a 1 in the units column (when in simplest form) might prove interesting. I do not know of any useful application for systems such as this, except as a mental exercise and pastime, though it may be of some service in algebraic number theory. For instance, the numbers expressible in the Tau System in terminating form consist of all the algebraic integers in $R(\sqrt{5})$, and some of the properties of numbers in this and other systems might correspond to facts about associated fields.

Definitions Invented for Work in the Tau System

Expand Alter three successive figures of a number by changing ...100... to ...011.... The result is the same in value as the original, because of rule (2).

This does not mean change zeros to ones and ones to zeros; just this specific change.

Simplify The reverse of expand; alter the figures thus: change ... 011 ... to ... 100 .... One speaks of simplifyng the 1s in the τ^{n-1} and τ^{n-2} columns into the τ^n column. Also, one speaks of expanding the 1 in the τ^n column into the τ^{n-1} and τ^{n-2} columns.

Simplest form That form of a number which has been simplified until no more simplification is possible. It therefore has no two 1s in succession. It also has the fewest 1s and is the easiest form to work with.

Columns Just as in our decimal system we speak of a units column, a tens column, a hundreds column, a tenths column, etc., in the Tau System we speak of a units column, a τ^1 column, a τ^{-1} column, etc.

Pair Two 1s in succession.

Cancellation A change of the form

$$\begin{vmatrix} 1 \\ \\ -1 \end{vmatrix} = \begin{vmatrix} \not{1} \\ \\ -\not{1} \end{vmatrix}$$

Set In division, two or more 1s arranged with the same spacing as the 1s in the divisor (regardless of intervening 1s). A set can be "exchanged" for a 1 in the quotient.

Part IV

The 1960s

Generalizations of Theorems about Triangles

Carl B. Allendoerfer

Mathematics Magazine
38 (1965), 253–59

Editors' Note: Carl Barnett Allendoerfer was president of the Mathematical Association of America (1959–60) and won the Lester R. Ford Award for this article, published a few years after his presidency. (That was before the MAA had established the Allendoerfer Award for papers in *Mathematics Magazine*.)

Allendoerfer was an undergraduate at Haverford College, then after a stay at Oxford University as a Rhodes Scholar, he received his PhD from Princeton in 1937. He spent most of his career at the University of Washington in Seattle, where he was department chair between 1951 and 1962. He was a member of the Institute for Advanced Study in Princeton, 1948–49, and a Fulbright lecturer at the University of Cambridge, 1957–58.

A topologist by profession he also wrote several elementary textbooks, the most successful of which was *Fundamentals of Freshman Mathematics* (McGraw-Hill, 1959), coauthored with Cletus O. Oakley of Haverford College.

Professor Allendoerfer died in 1974.

1. Introduction

Since one of the most powerful methods in mathematical research is the process of generalization, it is certainly desirable that young students be introduced to this process as early as possible. The purpose of this article is to call attention to the usually untapped possibilities for generalizing theorems on the triangle to theorems about the tetrahedron. Some of these, of course, do appear in our textbooks on solid geometry; but here I shall describe two situations where the appropriate generalizations seem to be generally unknown. The questions to be answered are: (1) What is the generalization to a tetrahedron of the angle-sum theorem for a triangle? (2) What is the corresponding generalization of the laws of sines and cosines for a triangle? Expressed in this form, the questions are certainly vague; for surely there are many generalizations. From these we are to select the ones which are most satisfying and which have a clear right to be called *the generalizations*. In attacking these problems we will need to reexamine the theorems as they are stated for a triangle, and perhaps to reformulate them so that the generalizations appear to be natural. Thus we have a bonus in that we learn additional ways of thinking about triangles.

2. The angle-sum theorem

Since this theorem is one of the most familiar in Euclidean geometry, it is strange that its three-dimensional generalization is not part of the classical literature on geometry. I ran across this generalization some years ago and have been putting the question to mathematicians wherever I find them. Only one of them, Professor Pólya, knew of it. He attributes it to Descartes [1].

The first question to be settled is that of the type of angles in a tetrahedron to be considered. It would be most natural to consider the inner solid angles and their sum. I remind you that the measure of a solid angle is the area of the region on the unit sphere which is the intersection of the sphere with the interior of the solid angle whose vertex is at the center. Thus the measure of the solid angle at a corner of a room is $4\pi/8 = \pi/2$, and the measure of a "straight" solid angle is $4\pi/2 = 2\pi$. By considering a few cases, we conclude that the sum of the measures of the inner solid angles of a tetrahedron is not a constant. For example consider the situation in Figure 1, where all the points lie in a plane.

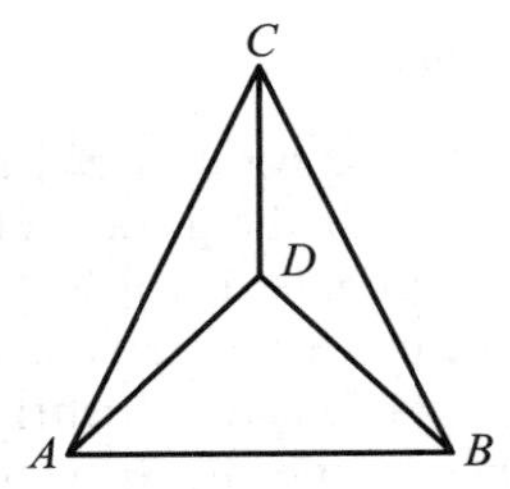

Figure 1.

If D is raised slightly, we have a tetrahedron the sum of whose interior solid angles is very near to 2π. On the other hand let us raise segment AB in the plane Figure 2 a small amount. Then we have a tetrahedron the sum of whose inner solid angles is very near to zero. Hence the obvious generalization is incorrect. As a matter of fact it has been proved [2] that the sum of the solid angles of a tetrahedron can take any value between 0 and 2π.

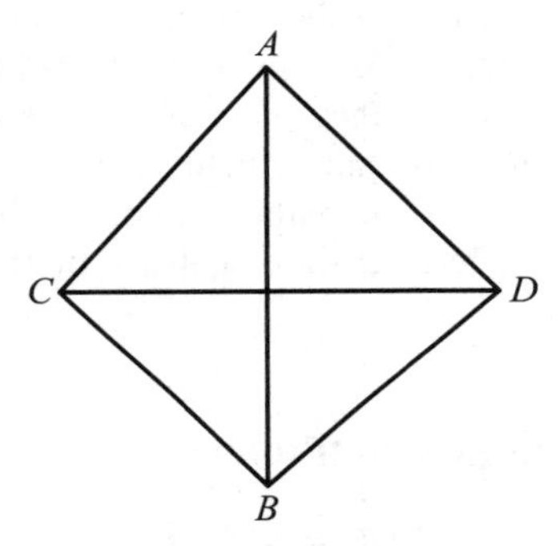

Figure 2.

In order to make a fresh start, let us reformulate the triangle theorem in the statement: The sum of the outer angles of a triangle equals 2π. There are two possible definitions of an outer angle. The usual one is that it is the angle between a pair of successive directed sides (Fig. 3). This clearly does not generalize to three dimensions. Less familiar is the definition that an outer angle at a vertex is the angle between the two outward drawn normals to the two edges which meet at this vertex (Fig. 4).

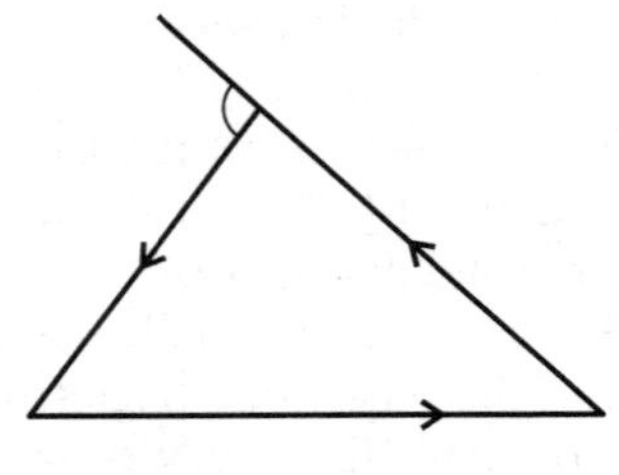

Figure 3.

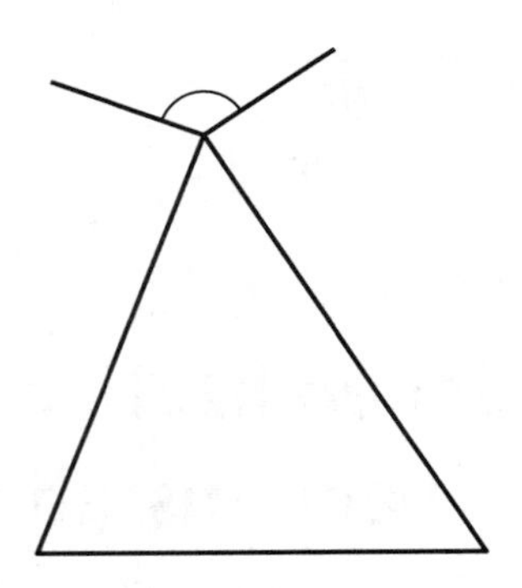

Figure 4.

Using this second definition, we can construct an elegant proof of the theorem. Choose any point P in the interior of the triangle and draw the perpendiculars from P to the three sides (Fig. 5). Then the outer angles α, β, and γ are equal to the three angles formed at P. Hence $\alpha + \beta + \gamma = 2\pi$.

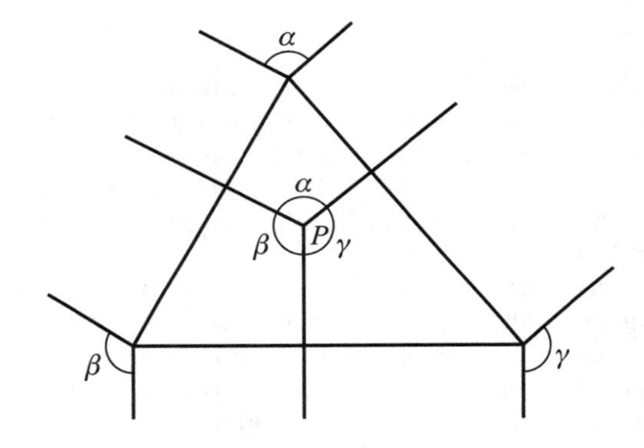

Figure 5.

Now we can generalize at once. To find the corresponding theorem on the tetrahedron, first define the outer angle at a vertex as the trihedral angle formed by the three outer normals to the three faces meeting at this vertex. Choose an interior point P and draw the perpendiculars from P to the four faces. By the same argument that we used for the triangle, we find that

Theorem 1 *The sum of the outer angles of a tetrahedron is* 4π.

By a straightforward generalization of the notion of an outer angle, we can similarly prove that

Theorem 2 *The sum of the outer angles of any convex polyhedron is equal to* 4π.

There is also an immediate generalization to higher dimensions.

3. The Laws of Sines and Cosines

Before considering the generalization of these laws to a tetrahedron, let me give unfamiliar proofs of them which will suggest the proper generalization.

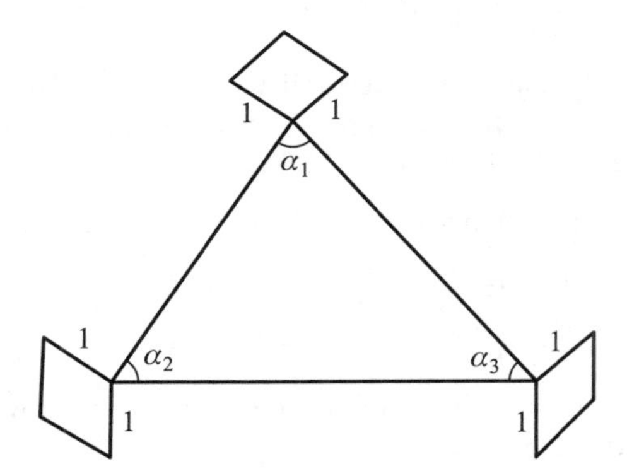

Figure 6.

First, consider the Law of Sines. At each vertex (Fig. 6) draw the unit outer normals to the sides meeting at that vertex and complete the parallelograms determined by these pairs. By a familiar theorem of trigonometry the areas of these parallelograms are respectively $\sin(\pi - \alpha_1) = \sin\alpha_1$, $\sin(\pi - \alpha_2) = \sin\alpha_2$, and $\sin(\pi - \alpha_3) = \sin\alpha_3$. We shall proceed to compute these areas in terms of the coordinates of the vertices of the triangle (Fig. 7), choosing the notation appropriately so that $A_1A_2A_3$ are labeled in a counterclockwise fashion.

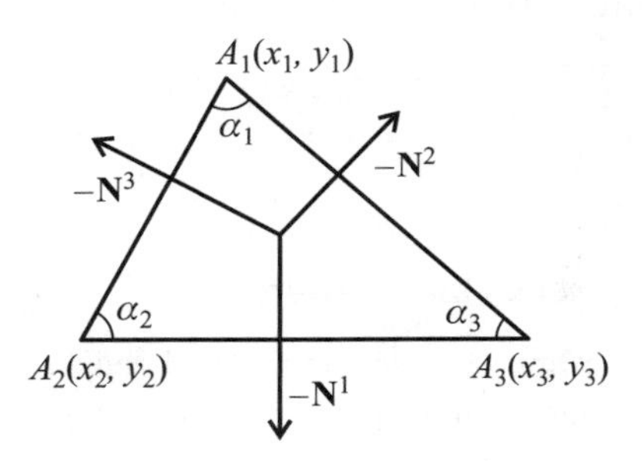

Figure 7.

The equation of side A_2A_3 is

$$xN_x^1 + yN_y^1 + (x_2y_3 - x_3y_2) = 0$$

where N_x^1 and N_y^1 are respectively the cofactors of x_1 and y_1 in the determinant

$$\Delta = \begin{vmatrix} x_1 & y_1 & 1 \\ x_2 & y_2 & 1 \\ x_3 & y_3 & 1 \end{vmatrix}$$

Thus the vector $\mathbf{N}^1$ with components (N_x^1, N_y^1) is normal to A_2A_3; $-\mathbf{N}^1$ is an outer normal; and $\mathbf{U}^1 = -\mathbf{N}^1/a_1$, is the unit outer normal (where a_1 is the length of A_2A_3). More generally $\mathbf{U}^i = -\mathbf{N}^i/a_i$ $(i = 1, 2, 3)$ are the three outer normals, where N_x^i and N_y^i are the cofactors of x_i and y_i respectively and a_i is the length of the side to which $\mathbf{U}^i$ is normal.

The area of the outer parallelogram at A_1 of which two sides are $\mathbf{U}^2$ and $\mathbf{U}^3$ is

$$\sin\alpha_1 = \begin{vmatrix} U_x^2 & U_y^2 \\ U_x^3 & U_y^3 \end{vmatrix} = \frac{1}{a_2a_3}\begin{vmatrix} N_x^2 & N_y^2 \\ N_x^3 & N_y^3 \end{vmatrix}.$$

By a classical theorem on determinants (Bôcher, *Introduction to Higher Algebra*, p. 31) it follows that

$$\begin{vmatrix} N_x^2 & N_y^2 \\ N_x^3 & N_y^3 \end{vmatrix} = \Delta \cdot 1.$$

Hence

$$\sin\alpha_1 = \frac{\Delta}{a_2a_3} \quad \text{and} \quad \frac{\sin\alpha_1}{a_1} = \frac{\Delta}{a_1a_2a_3}.$$

In a similar fashion we prove that

$$\frac{\sin\alpha_1}{a_1} = \frac{\sin\alpha_2}{a_2} = \frac{\sin\alpha_3}{a_3} = \frac{\Delta}{a_1a_2a_3}$$

which is the familiar Law of Sines.

To arrive at the Law of Cosines, we begin with a theorem of Möbius.

Theorem 3 $\mathbf{N}^1 + \mathbf{N}^2 + \mathbf{N}^3 = 0$.

This theorem follows from the facts that $N_x^1 + N_x^2 + N_x^3 = 0$ and $N_y^1 + N_y^2 + N_y^3 = 0$. These may be computed directly, or they maybe proved by expanding the determinants

$$\begin{vmatrix} 1 & y_1 & 1 \\ 1 & y_2 & 1 \\ 1 & y_3 & 1 \end{vmatrix} = 0 \quad \text{and} \quad \begin{vmatrix} x_1 & 1 & 1 \\ x_2 & 1 & 1 \\ x_3 & 1 & 1 \end{vmatrix} = 0.$$

This theorem can be rewritten in the form:

$$\mathbf{N}^1 = -\mathbf{N}^2 - \mathbf{N}^3.$$

Now take the scalar product of each side of this equation with itself. The result is

$$\mathbf{N}^1 \cdot \mathbf{N}^1 = \mathbf{N}^2 \cdot \mathbf{N}^2 + \mathbf{N}^3 \cdot \mathbf{N}^3 + 2\mathbf{N}^2 \cdot \mathbf{N}^3.$$

Since $\mathbf{N}^i \cdot \mathbf{N}^i = a_i^3$, and $\mathbf{N}^2\mathbf{N}^3 = -a_2a_3\cos\alpha_1$, this becomes $a_1^2 = a_2^2 + a_3^2 - 2a_2a_3\cos\alpha_1$.

4. The Generalized Laws of Sines and Cosines

These generalizations are due to Grassmann, but are relatively unfamiliar. Their proofs follow the lines just given in Section 3.

Consider a tetrahedron (Fig. 8) whose vertices are ordered so that

$$\Delta = \begin{vmatrix} x_1 & y_1 & z_1 & 1 \\ x_2 & y_2 & z_2 & 1 \\ x_3 & y_3 & z_3 & 1 \\ x_4 & y_4 & z_4 & 1 \end{vmatrix} > 0.$$

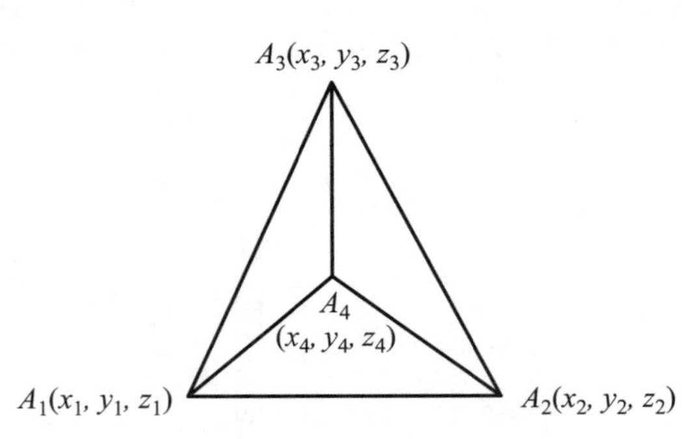

Figure 8.

Then the vector $\mathbf{N}^1$ whose components (N_x^1, N_y^1, N_z^1), are the cofactors of x_1, y_1, z_1 respectively in Δ, is normal to the face $A_2A_3A_4$. The length of $\mathbf{N}^1$, namely a_1, is equal to twice the area of this face. The vector $\mathbf{U}^1 = -\mathbf{N}^1/a_1$ is the unit outer normal to this face. Other normals $\mathbf{N}^i$ and $\mathbf{U}^i$ are defined in a similar fashion.

We now define the generalized sine ("G-sin") of the inner trihedral angle at A_1 to be the volume of the parallelopiped whose edges are $\mathbf{U}^2$, $\mathbf{U}^3$, and $\mathbf{U}^4$. Thus

$$G\text{-}\sin\alpha_1 = \begin{vmatrix} U_x^2 & U_y^2 & U_z^2 \\ U_x^3 & U_y^3 & U_z^3 \\ U_x^4 & U_y^4 & U_z^4 \end{vmatrix}$$

$$= \frac{-1}{a_2a_3a_4}\begin{vmatrix} N_x^2 & N_y^2 & N_z^2 \\ N_x^3 & N_y^3 & N_z^3 \\ N_x^4 & N_y^4 & N_z^4 \end{vmatrix}$$

$$= \frac{(-1)\Delta^2(-1)}{a_2a_3a_4} = \frac{\Delta^2}{a_2a_3a_4}.$$

By a continuation of this argument, we obtain the Generalized Law of Sines:

Theorem 4

$$\frac{G\text{-}\sin\alpha_1}{a_1} = \frac{G\text{-}\sin\alpha_2}{a_2} = \frac{G\text{-}\sin\alpha_3}{a_3} = \frac{G\text{-}\sin\alpha_4}{a_4} = \frac{\Delta^2}{a_1a_2a_3a_4}$$

To establish the Generalized Law of Cosines, we observe that we can prove the following generalization of the Theorem of Möbius.

Theorem 5 $\mathbf{N}^1 + \mathbf{N}^2 + \mathbf{N}^3 + \mathbf{N}^4 = 0.$

Then writing

$$\mathbf{N}^1 = -\mathbf{N}^2 - \mathbf{N}^3 - \mathbf{N}^4$$

and $f_i = a_i/2 =$ area of the ith face, we prove as above the result:

Theorem 6

$$f_1^2 = f_2^2 + f_3^2 + f_4^2 - 2\left[f_2f_3\cos(f_2, f_3) + f_2f_4\cos(f_2, f_4) + f_3f_4\cos(f_3, f_4)\right],$$

where (f_i, f_j) is the inner dihedral angle of the tetrahedron between the faces whose areas are f_i, and f_j respectively.

We also have another, rather novel, generalization if we start from $\mathbf{N}^1 + \mathbf{N}^2 = -\mathbf{N}^3 - \mathbf{N}^4$. The result is

Theorem 7

$$f_1^2 + f_2^2 - 2f_1f_2\cos(f_1, f_2) = f_3^2 + f_4^2 - 2f_3f_4\cos(f_3, f_4)$$

5. Supplementary matters

Another approach to the Generalized Law of Sines is to begin with a right tetrahedron (Fig. 9). Then it would be reasonable to define

$$G\text{-}\sin\alpha_1 = \frac{\text{Area}A_2A_3A_4}{\text{Area}A_1A_2A_3} = \frac{bc}{\{b^2c^2 + a^2c^2 + a^2b^2\}^{1/2}}.$$

Let us show that this agrees with our previous definition of G-$\sin\alpha_1$. We have:

$$\Delta = \begin{vmatrix} a & 0 & 0 & 1 \\ 0 & b & 0 & 1 \\ 0 & 0 & c & 1 \\ 0 & 0 & 0 & 1 \end{vmatrix}.$$

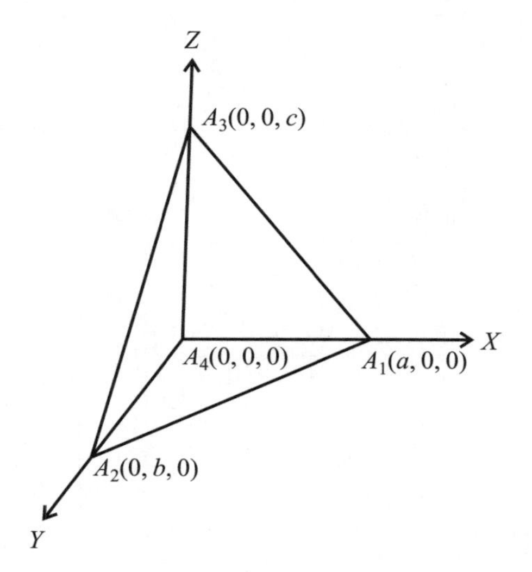

Figure 9.

Then

$$G\text{-}\sin\alpha_1 = \frac{\Delta^2}{a_2a_3a_4} = \frac{a^2b^2c^2}{(ac)(ab)\{b^2c^2+a^2c^2+a^2b^2\}^{1/2}} = \frac{bc}{\{b^2c^2+a^2c^2+a^2b^2\}^{1/2}}.$$

Also we have the reassuring result that for our right tetrahedron:

$$(G\text{-}\sin\alpha_1)^2 + (G\text{-}\sin\alpha_2)^2 + (G\text{-}\sin\alpha_3)^2 = 1.$$

It is natural to ask whether G-$\sin\alpha_1$ is actually the sine of the measure of the inner or the outer solid angle at A_1; the answer is "no". To give an elementary counterexample we consider the right tetrahedron with $a = b = c = 1$. Then G-$\sin\alpha_1 = 1/\sqrt{3}$; sine (measure of inner solid angle at A_1) $= 1/3$; and sine (measure of outer solid angle at A_1) $= \sin 7\pi/6 = -1/2$.

As a matter of fact, G-$\sin\alpha$ is not even a functon of either the inner or the outer solid angles at the given vertex. Rather it depends directly on the face angles of the outer trihedral angle. If these angles are λ, μ, ν and $s = (\lambda+\mu+\nu)/2$, then

$$G\text{-}\sin\alpha = \{\sin s \sin(s-\lambda)\sin(s-\mu)\sin(s-\nu)\}^{1/2}.$$

References

1. R. Descartes, *Œuvres*, vol. 10, pp. 257–276.
2. J. W. Gaddum, The sums of the dihedral and trihedral angles of a tetrahedron, *Amer. Math. Monthly*, 59 (1952) 370–375.

A Radical Suggestion

Roy J. Dowling

Mathematics Magazine
36 (1963), 59

Editor's Note: Professor Dowling, currently Professor Emeritus of Mathematics at the University of Manitoba, Winnipeg, contributed this clever one-paragraph proof to the *Magazine* when he was an Assistant Professor there. One can quickly conclude that this proof does not generalize to a general base b, to show that $\sqrt{b} = \sqrt{10_b}$ is irrational.

$\sqrt{10}$ is a useful number for illustrative purposes when one is discussing irrational numbers at an elementary level. It is the hypotenuse of a right triangle whose other sides are 1 and 3, and it can be shown to be irrational by a most unsophisticated argument.

When the nonzero integer p is squared, the resulting integer has exactly twice as many terminal zero digits as does p (even if p has none). Thus, if p is a nonzero integer, p^2 ends with an even number of zeros, and, if q is a nonzero integer, $10q^2$ ends with an odd number of zeros. It follows that p^2 cannot equal $10q^2$.

Topology and Analysis

R. C. Buck

Mathematics Magazine
40 (1967), 71–74

Editors' Note: Robert Creighton Buck received his undergraduate education at the University of Cincinnati and his PhD from Harvard in 1947, where his advisor was David Widder. After a short stay at Brown University he moved to the University of Wisconsin in Madison, where he stayed for the rest of his career, becoming department head in 1964.

He chaired the Committee on the Undergraduate Program, 1959–63, for the MAA and was active in the school reform movement, mainly the School Mathematics Study Group (SMSG), 1960–64. He is widely known for his very successful text *Advanced Calculus* (McGraw-Hill, 1978). Other books include *Studies in Modern Analysis* (MAA, 1962), for which he served as editor, and, with Ralph P. Boas, Jr., *Polynomial Expansions of Analytic Functions* (Academic Press, 1964).

Professor Buck died in 1998.

1. Introduction

In what follows, I speak as an analyst, not a topologist. In particular, I am not discussing topology as an axiomatic structure, nor yet as a touchstone to bring young minds to life, nor as an organized body of theorems. Rather, I shall confine my remarks largely to the role of topology as something which illuminates topics in analysis, and which provides a more geometric viewpoint. I will give a series of unconnected illustrations.

2. Elementary calculus

To "illuminate" can mean to clarify or simplify. Suppose we assume a background of the simplest anatomy of topology—acquaintance with the meaning of open, closed, connected, compact, etc., all applied chiefly in the context of sets in Euclidean space.

Here are three statements about sets and continuous functions:

(1) The set $\{\text{all } (x, y) \text{ with } x^2 - xy + y^3 \leq 1\}$ is a closed set.

(2) If f is a continuous real-valued function defined on $[a, b]$, then f is bounded there and achieves a maximum and minimum at points of $[a, b]$.

(3) If f is a continuous real-valued function defined on $[a, b]$, and $f(a) < 0$ and $f(b) > 0$, then for some choice of t, $a < t < b$, one has $f(t) = 0$.

Let us also consider three general statements of a topological nature:

(1)′ If E is a closed set, and F is a continuous mapping, then the inverse image $F^{-1}(E)$ is closed.

(2)′ Continuous mappings send compact sets into compact sets.

(3)′ Continuous mappings send connected sets into connected sets.

The connection between these triples should be very clear; each statement $(A)'$ implies the statement (A), and indeed, throws light upon it, supplying a clearer insight into the basic truth of these statements.

3. Intermediate calculus

Sometimes analysis can assist in putting over topological ideas. When one is first introducing topological concepts, a frequent tactic is to present a student with several different topologies on a space, and invite him to compare their properties. If the underlying set is the plane or a subset of the plane, then many of these alternate topologies may strike a student as contrived. He may be left with the feeling that the discussion of these topologies and their properties is of little permanent significance. This is especially true if he has already learned that there is a very definite sense in which the Euclidean topology is the *only* (normed) topology on the plane (or n-space) (see [2], pp. 191–192).

Perhaps this discussion of the comparative properties of several topologies on the same set could better be done in an intermediate analysis course. Suppose we select $\mathfrak{F}$ as the class of all real-valued functions defined on $X = (-\infty, \infty)$. Then, it is readily seen that many of the standard topics in analysis really deal with the comparison between two competing limit topologies defined on $\mathfrak{F}$:

(i) pointwise convergence: $\{f_n\} \to f$ iff $f_n(x) \to f(x)$ for each $x \in X$.

(ii) uniform convergence: $\{f_n\} \to f$ iff $f_n(x) \to f(x)$, uniformly for all $x \in X$.

Each of these is important in its own right, and each plays a role in analysis. Thus it is not an artificial question to ask for differences in the properties of these topologies.

As an illustration, let $\mathcal{C}$ be the subset of $\mathfrak{F}$ consisting of all the continuous functions. Then we can ask, in each case: Is $\mathcal{C}$ a closed set? Note that the answers differ, and that, furthermore, the results convey in different language certain central results about convergence of sequences of functions.

Other questions might be asked. What does a compact set look like? Does

$$F(\phi) = \int_0^1 \phi, \qquad \phi \in \mathcal{C}$$

define a continuous function on $\mathcal{C}$? on $\mathfrak{F}$? Is the set $\mathcal{P}$ of polynomials a dense subset of $\mathcal{C}$?

In each case, a natural topological question yields an important analytic result.

4. Advanced analysis

Another very important aspect of topology can also be discussed at this stage. (We also correct a possibly misleading inference in the preceding section.)

The notion of pointwise convergence (i) gives rise to the "pointwise" topology defined on $\mathfrak{F}$. A neighborhood $\mathfrak{N}$ of a function $f_0 \in \mathfrak{F}$ is specified by a finite number of points $x_1, \ldots, x_N$ and associated positive numbers $\delta_1, \delta_2, \ldots, \delta_N$, and $\mathfrak{N} = \mathfrak{N}(x_1, \ldots, x_N; \delta_1, \ldots, \delta_N)$ consists of all functions $f \in \mathfrak{F}$ such that $|f(x_j) - f_0(x_j)| < \delta_j$ for $j = 1, 2, \ldots, N$.

This can then be shown to be a topology which is *not* metrizable, even though it arises in a very natural manner; in it, the closure of a set *cannot* be obtained merely by adjoining all limits of converging sequences. [Thus, the notion of pointwise convergence, (i) does not define a true topology on $\mathfrak{F}$, although uniform convergence (ii) does!]

To show that this is the case, let $\mathcal{S}$ be the subset of $\mathfrak{F}$ consisting of all continuous functions f which satisfy the pair of conditions

$$\begin{aligned} &\int_0^1 f(x)dx = 1 \quad \text{and} \\ &0 \le f(x) \le 2, \quad \text{for all } x,\ -\infty < x < \infty. \end{aligned} \tag{4}$$

It is easy to see that the constant function 0 belongs to the pointwise closure of $\mathcal{S}$—i.e., every neighborhood $\mathfrak{N}(x_1, x_2, \ldots, x_N; \delta_1, \ldots, \delta_N)$ about 0 contains functions in $\mathcal{S}$. However, there cannot exist any sequence $f_n \in \mathcal{S}$ which converges pointwise to 0. For, if $f_n(x) \to 0$ for each $x \in [0, 1]$, then by the Lebesgue Bounded Convergence Theorem [1], we should have to have

$$\int_0^1 f_n \to \int_0^1 0 = 0 \tag{5}$$

which contradicts condition (4). (There is, of course, a *net* $\{f_\alpha\}$ in $\mathcal{S}$ which converges pointwise to 0.) (See [2], pp. 143–146.)

In a different direction, there are examples in which a topological viewpoint brings additional enlightenment. A standard theorem in elementary analysis is the continuity of the inverse of a monotonic continuous function on a bounded interval. Here is a simple alternate approach. We first investigate the relationship between the continuity of a function and the nature of its graph.

Theorem 1 *Let A and B be metric spaces, with A compact, and let $f : A \to B$ be a function. Let $G = \{all\ (a, f(a))\ for\ a \in A\}$ be its graph in $A \times B$. Then f is continuous iff G is a compact set.*

Proof. One direction is trivial. If f is continuous, so is the mapping $x \xrightarrow{T} (x, f(x))$, and $G = T(A)$ must be compact. Conversely, if G were compact but f not continuous, then there would exist a sequence $\{x_n\}$ in A convergent to $x_0 \in A$, but with $d\big(f(x_n), f(x_0)\big) \geq \delta > 0$ for all n. The sequence $p_n = \big(x_n, f(x_n)\big)$ lies in G, and must therefore have a subsequence $\{p_{n_j}\}$ converging to a point $(a, b) \in G$.

Clearly, $a = x_0$ so that $b = f(x_0)$, and therefore $f(x_{n_j}) \to f(x_0)$, which contradicts the assumed property of $\{x_n\}$.

[*Query*. Can you remove the hypothesis that A and B are *metric* spaces?]

An immediate trivial corollary of this is then the following basic result.

Corollary 1.1 *Let A and B be metric spaces, with A compact. Let $f : A \to B$ be continuous and one-to-one. Then f^{-1} is continuous.*

Proof. Let $B_0 = f(A)$, a compact set in B. Then, $f^{-1} : B_0 \to A$ is a function whose graph is homeomorphic to the graph of f. Since f is continuous, this set is compact, and f^{-1} is continuous.

At this point, one can point out the importance of the requirement that A be compact. Let $\mathcal{C}$ be the metrizable space consisting of all continuous real-valued functions on $[0, 1]$ with the uniform convergence topology. Define a function $F : \mathcal{C} \to \mathcal{C}$ by $F(\phi) = \psi$ where $\psi(x) = \int_0^x \phi$, $0 \leq x \leq 1$. It is easily checked that F is continuous. (Indefinite integration of a uniformly convergent sequence leaves it uniformly convergent, if we normalize at one point.) It is also one-to-one, for if $F(\phi_1) = F(\phi_2)$ then $\int_0^x (\phi_1 - \phi_2) = 0$ for all $x \in [0, 1]$, and $\phi_1 = \phi_2$. However, F^{-1}—which exists as a function—is *not* continuous! (One sees that F^{-1} is D, the differentiation operator, and if $\psi_n \to \psi_0$ uniformly, it does not follow that ψ_n' converges to ψ_0'.)

Here, then, is a convenient example of a function F which is continuous, and whose graph is therefore closed, but whose graph is not compact. Likewise, D is a function whose graph is closed, but which is not continuous on $\mathcal{C}$. (A simpler example is the function g on $[0, 1]$ defined by $g(x) = 1/x$, $g(0) = 1$.)

References

1. W. Rudin, *Principles of Mathematical Analysis*, McGraw-Hill, New York, 1964, pp. 246–247.

2. A. Wilansky, *Functional Analysis*, Blaisdell, New York, 1964.

An invited address at the MAA meeting in Denver, 1965.

The Sequence $\{\sin n\}$

C. Stanley Ogilvy

Mathematics Magazine
42 (1969), 94

Editors' Note: C. Stanley Ogilvy is Professor Emeritus of Mathematics at Hamilton College. Educated as an undergraduate at Williams College, he received his PhD from Syracuse University in 1954.

He is best known for his popular books *Through the Mathescope* (Oxford University Press, 1956), *Tomorrow's Math/Unsolved Problems for the Amateur* (Oxford, 1962), *Excursions in Number Theory* (with John T. Anderson) (Oxford,1966), and *Excursions in Geometry* (Oxford, 1969).

The following short proof is offered for the main result of the paper by John H. Staib and Miltiades S. Demos in Vol. 40, page 210 of this *Magazine*, namely that $\{\sin n\}$ is dense on $[-1, 1]$.

Let the points whose polar coordinates (r, θ) are $(1, 1), (1, 2), (1, 3), \ldots$ be called integral points. Then the integral points are dense on the unit circle. For, assume that there exists an arc-length of magnitude $\epsilon > 0$ such that between θ and $\theta + \epsilon$ there are no integral points. Then there is another such gap whose initial point is at $\theta + 1$, another whose initial point is at $\theta + 2$, and so on. Two initial points can never coincide, or π would be rational. But an infinite number (indeed any finite number $> (2\pi)/\epsilon$) of such gaps must cover the circle, so that the circle would contain *no* integral points, an absurdity.

The set $\{\sin n\}$ is a projection of the integral points onto the interval $[-1, 1]$ of the Y-axis. This set is dense in $[-1, 1]$ because every open interval is the projection of two open arcs of the unit circle which, in turn, contain integral points.

Probability Theory and the Lebesgue Integral

Truman Botts

Mathematics Magazine
42 (1969), 105–11

Editors' Note: After his undergraduate years at Stetson University, Botts went to the University of Virginia where he received his PhD in 1942. After a short stay at the University of Delaware, he took a faculty position at the University of Virginia, where he remained until he became Executive Director of the Conference Board of the Mathematical Sciences, a position he held from 1968 till 1982. He is now retired and living in Arlington, Virginia.

1. Random phenomena

Probability theory is the study of mathematical models for random phenomena. A random phenomenon is an empirical phenomenon whose observation under given circumstances leads to various different outcomes. When a coin is tossed, either heads or tails comes up; but on a given toss we can't predict which. When a die is tossed, one of the six numbered faces comes up; but again we can't predict which. In such simple random phenomena the various possible outcomes appear to occur with what is called "statistical regularity". This means that the relative frequencies of occurrence of the various possible outcomes appear to approach definite limiting values as the number of independent trials of the phenomenon increases indefinitely.

If we are trying to set up an abstract mathematical model for a random phenomenon, one of the features of the model should surely be a set E whose elements correspond to the various possible outcomes of this random phenomenon. For the case of tossing a coin this basic set might be the two-element set

$$E = \{H, T\}.$$

For the toss of a die it might be the six-element set

$$E = \{1, 2, 3, 4, 5, 6\}. \tag{1}$$

In the case of tossing a die we might also be interested in other "events," for instance, the event that an odd number comes up, or the event that the number which comes up is greater than 2, etc. We see that such events correspond, in our mathematical model, to subsets of the basic set E. In tossing a die, the event that an odd number comes up corresponds to the subset

$$\{1, 3, 5\}$$

of the set (1). The event that the number which comes up is greater than 2 corresponds to the subset

$$\{3, 4, 5, 6\},$$

and so forth.

The basic set E is not always finite. For instance, consider the random phenomenon of repeatedly tossing a coin until the first time heads comes up. Here the basic set E might be taken to be

$$E = \{H, TH, TTH, TTTH, \dots\},$$

or we might simply take E to be the set

$$E = \{1, 2, 3, \dots\}$$

of natural numbers, each natural number n corresponding to the outcome that n tosses are required.

Sometimes the basic set is not even countable. For example, consider the random phenomenon of spinning a pointer on a dial and, when it comes to rest, measuring—in radians, say—the angle θ it makes with some reference direction. Here it would be natural to take the basic set in our mathematical model to be

$$E = \{\theta : 0 \leq 0 < 2\pi\} = [0, 2\pi).$$

This is well known to be an *uncountably* infinite set.

2. Probability

Let us return to the simple example of tossing a coin. We said that in repeated tosses of a coin the relative frequency of heads (i.e., the number of times heads comes up divided by the number of tosses) appears to approach some definite limit p as the number of tosses increases indefinitely. We assign this number to the element H of the basic set $E = \{H, T\}$ and call it the *probability* of the outcome heads. When we choose $p = (\frac{1}{2})$, we say our mathematical model is for the random phenomenon of tossing a *fair* coin. When we choose $p \neq (\frac{1}{2})$, we say our model corresponds to tossing a biased coin. In any case, though, we choose $0 \leq p \leq 1$ (for otherwise p could not be a limit of relative frequencies). Since the relative frequencies of heads and tails always have sum 1, we see that we would then wish to assign to T, representing the outcome tails, a probability q such that $p + q = 1$.

In general, for a random phenomenon with n outcomes we would use a basic set

$$E = \{x_1, \ldots, x_n\}$$

of n elements, assigning to each element x_i a probability p_i such that $0 \leq p_i \leq 1$ and

$$p_1 + \cdots + p_n = 1.$$

And when the basic set E has a countable infinity of elements $x_1, x_2, \ldots, x_n, \ldots$, we assign to each x_i a probability p_i such that $0 \leq p_i \leq 1$ and

$$p_1 + p_2 + \cdots + p_n + \cdots = 1$$

in the sense of the sum of an infinite series.

Whenever E has a countable (i.e., finite or countably infinite) number of elements x_i, with respective probabilities p_i, we can assign to every *event*, i.e., to every subset A of E, a probability

$$P(A) = \sum_{x_i \in A} p_i.$$

That is, to get the probability of event A, we just add up the probabilities assigned to the various points of A. It is not hard to show that the resulting function P, defined on the class of all events (i.e., all subsets of E), has the following properties.

a. For every event A, $0 \leq P(A)$ (i.e., P is *nonnegative*). (2)

b. $P(E) = 1$.

c. If $A_1, A_2, \ldots$ is a sequence, finite or infinite, of mutually exclusive events, then

$$P(A_1 \cup A_2 \cup \cdots) = P(A_1) + P(A_2) + \cdots$$

(i.e., P is *countably additive*).

The development of probability theory has shown that the properties above are just the ones we need for a probability function in general.

3. The uncountably infinite case

Now how do we assign probabilities to events when the basic set E is uncountably infinite? Let us return to the example of spinning a pointer on a dial, where $E = [0, 2\pi)$. Here, thinking of the case of a "fair" pointer, it is natural to begin by assigning to each interval I in this set a probability proportional to its length:

$$P(I) = \frac{\text{length of } I}{2\pi} \tag{3}$$

Then

$$P(E) = \frac{2\pi}{2\pi} = 1.$$

But this only takes care of the *intervals* in E. What about *any* given subset A of E? It is natural to wish to consider the probability that the angle at which the pointer comes to rest will lie in such a set A. So the question now is: can we extend the domain of P to the class of *all* subsets of $E = [0, 2\pi)$, and in such a way that the essential requirements (2) for a probability function are satisfied? This is a very hard question, and unpleasantly enough, the answer turns out to be "No!" At least this is what can be proved if we permit the use of certain rather fancy tools of mathematical proof, called the axiom of choice and the continuum hypothesis. (If we *don't* permit use of the continuum hypothesis, the answer is as yet unknown!)

Our question becomes less difficult if we require in addition that the function P be "translation-invariant." This means that we require $P(A) = P(B)$ whenever the subsets A and B of $E = [0, 2\pi)$ are "rotations of each other," thought of as sets on the unit circle. This simpler question brings us to our first contact with the theory of measure and integration devised by the French mathematician Henri Lebesgue in the first years of this century. In effect, Lebesgue was able to answer this simpler question "No" without using the continuum hypothesis (though he still had to use the axiom of choice!).

We shall have to accept, then, that we *can't* assign probabilities to *all* the subsets of $E = [0, 2\pi)$, if we require properties (2) and (3) to hold. Well, what more modest class $\mathcal{S}$ of subsets of E, containing all the intervals of E, could we make do with? It would seem reasonable to require of any such class $\mathcal{S}$ that

Whenever A belongs to $\mathcal{S}$, then so does $E - A$. (4)

For our pointer-spinning example this just says that if we can consider the event that the pointer-angle *does* land in set A, then we certainly ought to be able to consider the event that it *doesn't*! Thinking of the countable additivity requirement in (2), it also seems reasonable to demand that

Whenever $A_1, A_2, \ldots$ is a (finite or infinite) sequence of sets belonging to $\mathcal{S}$, then also $A_1 \cup A_2 \cup \cdots$ belongs to $\mathcal{S}$. (5)

Mathematicians have a name (in fact, several names!) for any nonempty class $\mathcal{S}$ of subsets of a set E satisfying conditions (4) and (5). We call such a class a *σ-field* (or *Borel field* or *σ-algebra*) of subsets of E.

It was Lebesgue's theory of measure and integration that showed that P *could* be extended, so as to satisfy the requirements (2) and (3) and translation-invariance, from the class of intervals in $E = [0, 2\pi)$ to a σ-field of subsets of E, called the *Lebesgue measurable sets* of E. In the succeeding years this has led to a concept which lies at the foundation of modern probability theory, the concept of a probability space. A *probability space* is a triple $(E, \mathcal{S}, P)$ consisting of a set E (of "outcomes"), a σ-field $\mathcal{S}$ of subsets of E (the "events"), and a function P defined on $\mathcal{S}$ and having properties (2) (a "probability function"). We shall return to this concept after we have discussed certain ideas of integration, to which we now turn.

4. The Riemann integral

The integral of elementary calculus is sometimes called the *Riemann integral*, after the 19th century German mathematician who gave this integral a careful formulation. In order to compare it with the Lebesgue integral, let us recall briefly how it is defined.

If f is a bounded real-valued function on some interval $[a, b]$, then f is of course bounded, above and below, on every subset E of $[a, b]$. It follows from a basic property of the real numbers that f then has a *least upper bound*

$$\operatorname*{lub}_{x \in E} f(x)$$

on E and a *greatest lower bound*

$$\operatorname*{glb}_{x \in E} f(x)$$

on E. Corresponding to any partition λ of $[a, b]$ into pairwise-disjoint intervals $I_1, \ldots, I_n$, we may form—denoting the length of an interval I by $L(I)$—an "upper sum"

$$S(f; \lambda) = \sum_{k=1}^{n} \left(\operatorname*{lub}_{x \in I_k} f(x) \right) \cdot L(I_k)$$

and a "lower sum"

$$s(f; \lambda) = \sum_{k=1}^{n} \left(\operatorname*{glb}_{x \in I_k} f(x) \right) \cdot L(I_k).$$

It is easy to argue that as λ ranges over all such partitions of $[a, b]$ these upper sums form a collection of numbers having a lower bound and hence a greatest lower bound, which we define to be the *upper Riemann integral*

$$\operatorname*{glb}_{\lambda} S(f; \lambda) = \mathcal{R}\overline{\int_a^b} f(x)\, dx$$

of f over $[a, b]$. Similarly the lower sums form a collection having an upper bound and hence a least upper bound, which we define to be the *lower Riemann integral*

$$\operatorname*{lub}_{\lambda} s(f; \lambda) = \mathcal{R}\underline{\int_a^b} f(x)\, dx$$

of f over $[a, b]$. It is not hard to show that

$$\mathcal{R}\underline{\int}_{a}^{b} f(x)\,dx \leq \mathcal{R}\overline{\int}_{a}^{b} f(x)\,dx$$

When equality holds, we say that f is *Riemann-integrable* over $[a, b]$ and call this common value the *Riemann integral*

$$\mathcal{R}\int_{a}^{b} f(x)\,dx$$

of f over $[a, b]$. (Here the "R" for Riemann is just part of the integration sign.)

5. The Lebesgue integral

The Lebesgue integral of a bounded function f over $[a, b]$ may be defined in precisely the same manner as the Riemann integral, the only difference being that partitions of $[a, b]$ into intervals are replaced by partitions of $[a, b]$ into Lebesgue measurable sets.

Just as in the case of the particular interval $[0, 2\pi)$ already discussed earlier, Lebesgue's theory yields a σ-field of subsets of $[a, b]$ containing all subintervals of $[a, b]$ and called the *Lebesgue measurable sets* of $[a, b]$. This theory also yields a function L on this σ-field called *Lebesgue measure*. Like the probability function P already discussed, L is nonnegative and countably additive (and translation-invariant); and where P reduces to *relative length* for subintervals, L reduces to *actual length*; that is, for each subinterval I of $[a, b]$, $L(I) =$ length of I.

Now, corresponding to any partition μ of $[a, b]$ into pairwise-disjoint Lebesgue measurable sets $E_1, \ldots, E_n$, we form an upper sum

$$S(f;\mu) = \sum_{k=1}^{n}\left(\operatorname*{lub}_{x\in E_k} f(x)\right)\cdot L(E_k)$$

and a lower sum

$$s(f;\mu) = \sum_{k=1}^{n}\left(\operatorname*{glb}_{x\in E_K} f(x)\right)\cdot L(E_k).$$

As before, it is easy to argue that as μ ranges over all such partitions of $[a, b]$, these upper sums form a collection of numbers having a greatest lower bound, which we define to be the *upper Lebesgue integral* of f over $[a, b]$:

$$\operatorname*{glb}_{\mu} S(f;\mu) = \mathcal{L}\overline{\int}_{a}^{b} f(x)\,dx.$$

Analogously we define the *lower Lebesgue integral* of f over $[a, b]$ to be

$$\operatorname*{lub}_{\mu} s(f;\mu) = \mathcal{L}\underline{\int}_{a}^{b} f(x)\,dx.$$

As in the case of the Riemann integral, we have $\mathcal{L}\underline{\int}_{a}^{b} f(x)\,dx \leq \mathcal{L}\overline{\int}_{a}^{b} f(x)\,dx$. When equality holds we say that f is *Lebesgue integrable* over $[a, b]$ and that this common value is the *Lebesgue integral*

$$\mathcal{L}\int_{a}^{b} f(x)\,dx$$

of f over $[a, b]$.

One thing we can notice at once. Since every partition of $[a, b]$ into intervals is also a partition of $[a, b]$ into Lebesgue measurable sets, we see that

$$\begin{aligned}\mathcal{R}\underline{\int}_{a}^{b} f(x)\,dx &\leq \mathcal{L}\underline{\int}_{a}^{b} f(x)\,dx\\ &\leq \mathcal{L}\overline{\int}_{a}^{b} f(x)\,dx\\ &\leq \mathcal{R}\overline{\int}_{a}^{b} f(x)\,dx.\end{aligned}$$

It follows that whenever f is Riemann integrable over $[a, b]$, then f is also Lebesgue integrable over $[a, b]$, and $\mathcal{R}\int_a^b f(x)\,dx = \mathcal{L}\int_a^b f(x)\,dx$. Very simple examples show, though, that there are bounded functions f which are Lebesgue integrable over $[a, b]$ but not Riemann integrable over $[a, b]$. Thus the first advantage of the Lebesgue integral over the Riemann one is that it integrates more functions.

6. The Lebesgue integral in probability theory

The Lebesgue integral can be extended in a natural way to unbounded functions on unbounded domains. We shall indicate briefly how this extension of the Lebesgue integral is accomplished for the case of nonnegative functions only. These are of especial interest in probability theory.

First, there are Lebesgue measurable sets for the entire real line. These form a σ-field containing all the finite intervals. On this σ-field there is defined a Lebesgue measure. It has the properties that might be expected: that is, it is nonnegative, it is countably additive, it reduces to length for finite intervals, and it is translation-invariant. In fact, for every finite interval $[a, b]$ it reduces to the Lebesgue measure on the Lebesgue measurable subsets of $[a, b]$ already discussed above. (But of course it is no longer everywhere finite: it is easily seen that it couldn't be, since it reduces to length for intervals and is countably additive!) A not-necessarily-bounded function f on the real line is called *measurable* if for every real number c the set $\{x : f(x) \leq c\}$ is a Lebesgue measurable set.

When a measurable function f on the real line is nonnegative, we define its *Lebesgue integral*

$$\mathcal{L}\int_{-\infty}^{\infty} f(x)\,dx \tag{6}$$

(finite or infinite) over the whole real line to be the least upper bound of integrals $\mathcal{L}\int_a^b h(x)\,dx$ as $[a, b]$ ranges over all finite intervals and h ranges over all bounded Lebesgue integrable functions on $[a, b]$ for which $h(x) \leq f(x)$ on $[a, b]$. When the integral (6) is finite, we say f is *Lebesgue integrable* over the real line.

We are now in position to indicate briefly a place in probability theory where the Lebesgue integral is a natural tool, but where the Riemann integral is inadequate. There are numerical-valued random phenomena for which it is natural to take the set of outcomes to be the real line, and the events to be the sets of a σ-field $\mathcal{S}$ containing all the intervals and consisting entirely of Lebesgue measurable sets. To assign a probability to each of these events we can sometimes use a so-called *probability density function* f, that is, a nonnegative function whose integral over the whole real line exists and is $1 : \int_{-\infty}^{\infty} f(x)\,dx = 1$. (One very important such function is the so-called *normal* density function $f(x) = (1/\sqrt{2\pi})e^{-x^2/2}$.)

Using such a density function f we may try to define the probability $P(E)$ for each set E in the σ-field $\mathcal{S}$ by setting

$$P(E) = \int_{-\infty}^{\infty} f(x)K_E(x)\,dx \equiv \int_E f(x)\,dx, \tag{7}$$

where

$$K_E(x) = \begin{cases} 1 & \text{if } x \in E \\ 0 & \text{if } x \notin E \end{cases},$$

is termed the *characteristic function* (or *indicator function*) of E. It turns out that the sets E for which (7) exists as a Lebesgue integral always include all Lebesgue measurable sets and hence all sets of the σ-field $\mathcal{S}$ and that the resulting function P has the desired properties (2). But in general the sets E for which (7) exists as an (improper) Riemann integral don't form a σ-field or contain an adequate σ-field, at all.

As an indispensable setting and tool for probability, modern ideas of measure and integration really come into their own in the use of *multi-dimensional* measures and integrals to treat many related or independent random phenomena together. But this is a more complicated story and one we cannot go into here.

Suggestions for Further Reading

On probability as measure theory:

P. R. Halmos, The foundations of probability, *Amer. Math. Monthly*, 51 (1944) 493–510.

——, *Measure Theory*, Van Nostrand, Princeton, 1950, especially the "Heuristic Introduction" to Chapter IX.

On elementary modern accounts of the Lebesgue integral:

R. R. Goldberg, *Methods of Real Analysis*, Blaisdell, New York, 1964.

Edgar Asplund and Lutz Bungart, *A first course in integration*, Holt, Rinehart and Winston, New York, 1966.

On Round Pegs in Square Holes and Square Pegs in Round Holes

David Singmaster

Mathematics Magazine
37 (1964), 335–37

Editors' Note: David Breyer Singmaster, Professor Emeritus at the South Bank University in London, earned his PhD under the direction of D. H. Lehmer at the University of California, Berkeley, in 1966. He has written extensively on puzzles (especially Rubik's Cube), games, and general topics in recreational mathematics. He is the author (with Alexander Frey) of *Handbook of Cubik Mathematics* (Lutterworth, 1984).

He distributes privately a regularly updated list called *Mathematical Monuments* as an aid to the mathematical traveler. It's a list of places to visit anywhere in the world if one wants to see memorial plaques to mathematicians, places of birth or death, tombstones, the garage in Palo Alto, California, where William Hewlett and David Packard started Hewlett-Packard, and on and on. It's an invaluable resource for the right kind of tourist.

In this article Singmaster draws some interesting conclusions using the formula for the volume V_n of an n-ball with radius r,

$$V_n = \frac{\pi^{n/2}}{\Gamma(\frac{n}{2}+1)} r^n$$

where $\Gamma(x)$ is the gamma function. For balls with radius one, the sequence of volumes $V_1, V_2, V_3, \ldots$ also has surprises: $V_n \to 0$ as $n \to \infty$, with V_5 being the largest term in the sequence. The derivation of the above formula for V_n, by an argument "accessible to a multivariable calculus class," is given in Jeffrey Nunemacher's "The largest unit ball in any Euclidean space," 59 (1986), 170–71. By using the fact that the surface area of an n-ball is the derivative with respect to the radius of the volume, this article determines that the unit n-ball with $n = 7$ (not 5) has the greatest surface area.

Some time ago, the following problem occurred to me: which fits better, a round peg in a square hole or a square peg in a round hole? This can easily be solved once one arrives at the following mathematical formulation of the problem. Which is larger: the ratio of the area of a circle to the area of the circumscribed square or the ratio of the area of a square to the area of the circumscribed circle? One easily finds that the first ratio is $\pi/4$ and that the second is $2/\pi$. Since the first is larger, we may conclude that a round peg fits better in a square hole than a square peg fits in a round hole.

More recently, it occurred to me that the above question could be easily generalized to n dimensions. The remainder of this paper will be devoted to the following

Theorem 1 *The n-ball fits better in the n-cube than the n-cube fits in the n-ball if and only if $n \leq 8$.*

First, we take the following formula for the n-volume of the n-ball of radius r [1, p. 136]

$$V_n = \frac{\pi^{n/2} r^n}{\Gamma(\frac{n}{2}+1)}$$

where $\Gamma(x)$ is the well-known gamma function. (It is noteworthy that the volume of the unit n-ball decreases to zero with increasing n.)

Since we are interested only in ratios, we may, without loss of generality, assume that we have the unit n-ball in both ratios. Then the edge of the circumscribed cube is 2. Since the diagonal of an n-cube is $\sqrt{n}$ times its edge, we see that the edge of the n-cube which is inscribed in the unit n-ball is $2/\sqrt{n}$. Letting $V(n)$, $V_c(n)$, and $V_i(n)$ represent the n-volumes of the unit n-ball, its circumscribed n-cube, and its inscribed n-cube, respectively, we have:

$$V_n = \frac{\pi^{n/2}}{\Gamma(\frac{n}{2}+1)}, \quad V_c(n) = 2^n, \quad V_i(n) = \frac{2^n}{n^{n/2}}.$$

Now, the ratios under consideration are:

$$R_1(n) = \frac{V(n)}{V_c(n)} = \frac{\pi^{n/2}}{\Gamma\left(\frac{n+2}{2}\right) \cdot 2^n},$$

$$R_2(n) = \frac{V_i(n)}{V(n)} = \frac{2^n \Gamma\left(\frac{n+2}{2}\right)}{n^{n/2} \pi^{n/2}}$$

The ratio $R_1(n)$ measures how well an n-ball fits in an n-cube and the ratio $R_2(n)$ measures how well an n-cube fits in an n-ball. Our theorem can now be stated as: $R_1(n) \geq R_2(n)$ if and only if $n \leq 8$. We shall prove somewhat more.

n	$V(n)$	$R_1(n)$	$R_2(n)$
1	2.0	1.0	1.0
2	3.14159	.78540	.63662
3	4.18879	.52360	.36755
4	4.93479	.30842	.20264
5	5.26378	.16449	.10875
6	5.16770	.080745	.057336
7	4.72475	.036912	.029853
8	4.05870	.015854	.015399
9	3.29850	.0064423	.0078861
10	2.55015	.0024904	.0040154
11	1.88410	.00091997	.0020350
12	1.33526	.00032599	.0010273
13	.91062	.00011116	.00051691
14	.59926	.000036576	.00025936
15	.38144	.000011641	.00012982
16	.23533	.0000035908	.000064840
17	.14098	.0000010756	.000032325
18	.082145	.00000031336	.000016088
19	.046621	.000000088923	.0000079952
20	.025807	.000000024611	.0000039680
30	.000021915	2.0410×10^{-14}	3.4146×10^{-9}
40	3.6047×10^{-9}	3.2784×10^{-21}	2.7741×10^{-12}
50	1.7302×10^{-13}	1.5367×10^{-28}	2.1835×10^{-15}
60	3.0962×10^{-18}	2.6856×10^{-36}	1.6844×10^{-18}

Theorem 2 $R_1(n)/R_2(n) \to 0$ *as* $n \to \infty$.

Proof. From the definitions, we have:

$$\frac{R_1(n)}{R_2(n)} = \frac{\pi^n n^{n/2}}{2^{2n}\left[\Gamma\left(\frac{n+2}{2}\right)\right]^2}.$$

We apply Stirling's approximation:

$$\Gamma(z) \sim z^{z-1/2} e^{-z} \sqrt{2\pi},$$

thus obtaining

$$\left[\Gamma\left(\frac{n+2}{2}\right)\right]^2 \sim \left(\frac{n+2}{2}\right)^{n+1} e^{-n-2} 2\pi.$$

Hence, we have:

$$\frac{R_1(n)}{R_2(n)} \sim \frac{\pi^n n^{n/2}}{2^{2n}\left(\frac{n+2}{2}\right)^{n+1} e^{-n-2} 2\pi}$$

$$= \frac{e^2}{\pi(n+2)}\left[\frac{\pi e \sqrt{n}}{2(n+2)}\right]^n$$

$$< \frac{e^2}{\pi n}\left[\frac{\pi e}{2} \cdot \frac{1}{\sqrt{n}}\right]^n.$$

This last quantity is easily seen to approach zero as n increases, hence the theorem is proved.

Corollary 2.1 $R_1(n) < R_2(n)$ *for all large enough* n.

One may readily compute that the asymptotic approximation for $R_1(n)/R_2(n)$ has the value 1.06... for $n = 8$ and the value .84... for $n = 9$. Since Stirling's approximation has a relative error less than $1/12z$, we can say that the relative error in the asymptotic expression for $R_1(n)/R_2(n)$ is less than $2 \times 1/12 \cdot 5$ or 3.3% for $n \geq 8$. Hence we can be confident in stating that $R_1(n) < R_2(n)$ holds if $n \geq 9$, since the asymptotic approximation decreases with n, for $n \geq 5$. That the asymptotic approximation decreases with n is clear when $n \geq 14$ since $\sqrt{n}/(n+2)$ decreases with n for $n \geq 2$ and

$$\frac{\pi e \sqrt{n}}{2(n+2)} < 1$$

for $n \geq 14$. Further, one can compute that the asymptotic approximation decreases in the range $5 \leq n \leq 14$. Hence the theorem first stated has been half proven.

In order to check the theorem for small values of n, I programmed the IBM 7090 computer at Berkeley to compute $V(n)$, $R_1(n)$ and $R_2(n)$ for $1 \leq n \leq 100$. The results, which are partially reproduced above, show that $R_1(n) \geq R_2(n)$ holds if $n \leq 8$, with equality only for $n = 1$. The numerical results for small n, together with the asymptotic results for large n, show that $R_1(n) \geq R_2(n)$ if and only if $n \leq 8$, as originally claimed.

In closing, we remark that one can also show that $R_2(n)$ and $V(n)/R_2(n)$ each approach zero with $V(n) > R_2(n)$ if and only if $n \leq 61$.

Reference

1. D. M. Y. Sommerville, *An introduction to the geometry of N dimensions*, Dover, New York, 1958.

π_t: 1832–1879

Underwood Dudley

Mathematics Magazine
35 (1962), 153–54

Editors' Note: Professor Underwood Dudley wrote this "entertainment" when he was still a graduate student in mathematics at the University of Michigan, Ann Arbor. Currently Professor of Mathematics at Depauw University in Indiana, he is widely known for his number theory textbook, *Elementary Number Theory* (second edition), 1978, and several books on angle trisectors, numerologists, and mathematical cranks. Most recently he has been editor of the MAA's *College Mathematics Journal*.

When mathematicians are thought of, who remembers James Smith? Or Daniel West? Few people indeed. Yet these men (and 42 others) performed a valuable service in the middle of the last century: they kept track of π_t, the ratio of the circumference of a circle to its diameter at time t. See the table for the results of their calculations, rounded off to five decimal places. The data are mostly from DeMorgan [1] and Gould [2]. Lately, very little has been done in this field; we have let π_t get away from us.

t	π_t	Calculator
1832	3.06250	Parsey
1833	3.20222	Baddeley
1833	3.16483	Bouche
1835	3.20000	Oliveira
1836	3.12500	Lacomme
1837	3.23077	Bennett
1841	3.12019	McCook
1843	3.04862	Johnson
1844	3.17778	Dennison
1845	3.16667	Davis
1846	3.17480	Young
1848	3.20000	Peters
1848	3.12500	Merceron
1849	3.14159	deGelder
1850	3.14159	Parker
1851	3.14286	Adorno
1853	3.12381	"Futurus"
1854	3.17124	Bouche
1855	3.15532	Smith, A.
1858	3.20000	Anghera
1859	3.14159	Gee
1860	3.12500	Smith, J.
1860	3.14241	Hailes
1862	3.14159	Benson
1862	3.14214	Houlston
1862	3.20000	Pratt
1863	3.14063	Dean
1865	3.16049	Faber
1866	3.24000	May
1868	3.14214	Grosvenor
1868	3.14159	Harbord
1869	3.12500	Dircks
1871	3.15470	G. W. B.
1871	3.15544	Terry
1872	3.16667	"A. Finality"
1873	3.14286	Myers
1874	3.15208	Brower
1874	3.14270	Harris
1874	3.15300	Stacy
1875	3.14270	Goodsell

t	π_t	Calculator
1875	3.15333	Weatherby
1876	3.13397	Cart
1878	3.20000	"Durham"
1878	3.13514	Gidney
1879	3.14286	Crabb

But perhaps something can be saved. If we construct a least-squares line using the data of the table, we find that $\pi_t = .0000056060t + 3.14281$ where t is measured in years A.D.; in particular, $\pi_{1962} = 3.15381$. Consider the implications of our relation. For one thing, we see that the Biblical value $\pi_t = 3$ was an excellent approximation for those days. For another, schoolchildren in 10201 can look forward with more pleasure than usual to June, for in that month, $\pi_t = 3.20000$, and their calculations will be much simpler.

There are discrepancies, though. We find that π_t was 3.14 15926535... sometime around 10:54 P.M. on November 10, 219 B.C. (using the Gregorian calendar and Greenwich time). What happened then that fixed this erroneous approximation as an exact value? History is silent. Further, our expression gives the date of the creation—when π_t was zero—as 560,615 B.C., agreeing neither with astronomical theory nor with Archbishop Ussher's chronology. Clearly, more research is needed.

References

1. DeMorgan, Augustus, *A Budget of Paradoxes*, Chicago, 1915.
2. Gould, Sylvester C., *What is the Value of π?*, Manchester, NH, 1888.

Part V

The 1970s

Trigonometric Identities

Andy R. Magid

Mathematics Magazine
47 (1974), 226–27

Editors' Note: Here the author uses commutative ring theory to prove that every trigonometric identity is a consequence of $\sin^2 x + \cos^2 x = 1$. Here "trigonometric identity" means an identity in x that has been first simplified to a polynomial identity of the form $f(\sin x, \cos x) = 0$. In a letter to the editor Harry W. Hickey noted that this result could be proved by more elementary methods not requiring the power of commutative ring theory ([48] (1975), 4).

Magid, long a Professor of Mathematics at the University of Oklahoma, received his PhD in mathematics from Northwestern University, under the direction of Daniel Zelinsky. He has held various positions in the American Mathematical Society, principally as an Associate Secretary and currently as editor of *The Notices of the AMS*.

A *trigonometric identity* is an equation between two rational functions of trigonometric functions, e.g.:

$$\frac{\tan x}{\csc x - \cot x} - \frac{\sin x}{\csc x + \cot x} = \sec x + \cos x.$$

The *ad hoc* verification of such identities is a standard exercise in elementary trigonometry (see, for example, [1, Chapter 8], from which the above identity comes) which most students have done, usually not in any systematic way. Thus the following theorem, which fits nicely into an undergraduate abstract algebra course, is of some interest to them:

Theorem 1 *Every trigonometric identity is a consequence of* $\sin^2 x + \cos^2 x = 1$.

The proof of the theorem, which will be outlined here, uses only some elementary commutative ring theory, about at the level of [2, Chapter 3].

We begin the proof with some simplifications: first express all the trigonometric functions appearing in the identity in terms of $\sin x$ and $\cos x$, next clear denominators and finally subtract one side from the other. What remains is a polynomial identity of the form $f(\sin x, \cos x) = 0$. For convenience later on, assume that the polynomials have complex coefficients. Now let $\mathbb{C}$ denote the complex numbers and let A be the ring of all complex functions analytic in a neighborhood of zero. Then the theorem is equivalent to the following:

Proposition 1.1 *The kernel of the ring homomorphism* $\Phi\colon \mathbb{C}[X, Y] \to A$ *defined by* $\Phi(X) = \cos x$ *and* $\Phi(Y) = \sin x$ *is generated by* $X^2 + Y^2 - 1$.

Now A is an integral domain (using, for example, power series expansions) and so the kernel P of Φ is a prime ideal which contains the prime ideal P_0 generated by the irreducible polynomial $X^2 + Y^2 - 1$. The proposition will be proved by showing that $P = P_0$. For experts, this is immediate: $\mathbb{C}[X, Y]$ has Krull dimension 2, so every nonzero prime properly containing P_0 is maximal. So if $P \neq P_0$, P is a maximal ideal and the image of Φ is a field. But $\Phi(X) = \sin x$ is not even a unit in A, so this is impossible. The same line of reasoning will provide an elementary proof, once the following is established:

Lemma 1.1 *Every nonzero prime ideal of* $R = \mathbb{C}[X, Y]/P_0$ *is a maximal ideal.*

Here is a sketch of the proof: let x, y denote the images of X, Y in R. Then $x^2 + y^2 = 1$, so if

$u = x + iy$, $u^{-1} = x - iy$ and $R = \mathbb{C}[u, u^{-1}]$. If I is any ideal of R, then $I = (I \cap \mathbb{C}[u])R$ by the standard argument, but since $\mathbb{C}[u]$ is Euclidean this makes $I \cap \mathbb{C}[u]$, and hence I, principal. Thus R is a principal ideal domain. The usual arguments now show that every nonzero prime of R is maximal (see, for example, [2], p. 109): prime ideals are generated by prime elements and ideals generated by prime elements are maximal. This proves the lemma, hence also the proposition and the theorem.

References

1. A. W. Goodman, *The Mainstream of Algebra and Trigonometry*, Houghton Mifflin, Boston, 1973.
2. I. Herstein, *Topics in Algebra*, Ginn, Waltham, 1964.

A Property of 70

Paul Erdős

Mathematics Magazine
51 (1978), 238–40

Editors' Note: Paul Erdős is perhaps the best known of 20th century research mathematicians. Stories about his illustrious but unusual career, his unique way of doing mathematics, and his eccentricities abound. Since his death in 1996 two biographies of him have appeared—Paul Hoffman's *The Man Who Loved Only Numbers* (Hyperion, 1998) and Bruce Schechter's *My Brain Is Open* (Simon and Schuster, 1998). Further, an MAA film about him, "N Is a Number", has been widely distributed and shown on public television. He held only one regular academic appointment during his long life—he died at the age of 83—so he was able to travel continually to visit other mathematicians and to share with them news of outstanding problems. Extraordinarily prolific, he wrote, at last count, 1514 papers, very often with collaborators, mainly in number theory, combinatorics and graph theory. Because he had collaborators in so many countries, someone wrote the following limerick:

> A problem both deep and profound
> Is whether a circle is round.
> In a paper by Erdős,
> Written in Kurdish,
> A counterexample is found.

Here he finds an extension of a well-known theorem in number theory that says that 30 is the largest integer such that all smaller positive integers relatively prime to it are themselves primes.

It is well known (see, e.g., [3]) that 30 is the largest integer with the property that all smaller integers relatively prime to it are primes. In this note I will consider a related situation in which the corresponding special number turns out to be 70. (For a while I believed 30 to be the key figure in the new context, too, but E. G. Straus showed me that the correct value was indeed 70.) Following the proof of this special property of 70, I will mention a few related problems, some of which seem to me to be very difficult. I hope to convince the reader that there are very many interesting and new problems left in what is euphemistically called "elementary" number theory. Although these problems are easy to comprehend, their solutions will undoubtedly require either remarkable ingenuity or extensive application of known techniques.

Throughout this paper we will be studying sequences of positive integers related to a given integer n. The basic sequence $\{a_i\}_{i=0}^{\infty}$ begins with $a_0 = n$; once $a_0, a_1, \ldots, a_{k-1}$ are known, a_k is chosen to be the smallest integer greater than a_{k-1} that is relatively prime to the product $a_0 a_1 \cdots a_{k-1}$. Clearly each prime greater than n is an a_k. Moreover, each a_k greater than n^2 is a prime. Table 1 contains examples of the sequences $\{a_k\}$ corresponding to certain integers n.

Theorem 1 *70 is the largest integer for which all the a_k (for $k \geq 1$) are primes or powers of primes.*

Proof. I will try to make the proof as short as possible; thus it is not as elementary as it might be. We begin with the difficult but useful result [5] that for $x > 17/2$, there are at least *three* primes in the interval $(x, 2x)$. Hence, for $n > 17^2 = 289$ there are at least three primes in the interval $(\frac{1}{2}n^{1/2}, n^{1/2})$. Furthermore, at least one of these primes does not divide n since their product exceeds $n^{3/2}/8$ (which in turn is greater than n). Thus, if p_1 is the greatest prime satisfying $p_1 < n^{1/2}$ and $p_1 \nmid n$ we know that $p_1 > \frac{1}{2}n^{1/2}$. Also, for $n > 289$, there are at least three primes in $(2n^{1/2}, n/4)$ since $n > 16n^{1/2}$. At least one of these three primes does not divide n since their product exceeds $4n$. Hence, if q_1 is the least prime satisfying $q_1 \geq n^{1/2}$ and $q_1 \nmid n$, $q_1 < n/4 < p_1^2$.

Now consider $p_1 q_1$. If it is one of the a_k's, then the property stated in our theorem—namely, that all a_k are primes or powers of primes—is satisfied for $n > 289$. If not, then there must be an a_i with $n < a_i < p_1 q_1$ and $(a_i, p_1 q_1) > 1$. We only have to prove that this a_i must have at least two distinct

n	non-prime a_k	$f(n)$	a_k that are not prime powers
3	2^2	0	
4	3^2	0	
5	$2 \cdot 3$	1	6
6	5^2	0	
7	$2^3, 3^2, 5^2$	0	
8	$3^2, 5^2, 7^2$	0	
9	$2 \cdot 5, 7^2$	1	10
10	$3 \cdot 7$	1	21
11	$4 \cdot 3, 5^2, 7^2$	1	12
12	$5^2, 7^2, 11^2$	0	
15	$2^4, 7^2, 11^2, 13^2$	0	
18	$5^2, 7^2, 11^2, 13^2, 17^2$	0	
22	$5^2, 3^3, 7^2, 13^2, 17^2, 19^2$	0	
24	$5^2, 7^2, 11^2, 13^2, 17^2, 19^2, 23^2$	0	
30	$7^2, 11^2, 13^2, 17^2, 19^2, 23^2, 29^2$	0	
31	$2^5, 3 \cdot 11, 5 \cdot 7, 11^2, 13^2, 19^2, 23^2, 29^2$	2	33, 35
46	$7^2, 3 \cdot 17, 5 \cdot 11, 13^2, 19^2, 29^2, 31^2, 37^2, 41^2, 43^2$	2	51, 55
70	$9^2, 11^2, 13^2, 17^2, 19^2, 23^2, 29^2, 31^2, 37^2, 41^2, 43^2, 47^2, 53^2, 59^2, 61^2, 67^2$	0	
71	$2^3 \cdot 3^2, 7 \cdot 11, 5 \cdot 17, 13^2, 19^2, \ldots, 67^2$	3	72, 77, 85
97	$2 \cdot 7^2, 3^2 \cdot 11, 5 \cdot 23, 13^2, 17^2, 19^2, 29^2, \ldots, 97^2$	3	98, 99, 115
272	$3 \cdot 7 \cdot 13, 5^2 \cdot 11, 19^2, 23^2, 29^2, \ldots, 271^2$	2	273, 275

Table 1. Sample sequences generated from integers n by counting upwards from n, omitting every integer that contains a prime factor in common with any previous terms in the sequence. Since every prime larger than n will automatically be included, we record here only the non-prime numbers that occur in the sequences. (No non-primes occur beyond n^2—as observed in the text—so our record terminates before that point.) The column headed "$f(n)$" records the number of members of the sequence that are neither prime nor a power of a prime. Those numbers for which $f(n) = 0$ have the property that all members of the sequence are primes or powers of primes; they are 3, 4, 6, 7, 8, 12, 15, 18, 22, 24, 30, 70. It is proved in the accompanying article that no other numbers have this property.

prime factors. If this does *not* hold, then a_i would have to be a power of p_1 or a power of q_1. Clearly it cannot be a power of q_1 since $q_1^2 > p_1 q_1$. However, it cannot be a power of p_1, either, since $p_1^2 < n$ and $p_1^3 > p_1 q_1$ (because $p_1^2 > q_1$). Thus all a_k corresponding to $n > 289$ are primes or powers of primes. The same conclusion holds for $70 < n \leq 289$, and may be verified by direct computation.

By more complicated methods, we can prove the following related result:

Theorem 2 *For all sufficiently large n, at least one of the a_k's is the product of exactly two distinct primes.*

I shall not give the proof since it is fairly complicated and uses deep results in analytical number theory. Although I was fairly sure that this result held for every n greater than 70, and thus strengthened Theorem 1, I could not prove this. Recently C. Pomerance found a proof of Theorem 2 for n greater than 6000; he also observed that the result fails for $n = 272$ (see Table 1).

The following conjecture, related to Theorem 2, seems very difficult. Denote by $p(x)$, the least prime factor of x. Then for sufficiently large n, there are always composite numbers x satisfying

$$n < x < n + p(x) \tag{1}$$

The inequality (1) is a slight modification of an old conjecture that J. L. Selfridge and I proposed in [2]. In fact, I expect that for sufficiently large n there are squarefree x's satisfying (1) which have exactly k distinct prime factors. I am sure that this conjecture is very deep. It would of course imply Theorem 2 since the integers x satisfying (1) must be a_k's corresponding to the given value of n.

I wish now to state a few simple facts and pose some difficult problems about our a_k's. We have already noted that each prime greater than n is an a_k, and that each a_k greater than n^2 is prime. Let p

be a prime less than n and let $a(n, p)$ be the least a_k which is a multiple of p. It is easy to see that $a(n, p) \leq p^{\alpha+1}$ where $p^{\alpha} \leq n < p^{\alpha+1}$, for if none of the $a_k < p^{\alpha+1}$ are multiples of p, then $p^{\alpha+1}$ is an a_k.

Denote by $f(n)$ the number of those a_k which are not powers of primes. Clearly, $f(n) \leq \pi(n^{1/2})$ (where $\pi(x)$ denotes the number of primes $\leq x$) since each such a_k must have a prime factor exceeding $n^{1/2}$. (If $n^{1/2} < p$, then p^2 is an a_i and so no a_k can equal pt for $p \leq t$). I do not have any good upper or lower bounds for $f(n)$. I conjecture that $f(n) > n^{1/2-\epsilon}$ for $n > n_0(\epsilon)$. I am not sure whether $\lim_{n\to\infty} f(n)/\pi(n^{1/2}) = 0$.

Denote by $P(n)$ the largest prime which is less than n. It is not difficult to show that the largest a_k which is not a prime is just $P^2(n)$. On the other hand, I cannot determine the largest a_k which is not a power of a prime. In fact, I cannot even get an asymptotic formula for it and, in fact, have no guess as to its order of magnitude. It may be true that if $n \geq n_0(\epsilon)$ and a_k is not a power of a prime, then $a_k < (1+\epsilon)n$. Pomerance informs me that he can prove this, and in fact Penney, Pomerance and I are writing a longer joint paper on this subject.

References

1. P. Erdős, On the difference between consecutive primes, *Quart. J. Math.*, 6 (1935) 124–128.
2. P. Erdős and J. L. Selfridge, Some problems on the prime factors of consecutive integers, *Illinois J. Math.*, 11 (1967) 428–430. See also a forthcoming paper in *Utilitas Math* by R. Eggleton, P. Erdős and J. L. Selfridge.
3. H. Rademacher and O. Toeplitz, *The Enjoyment of Mathematics*, Princeton Univ. Press, (1957) 189–196.
4. R. A. Rankin, The difference between consecutive prime numbers, *J. London Math. Soc.*, 13 (1938) 242–247.
5. B. Rosser and L. Schoenfeld, Approximate formulas for some functions of prime numbers, *Illinois J. Math.*, 6 (1962) 64–94. (The result we need can of course be proved by completely elementary means.)

Hamilton's Discovery of Quaternions*

B. L. van der Waerden

Mathematics Magazine
49 (1976), 227–34

Editors' Note: In the mid-twentieth century there was a two-volume set of books that every graduate student in mathematics knew: *Moderne Algebra*, by B. L. Van der Waerden. This pioneering work, published by Springer in 1931 and appearing in English translation in 1949, brought to graduate education the work of Emil Artin and Emmy Noether that had revolutionized the approach to algebra. The English version was delayed in reaching American universities because of World War II, since the copyright was vested in the Alien Property Custodian in 1943 and the actual publication had to be authorized by the U.S. Attorney General!

Bartel Leendert van der Waerden studied at Amsterdam and Göttingen and in 1924, at the age of 22, he studied with Noether. This experience no doubt prompted his efforts to promulgate her ideas through writing the algebra text. After stays in Groningen, Leipzig, Johns Hopkins and Amsterdam, he took a position at the University of Zürich where he remained for the rest of his life.

His mathematical work ranged from algebraic geometry to various branches of algebra—Galois theory, ring theory, and Lie groups. He had been consistently interested in the history of mathematics throughout his career and in 1950 he published his *Ontwakende wetenschap*, published in 1954 in English as *Science Awakening*. This, along with his algebra text, is probably his best known work. In this article on Hamilton, he observes that Hamilton's work on quaternions would probably have moved more quickly had he read more Legendre!

* Originally published as "Hamiltons Entdeckung der Quarternionen" (Erweiterte Fassung eines Bortrages vor der Joachim Jungius-Gesellschaft der Wissenschaften). In *Veröffentlichungen der Joachim Jungius-Gesellschaft der Wissenschaften*, Vandenhoeck & Ruprecht, Göttingen, 1973. Reprinted with permission.

Introduction

The ordinary complex numbers $(a + ib)$ (or, as they were formerly written, $a + b\sqrt{-1}$) are added and multiplied according to definite rules. The rule for multiplication reads as follows:

First multiply according to the rules of high school algebra:

$$(a + ib)(c + id) = ac + adi + bci + bdi^2$$

and then replace i^2 by (-1):

$$(a + ib)(c + id) = (ac - bd) + (ad + bc)i.$$

Complex numbers can also be defined as couples (a, b). The product of two couples (a, b) and (c, d) is defined as the couple $(ac - bd, ad + bc)$. The couple $(1, 0)$ is called 1, the couple $(0, 1)$ is called i. Then we also have the result

$$i^2 = (0, 1)(0, 1) = (-1, 0) = -1.$$

By means of this definition the "imaginary unit" $i = \sqrt{-1}$ loses all of its mystery: i is simply the couple $(0, 1)$.

The **quaternions** $a + bi + cj + dk$ which William Rowan Hamilton discovered on the 16th of October, 1843, are multiplied according to fixed rules, in analogy to the complex numbers; that is to say:

$$\begin{aligned} &i^2 = j^2 = k^2 = -1, \\ &ij = k, \quad jk = i, \quad ki = j, \\ &ji = -k, \quad kj = -i, \quad ik = -j. \end{aligned}$$

They can also be defined as quadruples (a, b, c, d). Quaternions form a division algebra; that is, they cannot only be added, subtracted, and multiplied, but also divided (excluding division by zero). All rules of calculation of high school school algebra hold; only the commutative law $AB = BA$ does not hold since ij is not the same as ji.

How did Hamilton arrive at these multiplication rules? What was his problem and how did he find the solution? We are accurately informed about these matters in documents and papers which appear in the third volume of Hamilton's collected *Mathematical Papers* [3]:

First, through an entry in Hamilton's Note Book dated 16 October 1843 [3, pp. 103–105];

Second, through a letter to John Graves of the 17th of October 1843 [3, pp. 106–110];

Third, through a paper in the *Proceedings of the Royal Irish Academy* (2 (1844) 424–434) presented on the 13th of November 1843 [3, pp. 111–116];

Sir William Rowan Hamilton, a child prodigy whose maturity was all that his childhood promised, was born in Dublin, Ireland, in 1805. He was literate in seven languages and knowledgeable in half a dozen more. In 1827 while still an undergraduate Hamilton was appointed Andrews Professor of Astronomy and Superintendent of the Observatory, and soon afterwards Astronomer Royal, a position he held for the rest of his extraordinarily productive life. His work in dynamics is probably most well-known today. ``The Hamiltonian principle has become the cornerstone of modern physics," said Erwin Schrodinger, ``the thing with which a physicist expects every physical phenomenon to be in conformity." Hamilton's other major discovery is the system of quaternions. The flash of insight which produced this discovery occurred in 1843 and is described in the accompanying article. A century later the Irish government commemorated this achievement with the stamp pictured on the left.

Fourth, through the detailed Preface to Hamilton's "Lectures on Quaternions", dated June 1853 [3, pp. 117-155, in particular pp. 142–144];

Fifth, through a letter to his son Archibald which Hamilton wrote shortly before his death, that is shortly before the 2nd of September 1865 [3, pp. xv–xvi].

We can follow exactly each of Hamilton's steps of thought through all of these documents. This is a rare occurrence in which we can observe what flashed across the mind of a mathematician as he posed the problem, as he approached the solution step by step and then through a lightning stroke so modified the problem that it became solvable.

A brief history of complex numbers

Expressions of the form $A+\sqrt{-B}$ had already been encountered in the middle ages in the solution of quadratic equations. They were called "impossible solutions" or *numeri surdi*: absurd numbers. The negative numbers too were called "impossible." Cardan used numbers $A+\sqrt{-B}$ in the solution of equations of the third degree in the *casus irreducibilis* in which all three roots are real. Bombelli showed that it was possible to calculate with expressions such as $A+\sqrt{-B}$ without contradiction, but he did not like them: he called them "sophistical" and apparently without value. The expression "imaginary number" stems from Descartes.

Euler had no scruples about operating altogether freely with complex numbers. He proposed formulas such as $\cos\alpha = (1/2)(e^{i\alpha} + e^{-i\alpha})$. The geometric representation of complex numbers as vectors or as points in a plane stems from Argand (1813), Warren (1828) and Gauss (1832).

The first named, Argand, defined the complex numbers as directed segments in the plane. He took the basis vectors 1 and i as mutually perpendicular unit vectors. Addition is the usual vector addition, with which Newton made us familiar (the parallelogram law of velocities or of forces). The length of a vector was denoted at that time by the term "modulus", the angle of the vector with the positive x-axis as the "argument" of the complex number. Multiplication of complex numbers, according to Argand, then takes place so that the moduli are multiplied and the arguments are added. Independently of Argand, Warren and Gauss also represented complex numbers geometrically and interpreted their addition and multiplication geometrically.

"Papa, can you multiply triplets?"

Hamilton knew and used the geometric representation of complex numbers. In his published papers, however, he emphasized the definition of complex numbers as the couple (a, b) which followed definite rules for addition and multiplication. Related to that, Hamilton posed this problem to himself: *To find how number-triplets* (a, b, c) *are to be multiplied in analogy to couples* (a, b).

For a long time Hamilton had hoped to discover the multiplication rule for triplets, as he himself stated. But in October 1843 this hope became much stronger and more serious. He put it this way in a letter to his son Archibald [3, p. xv]:

> the desire to discover the law of multiplication of triplets regained with me a certain strength and earnestness,

In analogy to the complex numbers $(a + ib)$ Hamilton wrote his triplets as $(a + bi + cj)$. He represented his unit vectors 1, i, j as mutually perpendicular "directed segments" of unit length in space. Later Hamilton himself used the word vector, which I also shall use in the following. Hamilton then sought to represent products such as $(a + bi + cj)(x + yi + zj)$ again as vectors in the same space. He required, first, that it be possible to multiply out term by term; and second, that the length of the product of the vectors be equal to the product of the lengths. This latter rule was called the "law of the moduli" by Hamilton.

Today we know that the two requirements of Hamilton can be fulfilled only in spaces of dimensions 1, 2, 4 and 8. This was proved by Hurwitz [5]. Therefore Hamilton's attempt in three dimensions had to fail. His profound idea was to continue to 4 dimensions since all of his attempts in 3 dimensions failed to reach the goal.

In the previously mentioned letter to his son, Hamilton wrote about his first attempt:

> Every morning in the early part of the above-cited month [October 1843], on my coming down to breakfast, your brother William Edwin and yourself used to ask me, 'Well, Papa, can you multiply triplets?' Whereto I was always obliged to reply, with a sad shake of the head, 'No, I can only add and subtract them.'

From the other documents we learn more precisely about Hamilton's first attempts. To fulfill the "law of the moduli" at least for the complex numbers $(a + ib)$, Hamilton set $ii = -1$, as for ordinary complex numbers, and similarly so that the law would also hold for the numbers $(a + cj)$, $jj = -1$. But what was ij and what was ji? At first Hamilton assumed $ij = ji$ and calculated as follows:

$$\begin{aligned}(a + ib + jc)&(x + iy + jz)\\ &= (ax - by - cz) + i(ay + bx)\\ &\quad + j(az + cx) + ij(bz + cy).\end{aligned}$$

Now, he asked, what is one to do with ij? Will it have the form $\alpha + \beta i + \gamma j$?

First attempt. The square of ij had to be 1, since $i^2 = -1$ and $j^2 = -1$. Therefore, wrote Hamilton, in this attempt one would have to choose $ij = 1$ or $ij = -1$. But in neither of these two cases will the law of the moduli be fulfilled, as calculation shows.

Second attempt. Hamilton considered the simplest case

$$(a+ib+jc)^2 = a^2-b^2-c^2+2iab+2jac+2ijbc.$$

Then he calculated the sum of the squares of the coefficients of 1, i, and j on the right-hand side and found

$$(a^2-b^2-c^2)^2+(2ab)^2+(2ac)^2 = (a^2+b^2+c^2)^2.$$

Therefore, he said, the product rule is fulfilled if we set $ij = 0$. And further: if we pass a plane through the points 0, 1, and $a+ib+jc$, then the construction of the product according to Argand and Warren will hold in this plane: the vector $(a + bi + cj)^2$ lies in the same plane and the angle which this vector makes with the vector 1 is twice as large as the angle between the vectors $(a + bi + cj)$ and 1. Hamilton verified this by computing the tangents of the two angles.

Third attempt. Hamilton reports that the assumption $ij = 0$, which he made in the second attempt, subsequently did not appear to be quite right to him. He writes in the letter to Graves [3, p. 107]:

> Behold me therefore tempted for a moment to fancy that $ij = 0$. But this seemed odd and uncomfortable, and I perceived that the same suppression of the term which was *de trop* might be attained by assuming what seemed to me less harsh, namely that $ji = -ij$. I made therefore $ij = k$, $ji = -k$, reserving to myself to inquire whether k was 0 or not.

Hamilton was entirely right in giving up the assumption $ij = 0$ and taking instead $ij = -ji$. For example, if $ij = 0$ then the modulus of the product ij would be zero, which would contradict the law of the moduli.

Fourth attempt. Somewhat more generally, Hamilton multiplied $(a + ib + jc)$ and $(x + ib + jc)$. In this case the two segments which are to be multiplied also lie in one plane, that is, in the plane spanned by the points 0, 1, and $ib + jc$. The result of the multiplication was $ax - b^2 - c^2 + i(a + x)b + j(a + x)c + k(bc - bc)$. Hamilton concluded from this calculation [3, p. 107] that:

> the coefficient of k still vanishes; and $ax - b^2 - c^2$, $(a + x)b$, $(a + x)c$ are easily found to be the correct coordinates of the *product-point* in the sense that the rotation from the unit line to the radius vector of a, b, c being added in its own plane to the rotation from the same unit-line to the radius vector of the other factor-point x, b, c conducts to the radius vector of the lately mentioned product-point; and that this latter radius vector is in length the product of the two former. Confirmation of $ij = -ji$; but no information yet of the value of k.

The leap into the fourth dimension

After this encouraging result Hamilton ventured to attack the general case. ("Try boldly then the general product of two triplets, ..." [3, p. 107].) He calculated

$$(a + ib + jc)(x + iy + jz) = (ax - by - cz) + i(ay + bx) + j(az + cx) + k(bz - cy).$$

In an exploratory attempt he set $k = 0$ and asked: Is the law of the moduli satisfied? In other words, does the identity

$$(a^2 + b^2 + c^2)(x^2 + y + z^2) = (ax - by - cz)^2 + (ay + bx)^2 + (az + cx)^2$$

hold?

> No, the first member exceeds the second by $(bz - cy)^2$. But this is just the square of the coefficient of k, in the development of the product $(a + ib + ic)(x + iy + jz)$, if we grant that $ij = k$, $ji = -k$, as before.

And now comes the insight which gave the entire problem a new direction. In the letter to Graves [3, p. 108], Hamilton emphasized the insight:

> And here there dawned on me the notion that we must admit, in some sense, a *fourth dimension* of space for the purpose of calculating with triplets;

This fourth dimension appeared as a "paradox" to Hamilton himself and he hastened to transfer the paradox to algebra [3, p. 108]:

> or transferring the paradox to algebra, [we] must admit a *third* distinct imaginary symbol k, not to be confounded with either i or j, but equal to the product of the first as multiplier, and the second as multiplicand; and therefore [I] was led to introduce *quaternions* such as $a+ib+jc+kd$, or (a, b, c, d).

Hamilton was not the first to think about a multi-dimensional geometry. In a footnote to the letter to Graves he wrote:

> The writer has this moment been informed (in a letter from a friend) that in the Cambridge Mathematical Journal for May last [1843] a paper on Analytical Geometry of n dimensions has been published by Mr. Cayley, but regrets he does not yet know how far Mr. Cayley's views and his own may resemble or differ from each other.

"This moment" can in this connection only mean the same day in which he wrote the letter to Graves. In the Note Book of the 16th of October 1843 there is no mention of the paper by Cayley. Hamilton therefore appears to have arrived at the concept of a 4-dimensional space independently of Cayley.

After Hamilton had introduced $ij = -ji = k$ as a fourth independent basis vector, he continued the calculation [3, p. 108]:

> I saw that we had probably $ik = -j$, because $ik = iij$, and $i^2 = -1$; and that in like manner we might expect to find $kj = ijj = -i$;

From the use of the word "probably" it can be seen how cautiously Hamilton continued. He scarcely trusted himself to apply the associative law $i(ij) = (ii)j$ because he was not yet certain if the associative law held for quaternions. Likewise Hamilton could have used the associative law to determine ki:

$$ki = -(ji)i = -j(ii) = (-j)(-i) = j.$$

Instead he applied a conclusion by analogy. He wrote [3, p. 108]

> from which I thought it likely that $ki = j$, $jk = i$, because it seemed likely that if $ji = -ij$, we should have also $kj = -jk$, $ik = -ki$.

Finally k^2 had to be determined. Hamilton again proceeded cautiously:

> And since the order of multiplication of these imaginaries is not indifferent, we cannot infer that k^2, or $ijij$, is $= +1$, because $i^2 \times j^2 = (-1)(-1) = +1$. It is more likely that $k^2 = ijij = -iijj = -1$.

This last assumption $k^2 = -1$, asserts Hamilton, is also necessary if we wish to fulfill the "law of the moduli." He carried this out and concluded [3, p. 108]:

> My assumptions were now completed, namely,
>
> $$i^2 = j^2 = k^2 = -1$$
> $$ij = -ji = k$$
> $$jk = -kj = i$$
> $$ki = -ik = j.$$

And now Hamilton tested if the law of the moduli was actually satisfied.

> But I considered it essential to try whether these equations were consistent with the law of moduli,..., without which consistence being verified, I should have regarded the whole speculation as a failure.

He therefore multiplied two arbitrary quaternions according to the rules just formulated

$$(a, b, c, d)(a', b', c', d') = (a'', b''c'', d''),$$

calculated (a'', b'', c'', d'') and formed the sum of the squares

$$(a'')^2 + (b'')^2 + (c'')^2 + (d'')^2$$

and found to his great joy that this sum of squares actually was equal to the product

$$(a^2 + b^2 + c^2 + d^2)(a'^2 + b'^2 + c'^2 + d'^2).$$

In Hamilton's letter to his son we learn even more about the external circumstances which befell him at this flash of insight. Immediately after the previously cited words, "No, I can only add and subtract them." Hamilton continued [3, p. xx–xvi]:

> But on the 16th day of the same month [October 1843]—which happened to be a Monday and a Council day of the Royal Irish Academy—I was walking in to attend and preside, and your mother was walking with me, along the Royal Canal, to which she had perhaps been driven; and although she talked with me now and then, yet an under-current of thought was going on in my mind, which gave at last a result, whereof it is not too much to say that I felt at once the importance. An electric circuit seemed to close; and a spark flashed forth, the herald (as I foresaw immediately) of many long years to come of definitely directed thought and work, by myself if spared, and at all events on the part of others, if I should ever be allowed to live long enough distinctly to communicate the discovery. I pulled out on the spot a pocket-book, which still exists, and made an entry there and then. Nor could I resist the impulse—unphilosophical as it may have been—to cut with a knife on a stone of Brougham Bridge, as we passed it, the fundamental formula with the symbols i, j, k;
>
> $$i^2 = j^2 = k^2 = ijk = -1,$$
>
> which contains the solution of the Problem, but of course as an inscription, has long since mouldered away.

The entry in the pocket book is reproduced on the title page of [3]: it contains the formulas

$$\begin{aligned} &i^2 = j^2 = k^2 = -1, \\ &ij = k, \quad jk = i, \quad ki = j, \\ &ji = -k, \quad kj = -i, \quad ik = -j. \end{aligned}$$

I assume as likely that before his walk Hamilton had already written on a piece of paper the result of the somewhat tiresome calculation which showed that the sum of squares

$$(ax - by - cz)^2 + (ay + bx)^2 + (az + cx)2$$

still lacked $(bz - cy)^2$ compared with the product

$$(a^2 + b^2 + c^2)(x^2 + y^2 + z^2).$$

What then happened immediately before and during that remarkable walk along the Royal Canal, he described again on the same day in his Note Book, as follows:

> I believe that I now remember the order of my thought. The equation $ij = 0$ was recommended by the circumstances that
>
> $$(ax - y^2 - z^2)^2 + (a + x)^2(y^2 + z^2) = (a^2 + y^2 + z^2)(x^2 + y^2 + z^2).$$
>
> I therefore tried whether it might not turn out to be true that
>
> $$(a^2 + b^2 + c)(x^2 + y^2 + z^2) = (ax - by - cz)^2 + (ay + bx)^2 + (az + cx)^2,$$
>
> but found that this equation required, in order to make it true, the addition of $(bz - cy)^2$ to the second member. This *forced* on me the non-neglect of ij, and *suggested* that it might be equal to k, a new imaginary.

By underscoring the italicized words *forced* and *suggested* Hamilton emphasized that he was concerned with two entirely different facts. The first was a compelling logical conclusion, which came immediately out of the calculation: it was not possible to set ij equal to zero, since then the law of the moduli would not hold. The second fact was an insight which came over him in a flash at the canal ("an electric circuit seemed to close, and a spark flashed forth"), that is, that ij could be taken to be a new imaginary unit.

After the insight was once there, everything else was very simple. The calculations $ik = iij = -j$ and $kj = ijj = -i$ could be made easily enough by Hamilton in his head. The assumptions $ki = -ik = j$ and $jk = -kj = i$ were immediate. And k^2 could be easily calculated too: $k^2 = ijij = -iijj = -1$.

And so during his walk Hamilton also discovered the rules of calculation which he entered into the pocket book. The pocket book also contains the formulas for the coefficients of the product

$$(a + bi + cj + dk)(\alpha + \beta i + \gamma j + \delta k),$$

that is,

$$\begin{aligned} &a\alpha - b\beta - c\gamma - d\delta \\ &a\beta + b\alpha + c\delta - d\gamma \\ &a\gamma - b\delta + c\alpha + d\beta \\ &a\delta + b\gamma - c\beta + d\alpha \end{aligned}$$

as well as the sketch for the verification of the fact that in the sum of the squares of these coefficients all mixed terms (such as $ad\alpha\delta$) cancel and only $(a^2 + b^2 + c^2 + d^2)(\alpha^2 + \beta^2 + \gamma^2 + \delta^2)$ remains. In the Note Book of the same day everything was again completely restated.

Octonions

The letter to Graves in which Hamilton announced the discovery of quaternions was written on the 17th of October 1843, one day after the discovery. The seeds, which Hamilton sowed, fell upon fertile soil, since in December 1843 the recipient John T. Graves already found a linear algebra with 8 unit elements $1, i, j, k, l, m, n, o$, the algebra of octaves or octonions. Graves defined their multiplication as follows [3, p. 648]:

$$\begin{aligned} i^2 = j^2 = k^2 = l^2 = m^2 = n^2 = o^2 &= -1 \\ i = jk = lm = on &= -j = -ml = -no \\ j = ki = ln = mo &= -ik = -nl = -om \\ k = ij = lo = nm &= -ji = -ol = -mn \\ l = mi = nj = ok &= -im = -jn = -ko \\ m = il = oj = kn &= -li = -jo = -nk \\ n = jl = io = mk &= -lj = -oi = -km \\ o = ni = jm = kl &= -in = -mj = -lk. \end{aligned}$$

In this system the "law of the moduli" also holds:

$$(a_1^2 + \cdots + a_8^2)(b_1^2 + \cdots + b_8^2) = (c_1^2 + \cdots + c_8^2) \quad (1)$$

Hamilton answered on the 8th of July 1844 [3, p. 650]. He noted to Graves that the associative law $A \cdot BC = AB \cdot C$ clearly held for quaternions but not for octaves.

Octaves were rediscovered by Cayley in 1845; because of this they are also known as *Cayley numbers*. Graves also made an attempt with 16 unit elements but it was unsuccessful. It could not succeed since we know today that identities of the form (1) are only possible for sums of 1, 2, 4 and 8 squares. I should like to close with a brief comment about the history of these identities.

Product formulas for the sums of squares

It is likely that the "law of the moduli" for complex numbers was already known to Euler:

$$(a^2 + b^2)(c^2 + d^2) = (ac - bd)^2 + (ad + bc)^2.$$

A similar formula for the sum of 4 squares

$$(a_1^2 + \cdots + a_4^2)(b_1^2 + \cdots + b_4^2) = (c_1^2 + \cdots + c_4^2)$$

was discovered by Euler; the formula is stated in a letter from Euler to Goldbach on May 4th, 1748 [4]. The formula (1) for 8 squares, which Graves and Cayley proved by means of octonions, was previously found by Degen (1818) [6]. Degen erroneously thought that he could generalize the theorem to 2^n squares.

The problem, which started with Hamilton, reads: can two triplets (a, b, c) and (x, y, z) be so multiplied that the law of the moduli holds? In other words: is it possible so to define (u, v, w) as bilinear functions of (a, b, c) and (x, y, z) that the identity

$$(a^2+b^2+c^2)(x^2+y^2+z^2) = (u^2+v^2+w^2) \quad (2)$$

results?

The first to show the impossibility for this identity was Legendre. In his great work *Théorie des nombres* he remarked on page 198 that the numbers 3 and 21 can easily be represented rationally as sums of three squares:

$$\begin{aligned} 3 &= 1 + 1 + 1, \\ 21 &= 16 + 4 + 1, \end{aligned}$$

but the product $3 \times 21 = 63$ cannot be so represented, since 63 is an integer of the form $(8n + 7)$. It follows from this that an identity of the form (2) is impossible, to the extent that it is assumed that (u, v, w) are bilinear forms in (a, b, c) and (x, y, z) with rational coefficients. If Hamilton had known of this remark by Legendre he would probably have quickly given up the search to multiply triplets. Fortunately he did not read Legendre: he was self-taught.

The question for which values of n a formula of the kind

$$(a_1^2 + \cdots + a_n^2)(b_1^2 + \cdots + b_n^2) = (c_1^2 + \cdots + c_n^2)$$

is possible, was finally decided by Hurwitz in 1898. With the help of matrix multiplication he proved (in [5]) that $n = 1, 2, 4$ and 8 are the only possibilities. For further historical accounts the reader may refer to [1] or [2].

References

1. C. W. Curtis, The four and eight square problem and division algebras, *Studies in Mathematics*, vol. 2 (Ed. by A. A. Albert), p. 100, Mathematical Association of America, 1963.

2. L. E. Dickson, On quaternions and their generalizations and the history of the eight square theorem, *Ann. of Math.*, 20 (1919) 155.

3. *The Mathematical Papers of Sir William Rowan Hamilton*, vol. III, Algebra, Edited for the Royal Irish Academy, by H. Halberstam and R. E. Ingram, Cambridge University Press, 1967.

4. P. H. Fuss (Editor), *Correspondence Mathématique et Physique I*, St. Petersburg, 1843.

5. Hurwitz, Ueber die Composition der quadratischen Formen von beliebig vielen Variabeln, *Nachr. der königlichen Gesellschaft der Wiss. Göttingen* (1898) 309–316; *Mathematische Werke*, Bd. II, Basel, 1932, pp. 565–571.

6. C. P. Degen, Adumbratio Demonstrationis Theorematis Arithmeticae maxime generalis, *Mémoires de l'Académie de St. Petersbourg*, VIII, (1822) 207.

Geometric Extremum Problems

G. D. Chakerian and L. H. Lange

Mathematics Magazine
44 (1971), 57–69

Editors' Note: Gulbank D. Chakerian is Professor Emeritus of Mathematics at the University of California, Davis, where he joined the faculty after completing his PhD at Berkeley and after a short stay at the California Institute of Technology. His research is in geometric inequalities and convex sets.

Lester Henry Lange, a student of George Pólya's at Stanford, was for many years on the mathematics faculty at San Jose State University, where he was later Dean of the College of Science. Since retirement he has been active in the Moss Landing Marine Laboratories of the California State University system.

Both authors have received awards for expository writing from the MAA, Chakerian a Lester R. Ford Award for this paper in 1971, plus a Carl B. Allendoerfer Award in 1991 and a George Pólya Award in 1981; Lange the Ford Award for this paper and a Pólya Award in 1993. Lange is known for a popular text, *Linear Algebra* (Wiley, 1968). Recently Chakerian contributed to a collection, *Mathematical Adventures for Students and Amateurs* (MAA, 2004).

1. Introduction

A standard exercise for calculus classes reads: *Given a triangle of altitude a and base b, find the dimensions of the rectangle of maximum area which can be inscribed in this triangle with one side along the base.*

It at least broadens a student's perspective if he occasionally sees an alternate solution of such a problem avoiding the calculus. The above problem can be settled using an elementary inequality as follows:

Two essentially different possibilities face us, as shown in Figure 1. In case (i), the vertex C is "above" some point of the base; in case (ii), the vertex C is not so situated. We solve case (i), and then the solution of case (ii) is easily made to depend on that solution.

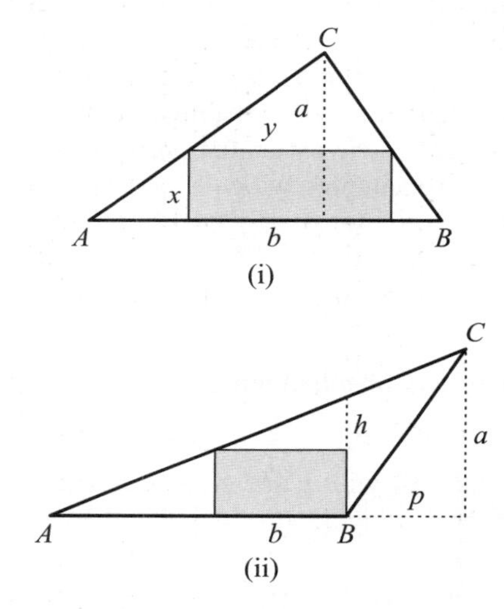

Figure 1.

Using the notation in Figure 1, we seek the maximum of the product xy, where, from similar triangles, we have $y = (b/a)(a-x)$. Thus, $xy = (b/a)(x)(a-x)$, and we need to find the value of x, $0 \leq x \leq a$, which maximizes this quantity. Rather than use the derivative, we can do this directly by observing that $x(a-x) = (a/2)^2 - \{x-(a/2)\}^2$, and this is a maximum if, and only if, $(x-a/2)^2 = 0$; i.e., iff $x = a/2$. Hence, $xy = (b/a)(x)(a-x) \leq \frac{1}{2}(ab/2) = \frac{1}{2}\,\text{area}\,(\triangle ABC)$, with equality holding iff $x = a/2$. Thus, the maximum rectangle has height $x = a/2$ and area exactly half that of the given triangle. In case (ii), the maximum rectangle will have height $h/2$ and area less than half that of the given triangle. (In this case, the maximum area is $\frac{1}{2}\{b/(b+p)\}\,\text{area}\,(\triangle ABC)$.)

Note that, in case (ii), if we choose AC as the base on which to place our rectangle, we would obtain a maximum rectangle with area half that of the given triangle. Hence, we observe that in any case, it is possible to inscribe some rectangle of area half that of the given triangle.

In calculus texts, it is common to treat, implicitly or explicitly, only those cases where the optimal figure is assumed to be in some special position. For example, in the problem above we considered only rectangles with a side lying on a base of the given triangle. But it is natural to inquire: of all rectangles contained in a given triangle, which yield the maximum area? Is this maximum area ever larger than half the area of the given triangle?

We shall answer this question in Section 2 (Theorem 3, below) and in later sections look at other extremum problems where the optimal figure is usually assumed to be in some restrictive special position. However, our main objective in this article is to provide an easily accessible account of a general class of geometric extremum problems of the above type, with examples which might be useful in the classroom (not necessarily only in calculus courses). An underlying theme, below, is the use of affine transformations to simplify problems of this type.

2. Polygons of minimum area circumscribed about a convex set

We shall deal with plane convex sets; that is, those plane sets having the property that the segment joining any two points of the set is contained in the set. By a *convex region* we shall mean a plane convex set which is also compact (i.e., closed and bounded), and has nonempty interior.

It is known that if K is a convex region and $n \geq 3$ a given integer, then there exists at least one convex n-gon of minimum area containing K. Obviously, such an n-gon must be circumscribed about K—that is, its sides must intersect the boundary of K. The following theorem features the property which interests us:

Theorem 1 *Let K be a convex region and $n \geq 3$ a given integer. Let P be a convex n-gon of minimum area containing K. Then the midpoints of the sides of P lie on the boundary of K.*

Remark. Although this is well known (see [3] p. 6), we know of no easily accessible published proof, so we feel justified in presenting an elementary proof in Section 4. The proof will illustrate the usefulness of the technique of affine transformation developed in Section 3.

As a simple application of Theorem 1, consider the case where K is a parallelogram, and let T be a triangle of minimum area containing K. According to Theorem 1, the midpoints of the sides of T meet K. The reader will readily convince himself that this is possible only if one or two sides of K lie on sides of T, and the relative positions are as depicted in Figure 2(a) or 2(b).

Notice, in any case, that area $(T) = 2$ area (K). Keeping in mind that T was a minimal triangle, we thus have the following result:

Theorem 2 *Let K be a given parallelogram, and let T be any triangle containing K. Then,* area $(T) \geq 2$ area (K), *and equality holds if, and only if, T is in a special position, as depicted in Figure 2.*

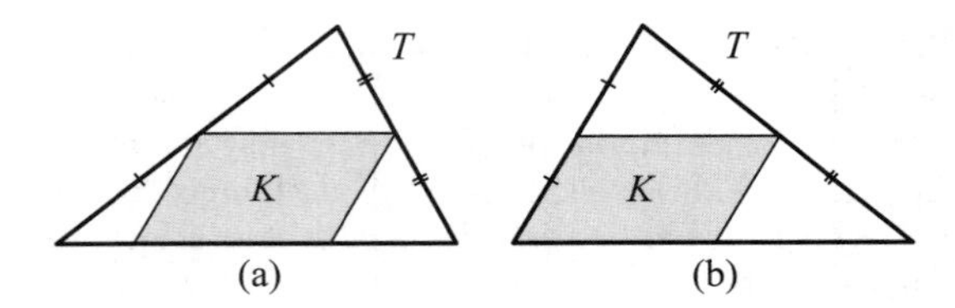

Figure 2.

We are now in a position to settle the question raised in Section 1. We have,

Theorem 3 *Let T be a given triangle, and let R_0 be a rectangle of maximum area contained in T. Then,* area $(R_0) = \frac{1}{2}$ area (T), *and R_0 is in a special position, with one side lying on a side of T and the midpoints of the other two sides of T are vertices of R_0.*

Proof. Suppose R is a rectangle contained in T but such that R is not in the special position described above. Then, Theorem 2 implies that T is not a triangle of minimum area containing R; hence, if T^* is such a minimum triangle, we have

$$\text{area}\,(R) = \frac{1}{2}\,\text{area}\,(T^*) < \frac{1}{2}\,\text{area}\,(T).$$

If, on the other hand, R_0 is a rectangle in the special position described above, then area $(R_0) = \frac{1}{2}$ area (T). It follows that R_0 is a rectangle of maximum area contained in T. (Note that there are three such rectangles if T is an acute triangle, two if T is a right triangle, and only one if T has an obtuse angle.) This completes the proof.

Remarks. The problem which leads to Theorem 3 was raised in a paper by one of the authors [7]. In the meantime, M. T. Bird [1] has given a simple direct proof of Theorem 3.

Theorem 2 is contained in a result of Fulton and Stein [4, Theorem 1].

Parallelograms are rather extreme in their behavior with regard to circumscribed triangles—one cannot keep them inside a triangle of less than twice their area. One is naturally led to ask two questions:

(a) Are parallelograms the only convex regions which behave in this (deplorable) fashion?
(b) Is every convex region K contained in some triangle whose area is less than or equal to twice the area of K?

We will return to these questions after developing some tools, namely affine transformations, to help us simplify these and related questions.

In relation to Theorem 3, C. Radziszewski [9] proved that every convex region K contains a rectangle of half its area.

3. Affine transformations

In this section we call attention to some useful properties of nonsingular affine transformations of the plane—that is, those transformations of the (x, y)-plane onto itself which send each point (x, y) to a point (x', y') such that

$$x' = ax + by + e, \quad y' = cx + dy - f,$$

where $a, b, \ldots, f$ are some given real numbers which satisfy the condition $ad - bc \neq 0$. Since we shall be interested only in nonsingular transformations, we shall consistently use the term "affine transformation" to mean nonsingular affine transformation.

Before listing those properties of affine transformations which we shall later find useful, we first look at an instructive example which involves a function of this type and the consideration of a certain kind of geometric extremum. (See [8] for an associated discussion.)

Letting a and b be given real numbers satisfying $0 < a \leq b$, and letting $\mathcal{P}$ be the set of all pairs (x, y) of real numbers—our *plane*—we consider the function μ which maps $\mathcal{P}$ into $\mathcal{P}$ by sending the point (x, y) into the point $(x', y') = (ax, by)$. If we now consider the set D of all points (x, y) such that $x^2 + y^2 \leq 1$, we see that the associated points (x', y') necessarily satisfy $(x'/a)^2 + (y'/b)^2 \leq 1$.

Our function μ is a very special affine transformation; it is, in fact, a simple example of a *linear transformation.* The equations $x' = ax$ and $y' = by$ tell us that, given any (x', y') in $\mathcal{P}$, there exists a unique (x, y) in $\mathcal{P}$ which is such that μ sends (x, y) into (x', y'); that is, μ is a *one-to-one, onto* function. In particular, we see that each point of the (closed) elliptical disc shown in Figure 3 is the image of exactly one point in the (closed) circular disc D. The function μ "picks up the circular sheet D, and, without any folding (since μ is one-to-one) covers completely (since μ is onto) and exactly the elliptical disc $\mu(D)$."

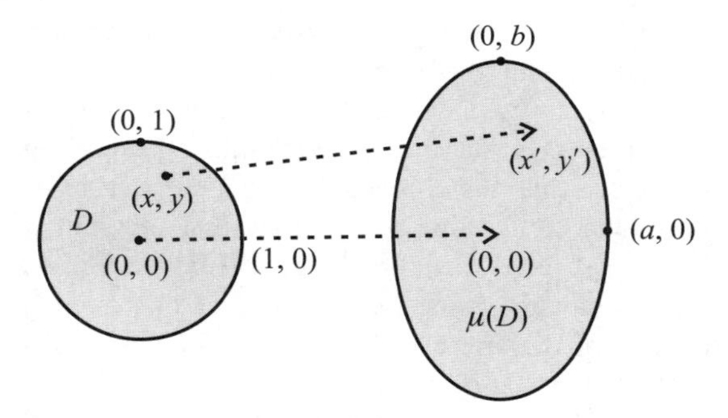

Figure 3.

Now suppose p, q, and r are given real numbers, and consider the set

$$\{(x, y) : px + qy + r = 0\},$$

which is a straight line. Then the function μ sends this set into the set

$$\left\{(x', y') : \frac{p}{a}x' + \frac{q}{b}y' + r = 0\right\}.$$

Thus, under μ, the image of any given straight line is again a straight line. We easily see that the image of the vertical line segment joining (x_1, y_1) and (x_1, y_2) is the vertical line segment which joins the image points (ax_1, by_1) and (ax_1, by_2). See Figure 4. We notice also that the length of this image segment is simply $(b) \cdot |y_2 - y_1|$, where $|y_2 - y_1|$ is, of course, the length of the original segment. Similarly, the length of an image horizontal segment is simply (a) times the length of the original horizontal segment.

Now, if we again look at the image of the circular disc D under the function μ, we see that if the *rectangle* A in Figure 4 is *inside* the disc D, then its image, $\mu(A)$ is a *rectangle* and it is *inside* the elliptical disc $\mu(D)$. Furthermore, if the area of A is α, then the area of the rectangle $\mu(A)$ is (ab) times α.

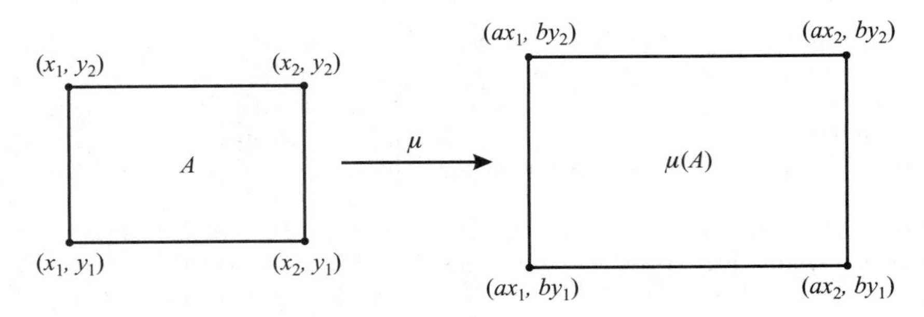

Figure 4.

It follows that, if $\sum \alpha_i$ is the combined area of any finite collection of such rectangles inside D, then $(ab)\sum \alpha_i$ is the area of the associated collection of rectangles inside the elliptical disc $\mu(D)$.

Consequently, if we "pack" any finite number of such rectangles (of various sizes) into D, requiring, of course, that these rectangles don't "overlap"—other than along edges—we know that the sum of their areas, $\sum \alpha_i$, satisfies $\sum \alpha_i < 4$, since 4 is the area of a square of side 2 enclosing D. Consequently, the sum of the areas of the associated rectangles in $\mu(D)$ satisfies $(ab)\sum \alpha_i < (ab)(4)$.

Now, the number 4 is not the *least* number which will serve as an upper bound for the collection of all possible numbers $\sum \alpha_i$ here. This distinction is, of course, held by the number π; and because π is the *least* upper bound of the collection of all such possible sums $\sum \alpha_i$, it is called the *area* of the disc D.

It follows that the number $(ab)\pi$ *must* be the area of the elliptical disc $(x/a)^2 + (y/b)^2 \leq 1$, for it is the least number which will serve as an upper bound to the collection of all numbers $(ab)\sum \alpha_i$.

Exercise 1. Show that the volume of the ellipsoid

$$(x/a)^2 + (y/b)^2 + (z/c)^2 \leq 1$$

is $\frac{4}{3}\pi abc$.

We now list—without proof—those *properties of affine transformations* of the plane which we shall be using.

(P-1) If l and m are parallel lines, then their images are parallel lines. ("Parallel lines are sent to parallel lines.")

(P-2) If u^* and v^* are the images of line segments u and v respectively, lying on parallel lines, then

$$\{\text{length}(u^*)/\text{length}(v^*)\} = \{\text{length}(u)/\text{length}(v)\}.$$

("The ratio of lengths of parallel segments is preserved.")

In particular, we have

(P-3) If u^* is the image of the line segment u, then the midpoint of u^* is the image of the midpoint of u. ("Midpoints are sent into midpoints.")

(P-4) The image of any convex region is again a convex region.

(P-5) If U^* and V^* are the respective images of the regions U and V, then

$$\{\text{area}\,(U^*)/\text{area}\,(V^*)\} = \{\text{area}\,(U)/\text{area}\,(V)\}.$$

("The ratio of areas is preserved.")

(P-6) Any given triangle can be mapped onto any other given triangle by some appropriate affine transformation. (Note that P-3 then implies that their centroids will correspond.)

(P-7) Any parallelogram can be mapped onto any other parallelogram by some affine transformation.

(P-8) The image of any ellipse is again an ellipse, and any ellipse can be mapped onto any other ellipse by some affine transformation. (Note that the center of an ellipse is sent to the center of the image. Indeed, P-3 implies that centrally symmetric figures are always sent into centrally symmetric figures, with centers corresponding.)

(P-9) Any (nonsingular) affine transformation is continuous and invertible, and the inverse is again a (nonsingular) affine transformation.

Here is a simple consequence of these properties:

Lemma 1 *There is contained in any triangle T one and only one ellipse E_0 tangent to the sides of T at their respective midpoints.*

Proof. At least one such E_0 exists. For we may affinely transform T onto an equilateral triangle T^* and consider the inscribed circle (call it E_0^*) of T^*. The image of E_0^* under the inverse transformation is the required ellipse E_0.

Suppose F_0 were another ellipse inside T tangent to the sides of T at their midpoints. Then, the image F_0^* of F_0 under the above transformation would be an ellipse tangent to the sides of the equilateral triangle T^* at their midpoints. But the center of F_0^* coincides with the centroid of T^*. (In order to see this, map F_0^* to a circle and note that T^* must map to an *equilateral* triangle circumscribed about that circle—then use properties P-6 and P-8.) Hence, central reflection through the centroid of T^* sends F_0^* onto itself. It follows that F_0^* contains not only the midpoints of the sides of T^*, but also the midpoints of the sides of the reflection of T^* through its centroid. Since these midpoints are the vertices of a regular hexagon inscribed in E_0^*, and since an ellipse is determined by five points, it follows that $F_0^* = E_0^*$; hence, $F_0 = E_0$. This completes the proof.

The last lemma enables us to establish an extremum property analogous to Theorem 3.

Theorem 4 *Any given triangle T contains a unique ellipse E_0 of maximum area. This ellipse is tangent to the sides of T at their midpoints, and* area $(E_0) = (\pi/3\sqrt{3})$ area (T).

Proof. Suppose E is an ellipse contained in T but such that the midpoints of the sides of T do not all meet E, and let E_0 be the unique ellipse inside T which is tangent to the sides of T at their midpoints. Further, let S be a triangle of minimum area containing E. Then, by Theorem 1, the midpoints of the sides of S lie on E, and area $(S) <$ area (T).

Now affinely map S onto T. Then E is mapped to an ellipse E^* in T tangent to the sides of T at their midpoints; hence, by the last lemma, $E^* = E_0$. By property P-5, we have then

$$\frac{\text{area}\,(E)}{\text{area}\,(S)} = \frac{\text{area}\,(E^*)}{\text{area}\,(T)} < \frac{\text{area}\,(E_0)}{\text{area}\,(S)},$$

hence, area $(E) <$ area (E_0). Thus, E_0 is the unique ellipse of maximum area contained in T. The ratio

$$\left\{\frac{\text{area}\,(E_0)}{\text{area}\,(T)}\right\} = \frac{\pi\sqrt{3}}{9}$$

is obtained by mapping T onto an equilateral triangle. This completes the proof.

Remark. Theorem 4 illustrates a special case of the following general result: *Every convex region contains a unique ellipse of maximum area, and also is contained in a unique ellipse of minimum area.* For a discussion of this result and the generalization to higher dimensional spaces, the reader may consult [2].

Exercise 2. Use the methods of this section to prove that any given triangle T is contained in a unique ellipse E_1 of minimum area, and area $(E_1) = 4\pi/3\sqrt{3}$ area (T). [Hint: affinely map T onto an equilateral triangle T^* and let E_1^* be the circumcircle of T^*. The required E_1 is the image of E_1^* under the inverse transformation.]

Exercise 3. Using Exercise 2 and Theorem 4, prove that the circumradius of any triangle is at least twice the inradius. (For an elegant proof of this property, by I. Ádám, see [3] p. 28.)

4. Proof of Theorem 1

For the proof we need the following lemma, which tells us how to cut off a triangle of minimum area with a line passing through a given point inside an angle.

Lemma 2 *Let XOY be a given angle (Figure 5). Then,* (a) *for each point M interior to the angle there exists one and only one line segment $\overline{AB}$ containing M, with endpoints A on $\overrightarrow{OX}$ and B on $\overrightarrow{OY}$, bisected by M. Moreover,* (b) *of all triangles cut from the angle by lines passing through M, $\triangle AOB$ is the unique one of minimum area.* (c) *Let $\triangle(Q)$ denote the minimum triangle associated with a point Q interior to our angle XOY. If $Q \to M$ along $\overline{AB}$, then $\triangle(Q) \to \triangle(M)$. By this we mean the vertices of $\triangle(Q)$ tend to the corresponding vertices of $\triangle(M)$.*

Proof. A simple continuity argument shows that there is always *some* segment $\overline{AB}$ bisected by M. Now transform the configuration with an affine transformation sending A to $A^* = (1, 0)$, O to

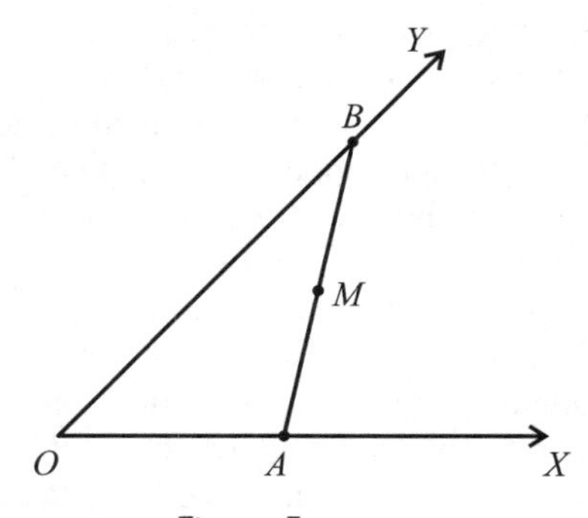

Figure 5.

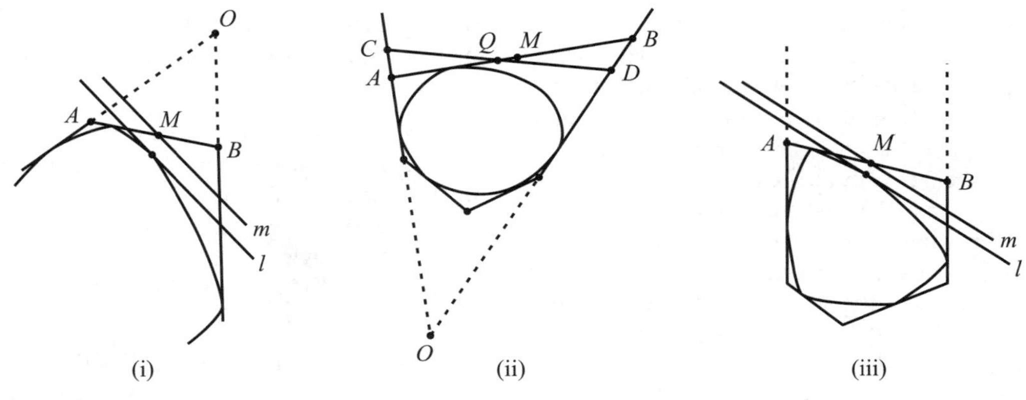

Figure 6.

$O^* = (0, 0)$, and B to $B^* = (0, 1)$. Then automatically M is sent to $M^* = (\frac{1}{2}, \frac{1}{2})$. We can now establish all the properties in this situation. For this purpose we revert to the notation in our original figure, omitting the asterisks.

Indeed, let $Q = (\xi, 1 - \xi)$, $0 < \xi < 1$, be any point in the interior of AB, and let $\triangle COD$ be any triangle cut off by a line passing through Q. Then, letting $C = (c, 0)$, we have

$$\text{area}\,(\triangle COD) = \frac{c^2(1-\xi)}{2(c-\xi)} \geq 2\xi(1-\xi).$$

(The inequality is a consequence of $(2\xi - c)^2 \geq 0$.) Equality holds if and only if $c = 2\xi$. Hence, there is a unique minimizing triangle $\Delta(Q)$ with a side passing through Q and bisected by Q, and,

$$\text{area}\,\big(\Delta\,(Q)\big) = 2\xi(1-\xi).$$

If $A'OB' = \Delta(Q)$, then note that $A' = (2\xi, 0)$ and $B' = (0, 2 - 2\xi)$. Thus, as $Q \to M$ along $\overline{AB}$, we have $A' \to A = (1, 0)$ and $B' \to B(0, 1)$; that is $\Delta(Q) \to \Delta(M)$. The properties (a), (b), and (c) asserted in the theorem are now easily verified using the properties P-3, P-5, and P-9 of affine transformations in Section 3, applied to the inverse of the transformation we used above.

We can now prove Theorem 1. Suppose P is an n-gon of minimum area circumscribed about the convex region K, and suppose the midpoint M of some side $\overline{AB}$ does not meet K. We consider the three possible cases depicted in Figure 6.

In case (i) the sides of P adjacent to $\overline{AB}$, when extended, meet in a point O such that $\triangle AOB$ does not contain K. In case (ii), $\triangle AOB$ contains K. In case (iii), the two sides adjacent to $\overline{AB}$ are parallel. We shall show that in each case it is possible to construct a polygon of area less than P containing K; this contradiction will then show that the midpoints of each side of the minimal n-gon must indeed meet K.

In case (i), since M is exterior to K, there exists a supporting line l of K separating M from K. (A supporting line of K is a line L such that K lies on one side of L and $K \cap L \neq \emptyset$.) Let m be the line through M parallel to l (Figure 6, (i)). Note that m cuts off a triangle of area strictly larger than area$\,(\triangle AOB)$, by the last lemma. But l cuts off an even larger triangle; hence, using l and P we can produce an n-gon containing K and having smaller area than P.

In case (ii), we can use the property (c) of the lemma to produce a line segment $\overline{CD}$ with its midpoint Q on $\overline{AB}$, $Q \neq M$, and $\overline{CD}$ not intersecting K (as depicted in Figure 6, (ii)). By the lemma again, $\triangle COD$ has strictly smaller area than all other triangles cut off by lines through Q; hence, area$\,(\triangle COD) <$ area$\,(\triangle AOB)$. It then is clear that we can produce an n-gon of smaller area than P containing K.

In case (iii), l is chosen to be a supporting line of K at the point (nearer M) where the midline (of the two parallel sides) intersects the boundary of K. Then m is the line through M parallel to l. It is easy to see that using m and P we obtain an n-gon containing K with the same area as P, and using l we obtain an n-gon of smaller area. This completes the proof.

5. The smallest triangle containing a convex region

We now answer a question raised at the end of Section 2.

Theorem 5 *Every convex region K is contained in some triangle of at most twice its area.*

Proof. It suffices to prove that if T_0 is a triangle of minimum area containing K, then $\text{area}(T_0) \leq 2\,\text{area}(K)$. In order to do this, let T_0 be a minimal triangle circumscribed about K, with the midpoints A, B, C of its sides on K. As indicated in Figure 7, let T be the triangle similar to T_0 formed by drawing supporting lines of K parallel to the sides of T_0, and let A', B', C' be points where the sides of T meet K.

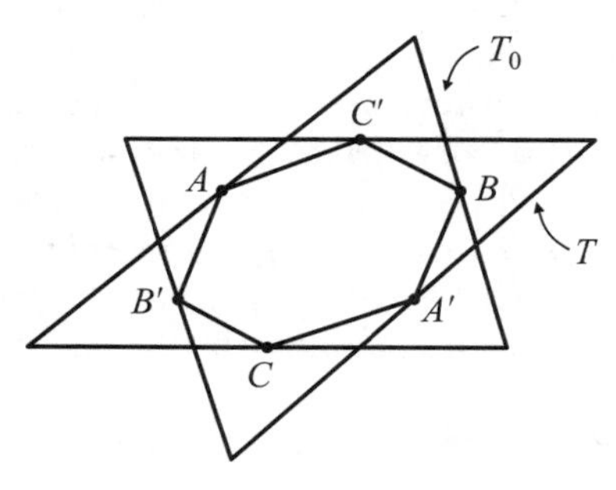

Figure 7.

We shall now show that the hexagon $AB'CA'BC'$ has area at least twice the area of $\triangle ABC$. It will then follow that

$$\begin{aligned}\text{area}(K) &\geq \text{area}(AB'CA'BC') \\ &\geq 2\,\text{area}(\triangle ABC) = \frac{1}{2}\,\text{area}(T_0),\end{aligned}$$

and this will prove the theorem.

We now affinely transform the configuration in Figure 7 so that T is mapped onto an equilateral triangle T^*. Then $\triangle ABC$ is mapped onto an equilateral triangle S^* inside T^* with sides parallel to those of T^*, and the hexagon $AB'CA'BC'$ is mapped to a hexagon like that dotted in Figure 8.

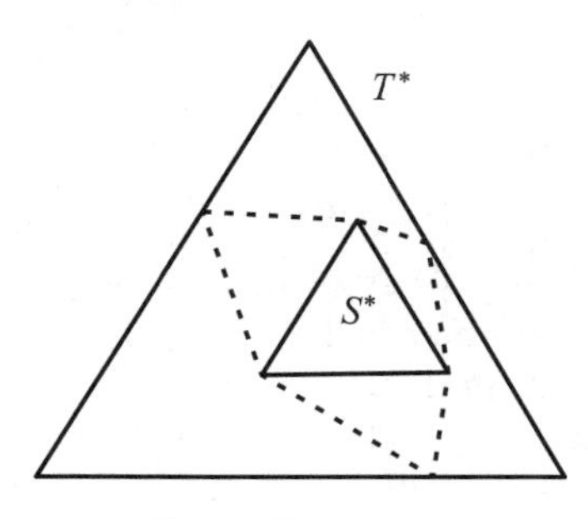

Figure 8.

Since T_0 is minimal and similar to T, each side of T is at least as long as the corresponding side of T_0, hence, at least twice as long as the parallel side of $\triangle ABC$. Thus, the sides of T^* are at least twice the length of the sides of S^*. Since affine transformations preserve ratios of areas, we need only show that under these conditions the dotted hexagon in Figure 8 has at least twice the area of S^*. In order to show this, form the dotted hexagon in Figure 9 by dropping perpendiculars to the sides of T^* from the centroid O of S^*.

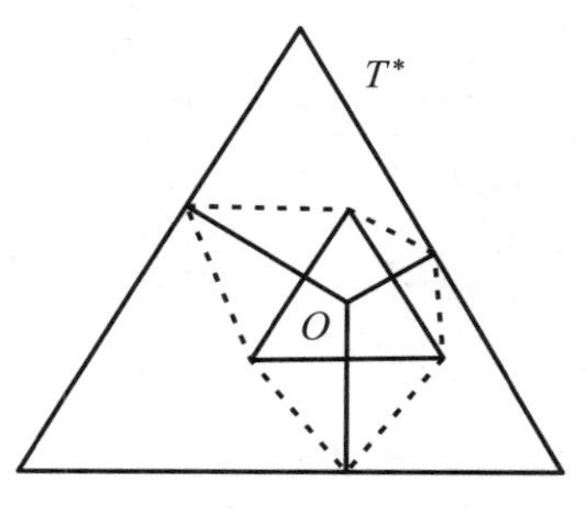

Figure 9.

The dotted hexagon in Figure 9 has the same area as that in Figure 8. We now leave it as an exercise to the reader to prove that this hexagon has at least twice the area of the small triangle S^*. [Hint: the sum of the lengths of the perpendiculars dropped to the sides of an equilateral triangle from any interior point is always equal to the altitude of the triangle. Keep in mind that the large triangle has sides at least twice the length of those of the small triangle.] With this, the reader completes the proof.

Remarks. Theorem 5 was first proved by Gross [5]. Gross also proves that if the minimal triangle containing K has exactly twice the area of K, then K must be a parallelogram. This answers affirmatively our question (a) raised at the end of Section 2.

It is natural to ask for analogues of Theorem 5 for n-gons with $n > 3$. The following partial result, pertaining to the case $n = 4$, appears to be new:

Theorem 6 *Every convex region K is contained in a quadrilateral Q_0 such that*

$$\text{area}(Q_0) \leq (\sqrt{2})\,\text{area}(K).$$

Proof. Let Q_0 be a quadrilateral of minimum area containing K, with the midpoints A, B, C, D of its sides on K. As is well known, $ABCD$ is a parallelogram with area half that of Q_0. In Figure 10 we have drawn with dotted lines the parallelogram Q circumscribed about K with sides parallel to those of $ABCD$ and meeting K in points $EFGH$.

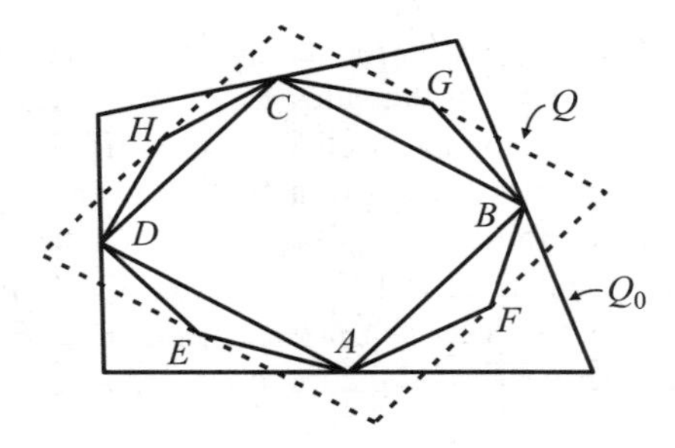

Figure 10.

If we could show that the octagon Z with vertices $AFBGCHDE$ satisfies

$$\text{area}\,(Z) \geq \sqrt{2}\,\text{area}\,(ABCD),$$

then we would have

$$\begin{aligned}\text{area}\,(K) &\geq \text{area}\,(Z)\\ &\geq \sqrt{2}\,\text{area}\,(ABCD) = \frac{\sqrt{2}}{2}\,\text{area}\,(Q_0),\end{aligned}$$

and the theorem would follow.

We note that the area of Z is unchanged if we let E, F, G, and H move on their respective sides of Q, and, moreover, it suffices to consider the case where Q and $ABCD$ are rectangles (using an appropriate affine transformation). In other words, it suffices to show that area $(Z) \geq \sqrt{2}$ area $(ABCD)$ in a situation like that depicted in Figure 11.

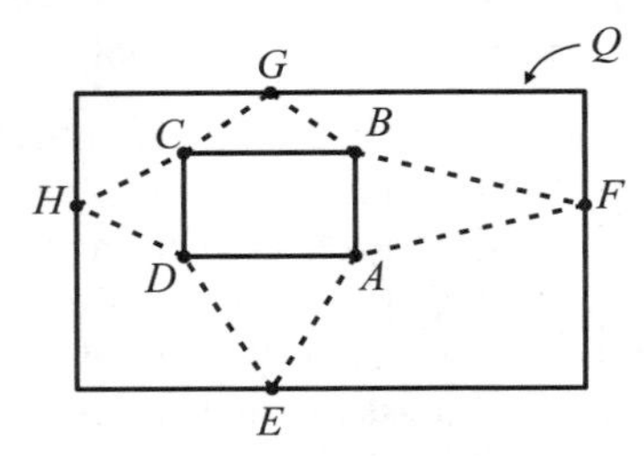

Figure 11.

In Figure 11, E, F, G, and H are the feet of the perpendiculars dropped from the center of $ABCD$ into the sides of Q. The dotted polygon is our new Z. If s and t are the lengths of the sides of $ABCD$, and s' and t' the lengths of corresponding parallel sides of Q, then it is easily checked that

$$\text{area}\,(Z) = \frac{1}{2}(st' + s't).$$

Recalling that

$$\text{area}\,(Q) \geq \text{area}\,(Q_0) = 2\,\text{area}\,(ABCD)$$

in our original configuration, we see that $s't' \geq 2st$. Thus, using the fact that the arithmetic mean of two numbers is always at least their geometric mean, we obtain

$$\begin{aligned}\text{area}\,(Z) &= \frac{1}{2}(st' + s't) \geq \sqrt{st's't} \geq \sqrt{2}st\\ &= \sqrt{2}\,\text{area}\,(ABCD),\end{aligned}$$

and the theorem follows.

Remark. We do not know if Theorem 6 is a best possible result. It is possible that every K is contained in some quadrilateral Q such that area $(Q) \leq \lambda$ area (K), where $\lambda < \sqrt{2}$. The question which needs to be answered is the following: *Is there a convex region K all of whose circumscribed quadrilaterals have area at least* $(\sqrt{2})$ area (K)?

Fejes Tóth [3, p. 38] remarks that the answers to corresponding questions for n-gons, $n > 3$, are unknown.

Theorem 6 is of interest in connection with problems of "packing" convex sets. A distribution of nonoverlapping congruent copies of K in the plane is called a packing. The basic question is: how large a fraction of the plane can be covered by nonoverlapping copies of K? In other words, what is the highest "density" of packing which can be achieved? If Q_0 is the minimal quadrilateral containing K, it is possible to cover the plane with nonoverlapping copies of Q_0. Then we obtain a packing with copies of K having density $\geq \sqrt{2}/2 > .707$. This packing even has a certain amount of regularity. It is the union of two "lattice" packings by K. The reference [3] contains a great deal of valuable information about packing problems.

6. Some familiar extremum problems

Another standard exercise in calculus texts is the following:

Given an ellipse E with semiaxes a and b, find the rectangle R_0 of maximum area inscribed in E.

In solving this problem, it is usually assumed that the sides of the rectangle are parallel to the axes of the ellipse. In order to justify this assumption, one needs to know that any rectangle R inscribed in E has its sides parallel to the axes (we are assuming E is not a circle). Let us show how to prove this fact using affine transformations.

Assume R is a rectangle inscribed in E. Affinely transform E to a circle E^*. Then, under the same transformation, R is sent to a parallelogram R^* inscribed in E^*. Now it is a trivial exercise to show that any parallelogram inscribed in a circle must be a rectangle. But the fact which interests us is that the center of R^* coincides with the center of E^*; hence, the center of R coincides with the center of E. Thus, the circumscribed circle C of R is centered at the center of E. Now it is obvious that such a circle C intersects E in four points which are the vertices of a rectangle with sides parallel to the axes; hence, R is such a rectangle.

Exercise 4. Use affine transformations to reduce the problem of finding R_0 to the problem of finding the maximum rectangle inscribed in a circle.

The result in the following exercise, intimately related to Exercise 2, can be established readily with an affine transformation.

Exercise 5. Prove that the maximum area of any triangle inside an ellipse E is $(3\sqrt{3}/4\pi)\,\text{area}\,(E)$.

The following theorem, proved in [3, p. 36] is complementary to our considerations concerning circumscribed n-gons of minimum area:

Theorem 7 *If P is an inscribed n-gon of maximum area in a convex region K, then*

$$\text{area}\,(P) \geq \frac{n}{2\pi} \sin \frac{2\pi}{n}\, \text{area}\,(K),$$

and equality holds only if K is an ellipse.

Remarks. Some of the examples of this article are also discussed in the paper of Klamkin and Newman [6]. The following interesting exercise is given there:

Exercise 6. Through a given point inside an ellipse, draw a line cutting off minimum area.

Although other examples, more or less familiar, do exist, we stop at this point.

References

1. M. T. Bird, Maximum rectangle inscribed in a triangle, to appear.
2. L. Danzer, D. Laugwitz, and H. Lenz, Über das Löwnersche Ellipsoid und sein Analogen unter den einem Eikörper einbeschriebenen Ellipsoiden, *Arch. Math.*, 8 (1957) 214–219.
3. L. Fejes Tóth, *Lagerungen in der Ebene, auf der Kugel, und in Raum*, Berlin, 1953.
4. C. M. Fulton and S. K. Stein, Parallelograms inscribed in convex curves, *Amer. Math. Monthly*, 67 (1960) 257–258.
5. W. Gross, Über affine Geometrie XIII: Eine Minimumeigenschaft der Ellipse und des Ellipsoids, *Leipziger Berichte*, 70 (1918) 38–54.
6. M. S. Klamkin and D. J. Newman, The philosophy and applications of transform theory, *SIAM Rev.*, 3 (1961) 10–36.
7. L. H. Lange, Some inequality problems, *The Math. Teacher*, 56 (1963) 490–494.
8. ——, *Elementary Linear Algebra*, Wiley, New York, 1968, pp. 140–147.
9. C. Radziszewski, Sur un problème extrémal relatif aux figures inscrites dans les figures convexes, *C. R. Acad. Sci. Paris*, 235 (1952) 771–773.

Pólya's Enumeration Theorem by Example

Alan Tucker

Mathematics Magazine
47 (1974), 248–56

Editors' Note: Alan Tucker was born into a mathematical family. His grandfather on his mother's side was David Raymond Curtiss, longtime Professor of Mathematics at Northwestern University and author of the second of the MAA's Carus Monographs, *Analytic Functions of a Complex Variable* (1926). Alan Tucker's father was Albert Tucker, an eminent mathematician at Princeton for many years, and his uncle was John Curtiss of Johns Hopkins and Miami, one-time executive director of the American Mathematical Society. His stepfather was E. F. Beckenbach at UCLA, and his brother is Thomas Tucker, on the faculty at Colgate University. Beckenbach played a role in the rescue of *Mathematics Magazine* in 1945.

Alan Tucker is a Distinguished Teaching Professor in the Department of Applied Mathematics and Statistics at the State University of New York in Stony Brook. He was an undergraduate at Harvard and took his PhD at Stanford University in 1969, writing his dissertation under the direction of George Dantzig and D. R. Fulkerson. He was also a teaching assistant for George Pólya in a course in combinatorics and was much influenced by him. Here he discusses Pólya's enumeration theorem. This effort may well have been part of planning that went into Tucker's writing his classic text, *Applied Combinatorics* (Wiley, 1980), which has been one of most popular combinatorics texts over the past couple of decades. His other books include *A Unified Introduction to Linear Algebra: Models, Methods, and Theory* (Macmillan, 1988) and *Linear Algebra* (Macmillan, 1994).

He has been active in the MAA and served as First Vice President between 1988 and 1990.

1. Introduction

One of the most important results in combinatorial mathematics is Pólya's enumeration formula. This formula constructs a generating function for the number of different ways to mark the corners of an unoriented figure using a given set of labels. For example, if the figure were an unoriented cube and one could color the corners black or white, Pólya's formula would yield the following generating function, called the *pattern inventory*:

$$b^8 + b^7w + 3b^6w^2 + 3b^5w^3 + 7b^4w^4 + 3b^3w^5 + 3b^2w^6 + bw^7 + w^8,$$

where the coefficient of $b^i w^j$ is the number of distinct colorings with i black corners and j white corners. More general nongeometric applications are also possible. Pólya's work was motivated by some problems of isomer enumeration in organic chemistry (half a century earlier, Kekulé had confirmed the structure of the benzene ring by comparing the observed number of isomers of $C_6H_3R_3$ against the theoretical number of isomers in competing hypothetical structures).

As a result of the recent surge of student interest in computer science and operations research on campuses, most universities now offer an upper-level undergraduate course in combinatorial mathematics. Indeed, the combinatorics course at Stony Brook has become very popular (this year its enrollment was the largest of all upper-level undergraduate mathematics courses), in large part, because it heavily emphasizes problem-solving and minimizes theory. In this spirit, the course develops Pólya's formula in an "experimental" fashion with several sets of small problems which lead the students to infer the formula's structure.

Fifteen years ago, there was no American textbook which presented Pólya's formula. Recent combinatorics texts which include the formula all present it in full generality. Unfortunately, the general proof of the formula is complicated and not intuitive. Indeed, when Professor Pólya taught part of the combinatorial mathematics course at Stanford (for which this author was a teaching assistant), he omitted a proof of his formula and gave only examples. A cookbook use of the formula is all that is needed for most applications. On the other hand, the derivation of the formula has much to recommend it: the basis of the formula is Burnside's theorem whose proof is a marriage of elementary group theory and a standard combinatorial argument; and it makes the most

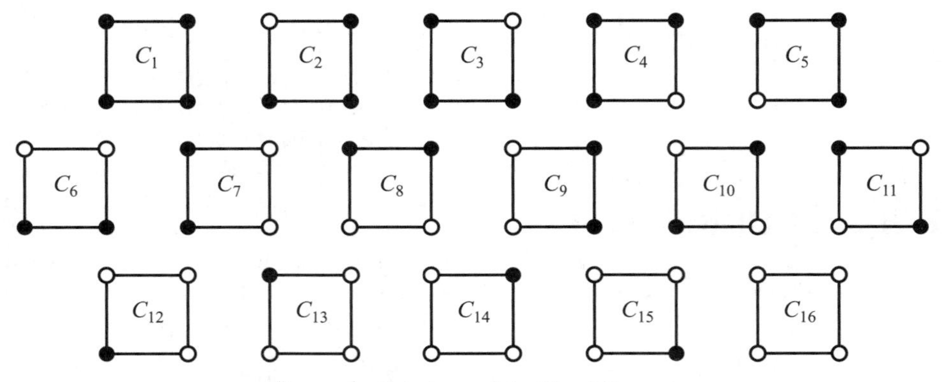

Figure 1. Colorings of the Fixed Square

elegant and powerful use of generating functions to be found anywhere in basic combinatorial mathematics. After several unrewarding attempts at a standard presentation of Pólya's formula, this author evolved a presentation of a slightly simplified form of the formula that retains the pedagogical aspects of the standard approach but replaces the usual complexities with a score of simple, suggestive exercises.

2. Basic concepts

We wish to find a formula for the pattern inventory of the number of ways to color the corners (or sides or faces) of an unoriented geometric figure choosing from n colors. We develop our theory around the sample problem of coloring corners of an unoriented (floating in three dimensions) square with black or white. By observation, we see that the pattern inventory for this problem is $b^4 + b^3w + 2b^2w^2 + bw^3 + w^4$ (the computations would be unduly messy if a less obvious sample problem were chosen). Since the inventory polynomial cannot be factored, we cannot sneak up upon a mathematical formulation by working backward. Thus we must start from scratch; but what is there to work with? The most obvious thing is the set of 16 colorings of the fixed square (see Figure 1). As a forerunner of the pattern inventory formula of the unoriented square, we observe that

$$(b + w)^4 = b^4 + 4b^3w + 6b^2w^2 + 4bw^3 + w^4$$

is the pattern inventory of the 2-colorings of the fixed square (each $(b + w)$ term corresponds to the color choices at a given corner). We can partition these 16 colorings into subsets of colorings that are all equivalent when the square is unoriented, i.e., partition into equivalence classes. As yet, the underlying equivalence relation has no mathematical foundation. Note, as a measure of the difficulty of our whole problem, that the equivalence classes vary greatly in size.

For information about this equivalence, we turn our attention to the motions π that map the square into itself, called the *symmetries* of the square (see Figure 2). Clearly, two colorings are equivalent when there exists a motion which acts to transform one into the other. The problem of finding all symmetries is often difficult, but with the square they can be enumerated as the ways of moving corner a to any other corner and simultaneously placing corner b on the clockwise or counterclockwise side of a. For any regular n-gon, it turns out that the motions are the set of all possible rotations and reflections (in Figure 2, π_1, π_2, π_3, π_4 are rotations of 0°, 90°, 180°, 270°, respectively, π_5, π_6, reflections about opposite sides, and π_7, π_8, reflections about opposite corners). The motions are naturally characterized by the way they permute the corners. Thus the motion π_3 can be defined as the corner permutation $(ac)(bd)$ (the representation of permutations as a product of disjoint cycles is central to our development). In permuting the corners, the motions induce permutations of the colorings of the corners. We let π_i^* denote the permutation of the 16 2-colorings induced by the motion π_i. Thus colorings C and C' are equivalent if there exists a motion π_i such that $\pi_i^*(C) = C'$.

It is well known that the symmetries of a geometric figure form a group under the associative operation of composition. Furthermore, it is exactly the properties of a group that induce the equivalence relation on the colorings, i.e., closure in the group makes the relation transitive, existence of an identity makes it reflexive, and existence of inverses makes it symmetric. (This correspondence is no coincidence—historically, the concept of a group

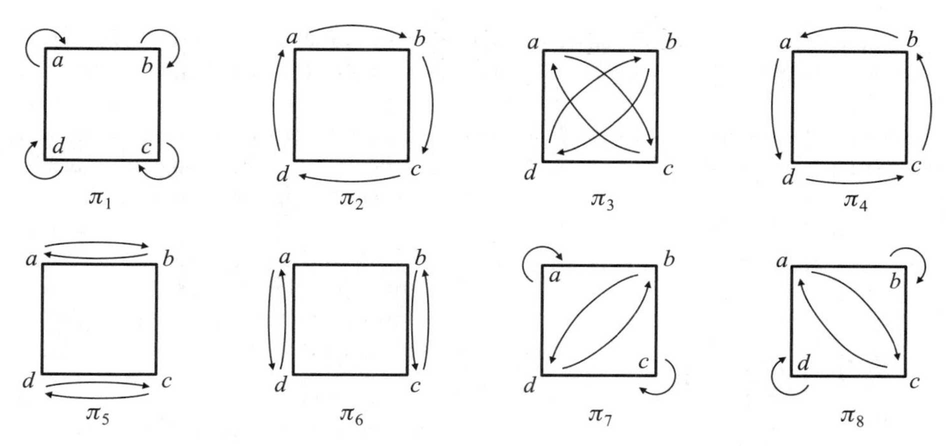

Figure 2. Symmetries of the Square; the use of arrows indicating the mapping at each corner makes it easier for the student to see the fundamental role which the cyclic decomposition will later play.

grew out of the study of various symmetries and their induced equivalences.)

Now we formalize the relation between π_i and π_i^*. We say a group G *acts* on a set S if each element $\pi \in G$ induces a permutation π_i^* on S, called the *action* of π_i, and if $(\pi_i \circ \pi_j)^* = \pi_i^* \circ \pi_j^*$. The second condition implies that the set of π_i^*'s also forms a group, but the π_i's are easier to work with than the π_i^*'s; for example in our current problem, a permutation on 4 elements is preferable to a permutation on 16 elements. Note that two different π's could have the same action—such a situation arises naturally when dealing with subsets of the 2-colorings; for example, all motions have the same action on the subset $\{C_1, C_{16}\}$. The following summarizes our discussion:

Theorem 1 *Suppose the group G (symmetries of the square) acts on a set S (2-colorings of the fixed square). Then this action induces a partition of S into equivalence classes with the equivalence relation defined by $C \sim C'$ $\Leftrightarrow$, there exists $\pi \in G$ whose action takes C to C' (i.e., $\pi^*(C) = C'$).*

Although the concept of the action of a group seems to portend much abstract mathematics, we use it only in the intuitive setting of symmetries of a square acting on colorings of the square. It will eliminate the need for always having a separate group of π^*'s.

3. Burnside's Lemma

Instead of seeking a formula for the pattern inventory for the different 2-colorings of the unoriented square, let us simplify the problem and ask for a formula for just the total number of different 2-colorings, i.e., the size of the pattern inventory. The following modified form of Burnside's Lemma answers this question in the generality of Theorem 1.

Theorem 2 (Burnside, 1897) *The number N of equivalence classes in the equivalence relation on S (the 2-colorings) induced by the action on S of a group G (symmetries of the square) is given by the formula:*

$$N = \sum_{\pi \in G} \psi(n) \tag{1}$$

where $\psi(\pi)$ is the number of elements of S left fixed by π^*, the action of π.

Although in general, $\psi(\pi)$ is only indirectly a function of π, in our square coloring problem, this dependence is quite obvious. Since the result of the theorem is basically combinatorial, it is best to extract all the algebra in the proof and place it in a separate lemma.

Lemma 1 *Let G be a group which acts on a set S. For $a \in S$, let $\eta(a)$ be the number of π's whose action fixes a (i.e., $\pi^*(a) = a$) and let S_a be the equivalence class containing a in the equivalence relation on S induced by G. Then $\sum_{b \in S_a} \eta(b) = |G|$.*

Proof. Let $T(a,b)$ be the set of distinct π's such that $\pi^*(a) = b$. Let $T(a,a) = \pi_1, \pi_2, \ldots, \pi_k$ and let $\pi_x(a) = b$. Then for $1 \le i \le k$, $\pi_i \circ \pi_x \in T(a,b)$; moreover

$$\pi_i \circ \pi_x = \pi_j \circ \pi_x \Rightarrow \pi_i \circ \pi_x \circ \pi_x^{-1} = \pi_j \circ \pi_x \circ \pi_x^{-1} \Rightarrow \pi_i = \pi_j,$$

i.e., $i = j$. Suppose $\pi_y \in T(a,b)$. Then $\pi_y \circ \pi_x^{-1} \in T(a,a)$ and so for some j, $\pi_j = \pi_y \circ \pi_x^{-1}$; then $\pi_j \circ \pi_x = \pi_y \circ \pi_x^{-1} \circ \pi_x = \pi_y$. Thus $T(a,b) = \{\pi_1 \circ \pi_x, \ldots, \pi_k \circ \pi_x\}$. We can also show that $T(b,a) = \{\pi^{-1} : \pi \in T(a,b)\}$. Thus

$$\eta(a) = |T(a,a)| = |T(a,b)| = |T(b,a)| = |T(b,b)| = \eta(b).$$

Since the sets $T(a,b)$, $b \in S_a$, exhaust G,

$$\sum_{b\in S_a} \eta(b) = \sum_{b\in S_a} |T(a,b)| = |G|.$$

(QED)

Note that we additionally have proved that $|T(a,a)| \cdot |S_a| = |G|$. Moreover, $T(a,a)$ is a subgroup and the $T(a,b)$, $b \in S_a$, are its cosets. Thus the equation $|T(a,a)| \cdot |S_a| = |G|$ is a special case of Lagrange's theorem for subgroups. Now we complete the proof of Theorem 2.

Proof of Theorem 2. By Lemma 1, $\sum_{b\in S_a} \eta(b) = |G|$. Then

$$\sum_{\text{all } c\in S} \eta(c) = \sum_{\substack{\text{each eq} \\ \text{class } S_a}} \left(\sum_{b\in S_a} \eta(b) \right)$$

$$= N \cdot |G|; \quad \text{or} \quad N = \frac{1}{|G|} \sum_{c\in S} \eta(c).$$

It remains to show that $\sum_{c\in S} \eta(c) = \sum_{\pi\in G} \psi(\pi)$. Define

$$\gamma(a,\pi) = \begin{cases} 1, & \pi^*(a) = a \\ 0, & \pi^*(a) \neq a \end{cases}.$$

So $\eta(c) = \sum_{\pi\in G} \gamma(c,\pi)$ and $\psi(\pi) = \sum_{c\in S} \gamma(c,\pi)$. Thus

$$\sum_{c\in S} \eta(c) = \sum_{c\in S} a \sum_{\pi\in G} \gamma(c,\pi)$$

$$= \sum_{\pi\in G} \sum_{c\in S} \gamma(c,\pi) = \sum_{\pi\in G} \psi(\pi).$$

Now (1) follows. In words, $\sum_{c\in S} \eta(c)$ and $\sum_{\pi\in G} \psi(\pi)$ both count up all instances of some c fixed by the action of some π: the first expression sums over the c's and the second over the π's.

(QED)

Several examples would now be in order. A typical application involves counting the number of different necklaces consisting of a circular string of 5 beads that can be formed by beads of 3 colors (S is the set of all 3^5 strings with beads in fixed positions and G is the group of different rotations of the string). At this point, the necklace question would be very difficult if the string had 6 beads (the problem is in determining the $\psi(\pi_i)$—if the size is not a prime number, it is extremely difficult to count directly the number of strings left fixed by some π_i^*).

4. Discovering the cycle index

Without further theory, we would find it very difficult to apply formula (1) to most coloring problems. When the set S in Theorem 2 is large, such as 3-colorings of a cube, it is impractical to determine the coloring permutations π_i^*, and hence to determine the $\psi(\pi_i)$, explicitly. We shall show that $\psi(\pi)$ can be calculated directly from the simpler π (thus justifying the notation $\psi(\pi)$). This approach leads to greatly shortened computations. The steps leading to the simplified calculation of $\psi(\pi)$ can be motivated by many mini-examples which allow the student to anticipate the final formula.

Let us apply formula (1) to the 2-colorings of the square. We shall determine the number of 2-colorings left fixed by π_i empirically. (For simplicity we shall write "left fixed by π_i" in place of "left fixed by the action of π_i"—most students naturally think of a coloring being left fixed by a motion, rather than by its induced action.) As we determine $\psi(\pi)$ for successive motions π_i in Figure 2, we look for a pattern that would enable us mathematically to predict which colorings (and implicitly, how many colorings) will be left fixed by the π_i. It is helpful to make a table of the π_i's and the colorings that they leave fixed; see columns i and ii in Figure 3 (columns iii and iv are used later). It will become evident as we collect this data that the number of colorings left fixed by π_i can be determined from the cyclic decomposition of π_i. The 0° rotation π_1 leaves each corner fixed and hence it leaves each coloring fixed. Next consider the 90°° motion π_2 which cyclicly permutes corners a, b, c, d (throughout the following discussion, it is assumed that for each motion the instructor asks the class which colorings are left fixed and why—and only afterwards gives his own explanation; this discussion can become dull and tedious for students if they are not involved in the development). In terms of corners, a coloring is

(iv) Inventory of Colorings Left Fixed by π_i^*	(i) Motion π_i	(ii) Colorings left Fixed by π_i^*	(iii) Cycle Structure Representation
$(b+w)^4 = b^4 + 4b^3w + 6b^2w^2 + 4bw^3 + w^4$	π_1	16: all colorings	x_1^4
$(b^4+w^4)^1 = b^4 + w^4$	π_2	2: C_1, C_{16}	x_4^1
$(b^2+w^2)^2 = b^4 + 2b^2w^2 + w^4$	π_3	4: $C_1, C_{10}, C_{11}, C_{16}$	x_2^2
$(b^4+w^4)^1 = b^4 + w^4$	π_4	2: C_1, C_{16}	x_4^1
$(b^2+w^2)^2 = b^4 + 2b^2w^2 + w^4$	π_5	4: C_1, C_6, C_8, C_{16}	x_2^2
$(b^2+w^2)^2 = b^4 + 2b^2w^2 + w^4$	π_6	4: C_1, C_7, C_9, C_{16}	x_2^2
$(b+w)^2(b^2+w^2) = b^4 + 2b^3w + 2b^2w^2 + 2bw^3 + w^4$	π_7	8: C_1, C_2, C_4, C_{10} $C_{11}, C_{13}, C_{15}, C_{16}$	$x_1^2x_2^1$
$(b+w)^2(b^2+w^2) = b^4 + 2b^3w + 2b^2w^2 + 2bw^3 + w^4$	π_8	8: C_1, C_3, C_5, C_{10} $C_{11}, C_{12}, C_{14}, C_{16}$	$x_1^2x_2^1$
Total $= 8b^4 + 8b^3w + 16b^2w^2 + 8bw^3 + 8w^4$			$P_G = \frac{1}{8}(x_1^4 + 2x_4^1 + 3x_2^2 + 2x_1^2x_2^1)$

Figure 3.

left fixed if each corner has the same color after the motion as it did before. Since π_2 takes a to b, then a coloring left fixed by π_2 must have the same color at a as at b. Similarly, such a coloring must have the same color at b as at c, the same color at c as at d, and the same at d as at a. Taken together these conditions imply that only the colorings of all white or all black corners, C_1 and C_{16}, are left fixed. In general, a coloring C will be left fixed by π_i if and only if for each corner s, the color at s is the same as the color at $\pi(s)$ (thus keeping the color at $\pi(s)$ unchanged). Next we consider the 180° motion π_3. Looking at the depiction of π_3 in Figure 2, we see that π_3 causes corners a and c to interchange and corners b and d to interchange. It follows that a coloring left fixed by the action of π_3 must have the same color at corners a and c and the same color at b and d (no further conditions are needed). With two color choices for a, c and with two color choices for b, d, we can construct $2 \cdot 2 = 4$ colorings which will be left fixed, namely C_1, C_{10}, C_{11}, C_{16}. The motion π_4 is similar to π_2 and only C_1 and C_{16} are left fixed. The horizontal flip π_5 interchanges a and b and interchanges c and d. Again we must have two pairs of like-colored corners when constructing the colorings that will be left fixed by π_5. Like π_3, the motion π_5 will leave $2 \cdot 2 = 4$ colorings fixed.

A pattern is becoming clear. The student should now be able quickly to predict that π_6 will also leave $2 \cdot 2 = 4$ colorings fixed, for again the motion interchanges two pairs of corners. Formally, an interchange is a cyclic permutation on 2 elements. All our enumeration of fixed colorings has been based on the fact that if π_i cyclicly permutes a subset of corners (that is, the corners form a cycle of π_i), then those corners must be the same color in any coloring left fixed. Thus all we need to do is get a disjoint-cycle representation of a π_i and use the number of cycles. For future use, let us also classify the cycles by their length. It will prove convenient to encode a motion's cycle information in an expression of the form

$$x_1^{n_1} \cdot x_2^{n_2} \cdots x_r^{n_r},$$

where n_i is the number of i-cycles, cycles of length i. This expression is called the *cycle structure representation* of a motion. Observe that

$$n_1 + 2n_2 + \cdots + kn_k = \left| \begin{matrix} \text{number of} \\ \text{corners} \end{matrix} \right|.$$

Now we add column iii to Figure 3 in which we write the cycle structure representation of each motion. So for π_2 and π_4 we enter x_4^1 and for π_3, π_5 and π_6 we enter x_2^2, but what about π_i? Previously, it sufficed to say that π_1 leaves all colorings fixed. Now it is time to point out that an element left fixed by a permutation is classified as a 1-cycle. Thus π_1 is really four 1-cycles. Its cycle structure representation is x_1^4. Since all the corners in each cycle must be one of the two colors in a fixed coloring, we predict *a posteriori* that π_1 leaves $2^4 = 16$ colorings fixed. For any π_i, the number of colorings left fixed will be given by setting each x_j equal to 2 (or, in general, the number of colors available) in the cycle structure representation of π_i. Finally we turn to π_7 and π_8. For each motion, the cycle structure representation is seen to be $x_1^2 \cdot x_2^1$ and thus for each we can find $2^2 \cdot 2 = 8$ colorings that are left fixed.

According to Theorem 2, we sum the numbers in column ii of Figure 3 and divide by 8 to obtain the number of 2-colorings of the unoriented square. There is a simpler way: algebraically sum the cycle structure representations of each motion, collecting like terms, and then divide by 8. From column iii, we obtain

$$\frac{1}{8}(x_1^4 + 2x_4^1 + 3x_2^2 + 2x_1^2x_2^1).$$

This expression is called the *cycle index* $P_G(x_1, x_2, \ldots, x_k)$ for our group G of motions. By setting each $x_i = 2$ in P_G, i.e., $P_G(2, 2, \ldots, 2)$, we get the same answer as before (before, the steps were reversed: we set $x_i = 2$ in each cycle structure representation and then added). Another advantage to the latter approach is that for any m, $P_G(m, m, \ldots, m)$ is the number of m-colorings of an unoriented square. The argument used to derive this coloring counting formula with the cycle index is valid for colorings of any set with associated symmetries (as an example, we can use the cycle index formula to rework a necklace counting problem for any number of colors). We shall state this as a corollary of Theorem 2. We emphasize again that although we were looking at properties of the π_i^*'s, the actions of π_i's, the new formula involves only the π_i's.

Corollary 2.1 *Let T be a set of elements and G be a group of permutations of T which acts to induce an equivalence relation on the colorings of T. Then the number of nonequivalent m-colorings of T is given by $P_G(m, m, \ldots, m)$.*

5. Discovering Pólya's formula

We are now ready to return to our original goal of a formula for the pattern inventory. The pattern inventory can be considered as giving the results of several formula (1)-type counting subproblems. In the case of 2-colorings of the unoriented square, we divide the colorings in Figure 1 into sets based on the number of black and white corners: $S_0 = \{C_1\}$, $S_1 = \{C_2, C_3, C_4, C_5\}$, $S_2 = \{C_6, C_7, C_8, C_9, C_{10}, C_{11}\}$, $S_3 = \{C_{12}, C_{13}, C_{14}, C_{15}\}$ and $S_4 = \{C_{16}\}$. In the pattern inventory, the coefficient of b^3w is the number of different colorings with three blacks. This is the result obtained by (1) for the counting problem where G, the set of motions of the square, acts on the set S_1. The coefficient of b^4 is the result when S_0 is the set. In general, the coefficient of $b^{4-i}w^i$ is the result of (1) when S_i is the set on which G acts. (One must check that G does indeed act on each S_i, i.e., show each motion induces an action (permutation) on S_i; as noted before, different π's may induce the same action on S_i.) Let us try to solve these five subproblems simultaneously. That is, we will list in a row the numbers of 2-colorings left fixed in each subproblem by the action of π_1, then below this row we will list the numbers left fixed in each subproblem by π_2, then by π_3, etc. (see column iv in Figure 7); then we total up the first column (the first number in each row), and divide by 8, total up the second column and divide by 8, etc. Since the action of π_1 leaves all C's fixed, the first row is 1, 4, 6, 4, 1. Let us put this data in the same form as the pattern inventory. We write: $b^4 + 4b^3w + 6b^2w^2 + 4bw^3 + w^4$; this is an inventory of fixed colorings. Note that to get the pattern inventory we do not have to add columns separately; we can simply add up all these inventories, collect terms, and divide by 8. For π_1, the inventory of fixed colorings is an inventory of all colorings. As observed at the beginning, this inventory is simply $(b + w)^4 = (b + w)(b + w)(b + w)(b + w)$, one $(b + w)$ for each corner. For π_2, the inventory is $b^4 + w^4$ (or suggestively $(b^4 + w^4)$). For π_3, we find by observation that the inventory is $b^4 + 2b^2w^2 + w^4$. This expression factors into $(b^2 + w^2)^2$. Just as we did before when counting the total number of colorings fixed by the action of some π, let us look for a "pattern" in the inventories of fixed colorings. Again the key to the "pattern" is the fact that in a coloring fixed by π, all corners in a cycle of π must have the same color. Since π_2 has one cycle involving all four corners, the possibilities are thus all corners black or all corners white; hence the inventory is $b^4 + w^4$. The motion π_3 has two 2-cycles (ac) and (bd). Each 2-cycle uses two blacks or two whites in a fixed coloring; hence the inventory of a cycle of size two is $b^2 + w^2$. The possibilities with two such cycles have the inventory $(b^2 + w^2)(b^2 + w^2)$. Thus the inventory of fixed colorings for π_i is a product of factors $(b^j + w^j)$, one factor for each cycle of π_i with j equal to the size of the cycle. So for each π_i, we need to know the number of cycles in π_i of each length. But this is exactly the information encoded in the cycle structure representation. Indeed, setting $x_j = (b^j + w^j)$ in the representation yields precisely the inventory of fixed colorings for π_i. By this method we compute the rest of the inventories of fixed colorings. For π_7 especially, the inventory should be checked against the list of colorings in column ii. As noted above, the pattern inventory is obtained by adding together the inventories of fixed colorings, collecting like-power terms and dividing by 8. In turn, the inventories are obtained by setting $x_j = (b^j + w^j)$ in each cycle structure representation and expanding the resulting expressions. As before in section 4, we get a more compact formula and save some computations by first adding together the cycle structure representations, then setting each $x_j = (b^j + w^j)$, and doing the polynomial algebra all at once. Again the first step in this approach (followed by division by 8) yields the cycle index $P_G(x_1, x_2, \ldots, x_k)$. Thus by setting $x_j = (b^j + w^j)$ in P_G, i.e., $P_G\big((b+w), (b^2+w^2), \ldots, (b^k+w^k)\big)$, we obtain the pattern inventory.

If three colors, black, white and green, were permitted, each cycle of size j would have an inventory of $(b^j + w^j + g^j)$ in a fixed coloring; so we would set $x_j = (b^j + w^j + g^j)$ in P_G. The preceding argument applies for any number of colors and any figure. In greater generality we have the following theorem (note again that only π_i's are involved in our formula):

Theorem 3 (Pólya's Enumeration Formula)
Let T be a set of elements and G be a group of permutations of T which acts to induce an equivalence relation on the colorings of T. The inventory of nonequivalent colorings on T using colors $c_1, c_2, \ldots, c_m$ is given by the generating function

$$P_G\left(\sum_{j=1}^{m} c_j, \sum_{j=1}^{m} c_j^2, \ldots, \sum_{j=1}^{m} c_j^r\right).$$

For a moment, let us return briefly to the problem of counting the total number of 2-colorings left fixed. This number is simply the sum of the coefficients in the pattern inventory. To sum coefficients, we simply set the indeterminants, b and w (and hence their powers), equal to 1, or equivalently, set $x_i = 2$ in P_G. If m colors were allowed, we would set $x_i = m$ in P_G, obtaining the same formula as in Corollary 1.

Finding the pattern inventories for the corner or face colorings of other unoriented regular m-gons is a good exercise, but practice has shown that only the most trivial 3-dimensional figures, such as a tetrahedron, are appropriate exercises. The cube can be treated in class with a tinker-toy model (without such a model, it is almost impossible to visualize the actions of all the motions). In doing such problems, there are two steps required before the formula can be used: first, a list of all the motions must be made (after some specific examples, the students should be able to deduce the size and contents of this list for an arbitrary regular m-gon); and second, the cycle structure representation of each motion must be determined (for a regular m-gon, this is equivalent to determining all the subgroups of the cyclic group of order m).

While the presentation here may seem fairly straightforward, it reads very differently from the standard treatment of Pólya's formula. The importance of our approach is that theory and precise mathematical statements have been avoided in Sections 4 and 5 in favor of the underlying ideas that motivated Pólya. For example, formally a coloring is a function $f : V \to C$ that assigns to each element $v \in V$ a color $c \in C$ and a permutation π leaves the coloring f fixed if $f\big(\pi^{-1}(v)\big) = f(v)$ for every v. For most students, such formalisms (and many are possible with this topic) make a derivation of this formula impossible to follow.

Logic from A to G

Paul R. Halmos

Mathematics Magazine
50 (1977), 5–11

Editors' Note: Paul R. Halmos was one of the great mathematical expositors of our time. Born in Hungary he came to the United States at age 13. He grew up in Chicago and took all of his degrees at the University of Illinois at Champaign-Urbana, where he received his PhD in 1938 under the direction of Joseph Doob. He then moved to the Institute for Advanced Study in Princeton where he was assistant to John von Neumann. He taught at Syracuse University, the University of Chicago (during the famous "Stone Age"), the University of Michigan at Ann Arbor, the University of Hawaii, the University of California at Santa Barbara, Indiana University at Bloomington, and Santa Clara University, from which he retired in 1995.

Many students learned their linear algebra from his *Finite-Dimensional Vector Spaces* (Van Nostrand, 1942; Springer, 1974). Others remember fondly his *Measure Theory* (Van Nostrand, 1950; Springer, 1974), *Naive Set Theory* (Van Nostrand, 1960; Springer, 1974), and his *Hilbert Space Problem Book* (Van Nostrand, 1967; Springer, 1974, 1982). The MAA has available in its catalogue his *I Want To Be a Mathematician/An Automathography* (1985), *Problems for Mathematicians Young and Old* (1991); *Linear Algebra Problem Book* (1995); and, with Steven Givant, *Logic as Algebra* (1998). In all he wrote over 120 papers, 25 reviews, and 17 books. Besides his writing on a great variety of topics—his fields of research included operator theory, ergodic theory, and algebraic logic—he often wrote about how to write. For the excellence of his own expository writing he was recognized by the MAA with its Lester R. Ford Award twice (1971, 1977), the Pólya Award (1983), and the Chauvenet Prize (1947), and by the American Mathematical Society with the Leroy P. Steele Prize for expository writing (1983). He died in 2006.

Here he informs us about logic with the expected Halmos charm and clarity.

What logic is and is not

Originally "logic" meant the same as "the laws of thought" and logicians studied the subject in the hope that they could discover better ways of thinking and surer ways of avoiding error than their forefathers knew, and in the hope that they could teach these arts to all mankind. Experience has shown, however, that this is a wild-goose chase. A normal healthy human being has built in him all the "laws of thought" anybody has ever invented, and there is nothing that logicians can teach him about thinking and avoiding error. This is not to say that he knows *how* he thinks and it is not to say that he never makes errors. The situation is analogous to the walking equipment all normal healthy human beings are born with. I don't know how I walk, but I do it. Sometimes I stumble. The laws of walking might be of interest to physiologists and physicists; all I want to do is to keep on walking.

The subject of mathematical logic, which is the subject of this paper, makes no pretense about discovering and teaching the laws of thought. It is called *mathematical* logic for two reasons. One reason is that it is concerned with the kind of activity that mathematicians engage in when they prove things. Mathematical logic studies the nature of a proof and tries to forecast in a general way all possible types of things that mathematicians ever will prove, and all that they never can. Another reason for calling the subject *mathematical* logic is that it itself is a part of mathematics. It attacks its subject in a mathematical way and proves things exactly the same way as do the other parts of mathematics whose methods it is concerned with. The situation is like that of a factory that makes machines whose purpose is to make machines. A worker at such an establishment is no different from a worker at any other machine factory, except perhaps that he understands a little better what makes machines in general work as they do (and a little less well how any particular machine works).

The history of logic, like the history of most subjects, developed all in the wrong order. If I told it to you straight, you'd get completely confused. I propose to tell you a little bit of the history of logic in the "right" order, that is, in the order in which it *should* have happened.

First Boole and propositions

According to my version of history it all begins with George Boole about 100 years ago. Boole's contribution was a systematic study of the innocuous little words that we all use every day to tie propositions together, the so-called propositional connectives. These connectives are, in English,

and, or, not, implies, and *if-and-only-if.*

It is convenient to have abbreviations for these words. The customary mathematical symbols used to abbreviate them are

$$\&,\ \vee,\ \neg,\ \Rightarrow,\quad \text{and}\quad \Leftrightarrow .$$

Thus, for example, if P and Q are propositions, then so also is $P\&Q$. If P says "the sun's shining" and Q says "it's hot," then $P\&Q$ says "the sun's shining and it's hot."

Next Peano and numbers

The next figure in our revised history is the 19th-century Italian mathematician Peano who studied the foundations of arithmetic. He studied, in particular, the properties of the basic numbers *zero* and *one*, the basic operations of *addition* and *multiplication*, and the basic relation of *equality*. The symbols for these things are, of course, known to everybody: they are

$$0,\ 1,\ +,\ \times,\quad \text{and}\quad = .$$

Thus, for example, the popular proposition that "two and two make four" can be written in the unabbreviated form

$$(1+1)+(1+1)=1+(1+(1+1)).$$

Then Aristotle and a quantifier

Immediately after Peano comes Aristotle, who lived well over 2000 years ago, and to whom we are indebted for the first analysis of the crucial words *all* and *some*. (Incidentally, we have now also reached the beginning of the alphabet: the "A" in "Logic from A to G" is, of course, Aristotle.) The abbreviations are $\forall$ and $\exists$. To illustrate the use of these symbols, consider a well-known sentence such as "He who hesitates is lost." In pedantic mathematese this can be said as follows: "For all X, if X hesitates, then X is lost." Using $H(X)$ and $L(X)$ as abbreviations for "X hesitates" and "X is lost," we may write

$$(\forall X)(H(X) \Rightarrow L(X)).$$

If we doubt this assertion, if, that is, we are inclined to believe it quite likely that hesitation is possible without subsequent perdition, we may express our skepticism in the form

$$(\exists X)(H(X)\&(\neg L(X)).$$

Finally Frege and many quantifiers

An essential part of such Aristotelean abbreviations is the use of auxiliary symbols such as the "X" above; symbols such as that play the role of pronouns in the language of logic. Recall that, in the example, "X" took the place of "he." Symbols used in this way are called *individual variables*, and the next historical figure to be mentioned is the first one to face them courageously. His name is Frege, and he too lived in the 19th century. His role for us today is to emphasize that *one* variable, that is one pronoun, is much too meager equipment for most scientific and mathematical purposes. Thus, for instance, if we want to express the very modest assertion that there are more than two numbers, that is, that there exist at least three different ones, we would do it this way:

$$(\exists X)(\exists Y)(\exists Z)$$
$$(\neg(X=Y)\&\neg(Y=Z)\&\neg(X=Z)).$$

For this purpose, pretty clearly, we need at least three variables. To say that there are more than ten numbers, we need at least eleven variables. To do any non-trivial amount of mathematics, we need an (at least potentially) infinite supply of variables. An efficient way of getting such a supply (since ordinary alphabets are much too finite) is to use one letter, say "X", and one extra symbol, say a dash $'$, and then use in the role of variables the symbols

$$X, X', X'', X''', X'''',\ \text{etc.}$$

In addition to all the symbols I have mentioned so far, there are two others that I have already used, and compulsive honesty now forces me to adjoin them to the list. The symbols I mean are the *left parenthesis* and *right parenthesis* denoted, of course, by

(

and

).

How many symbols?

It turns out that a considerably large body of mathematics, namely, all arithmetic, can be expressed by means of the symbols I have listed so far. (Thus, for instance, Euler's celebrated theorem that every positive integer is the sum of four squares can be written as follows:

$$(\forall X)(\exists X')(\exists X'')(\exists X''')(\exists X'''')(X = (X' \times X') + ((X'' \times X'') + ((X''' \times X''') + (X'''' \times X''''))))).)$$

In fact certain savings can be made; some symbols are naturally and easily definable in terms of others. Our list can easily be cut down to an even dozen, namely to

$$\& \neg + \times 0\ 1 \exists X\ '\ = (\,).$$

To recapture $\vee$ observe that $P \vee Q$ is the same as $\neg(\neg P \& \neg Q)$. To recapture $\Rightarrow$ observe that $P \Rightarrow Q$ is the same as $\neg P \vee Q$. To get $\forall$ note that $(\forall X)P(X)$ is the same as $\neg(\exists X)(\neg P(X))$. In words: to say that everybody hates spinach is the same as to deny that there is somebody who loves it. There is some technical advantage in such abbreviations, but there's no sense in tying ourselves down. Whenever convenient I'll still use $\vee$ and $\Rightarrow$ and $\forall$, and I'll even use Y, Z, U, V, and such, as variables. The proper way to interpret these irregularities is clear by now: replace $\vee$, $\Rightarrow$, $\forall$, and the like by their definitions, and replace Y, Z, U, V, and such, by X', X'', X''', X'''', etc.

It might be fun to make a couple of side observations here. One of them is that what we are doing for elementary arithmetic can just as well be done for all extant mathematics. The technical machinery, that is the symbolism and the rules governing it, doesn't even get more complicated: the only difference is that we have to think a little harder. Since that is clearly undesirable, I am going to stick to elementary arithmetic. The second observation is that there is nothing magic about the number twelve. A dozen symbols are adequate for arithmetic, and in fact for all mathematics (though we might have to find a different dozen for that). A little care and stinginess, however, can easily reduce the dozen to still fewer, and the best possible result that you would hope for in your wildest dreams is true, namely, that two symbols are enough. A complete exposition of all mathematics written with, say, the dots and dashes of the Morse code wouldn't make particularly thrilling reading, but in principle it is perfectly feasible.

The mechanical mathematician

Let us now set about designing the mechanical mathematician to end all mathematicians. The virtues of this imaginary machine are purely conceptual; it will not be claimed that there is any practical advantage in building it. It will never replace the live mathematician.

The first step is to build into the machine a typewriter ribbon, a dozen typewriter keys (one bearing each one of the dozen basic symbols), and a potentially infinitely long piece of typing paper. The idea is that when a certain crucial button is pushed the machine is to start printing, and to print one after another all things that it could conceivably ever print. One way to program this would be to arrange the dozen symbols in an arbitrary order (call it alphabetical), and then to direct the machine to print the first twelve symbols ("letters") of the alphabet, then to print the 144 two-letter "words," next the 1728 three-letter "words," and so on *ad infinitum*. A machine so designed would print a lot of nonsense (for example

$$= = = \ (((0 + '),$$

and it would print a lot of lies (for example

$$(\exists X)((0 \times X) = 1)),$$

but it would also, sooner or later, get around to printing all arithmetical statements.

Insist on grammar

In order to make the machine more like a flesh-and-blood mathematician, the next thing we must do is to arrange matters so that the output of the machine, while possibly false, should never be arrant nonsense. This is in principle quite easy. There is no point in listing here all the restrictions to which the machine must be subjected, but let us look at a few samples. First, teach the machine some "grammar." Let us say that a noun is any sequence of symbols built up from 0's and 1's by successively sandwiching $+$'s and $\times$'s between them and separating the results by the appropriate parentheses. Examples:

$$0,$$
$$1 + 1,$$
$$((1 + 1) \times (1 + (1 + (1 + 1))))$$

are nouns in this sense. Let us, similarly, say that a pronoun is anything obtained from X by successively appending dashes. Thus the pronouns are

$$X,\ X',\ X'', X''',\ \text{etc.}$$

To go on in the same spirit, a *substantive* shall be a string of nouns and pronouns put together by means of addition and multiplication (and the auxiliary use of parentheses) the same way as nouns were originally put together from 0's and 1's. It is not at all difficult to design the machine so that it is able to recognize a substantive when it sees one. Once that is done, we can direct the machine as follows: "Start printing substantives (in some systematic order). After printing one, print an equal sign and then print another substantive. Learn to recognize the strings you have printed in this manner (substantive, equal, substantive), and call each such string a *clause*." Our machine can now print sensible clauses, and recognize them as such. It is just one step from here to teach the machine to print (and to recognize) compound *phrases*. The idea is to put clauses together by suitably restricted use of the logical operators *or*, *not*, and *some*. A machine that prints only phrases is thus within conceptual reach. Such a machine might still print incomplete phrases (for example $X = 0$) and lies (for example $1 = 0$), but it will no longer print gibberish.

The incidental mention of "incomplete phrases" suggests another look at what we want the mechanical mathematician to do. An incomplete phrase, in the sense I want that expression to have now, is something like "he is lost." The natural reaction upon seeing or hearing those words is to ask "Who is lost?" Similarly, "$X = 0$" should evoke the reaction "What is X?". Phrases with "dangling pronouns" like the "he" in "he is lost" and like the "X" in "$X = 0$" are the ones I mean to call incomplete here. A phrase that has no such dangling pronouns shall be called a *sentence*. The next step in perfecting the mechanical mathematician is to teach it to recognize a sentence when it sees one, and to instill in it an inhibition that permits it to print complete sentences only (that is, no incomplete phrases). Now if the button is pushed, the machine will start printing sensible sentences. It will never present the machine operator with "$X = 0$". It might say something uninteresting (for example $(\exists X)(X = 0)$) or false (for example $(\forall X)(X = 0)$), but at any rate it will always say something.

Establish the axioms,

The machine now knows how to talk; the next step is to teach it how to prove things. Neither the machine nor its flesh-and-blood prototype, however, can prove something from nothing. A live mathematician has his axioms; the machine must have built into it certain sentences that it is instructed to start from. Let us call those sentences *axioms*. Once again I shall not define here exactly which sentences are to be the axioms of arithmetic, but I shall indicate some examples. (The complete definition of "axiom" for elementary arithmetic is not even very long or complicated. This is just not the time and place to get technical.) Very well then: the machine might be taught that whenever P and Q are sentences, then the sentence

(A) $$P \Rightarrow (P \vee Q)$$

shall be printed in red ink. The idea, of course, is that every such sentence is to be regarded as an axiom. Similarly, we might say that if $P(X)$ and $Q(X)$ are phrases (containing the dangling pronoun "X", but no other), then the sentence

(B) $$(\exists X)(P(X) \vee Q(X)) \Rightarrow (\exists X)P(X) \vee (\exists X)Q(X)$$

shall be printed in red. Final sample: a sentence such as

(C) $$(\forall X)(\forall Y)(X + Y = Y + X)$$

might be an axiom of elementary arithmetic. (Interpretive examples: (A) If it's hot, then either the sun's shining or else it's hot or both. (B) If somebody likes either spinach or broccoli, then either somebody likes spinach or somebody likes broccoli. (C) $2 + 3 = 3 + 2$.)

The main point is this: in some sensible and systematic manner a certain collection of sentences is singled out from among all the sentences that the machine can print, and the machine is directed to print those special sentences in red. The red sentences are called the axioms for the machine.

Program the procedure

I said before that nobody can prove something from nothing, and, for that reason, I endowed the mechanical mathematician with some starting sentences. But it is just as true that nobody can prove something *with* nothing. The machine can print black and red sentences, and, if the axioms were chosen in accordance with the wishes of a reasonable being, the red sentences (the axioms) will be true. The machine might nevertheless still print a lot of false stuff, and it still has no means of producing new true sentences out of old ones.

The process of educating the machine has now reached its final step. In that step the machine must learn to recognize certain patterns of sentences and is rewarded by being permitted to use more red ink. The total list of such *rules of procedure* is not long, and the list of two examples from among them, that I propose actually to give, is even shorter. Possible rule number one: if P is a red sentence, and if $P \Rightarrow Q$ is a red sentence, then print Q in red. (Interpretation: if "The sun's shining" is an axiom or has already been proved, and if the same is true of "If the sun's shining, then it's hot", then we may consider to have proved "It's hot.") Possible rule number two: if an incomplete phrase $P(X)$ containing the dangling pronoun "X" (and no others) is such that the sentence $P(0)$ is red ($P(0)$ is the sentence obtained from $P(X)$ by substituting 0 for X), then print the sentence $(\exists X)P(X)$ in red. (Interpretation: if we substitute 0 for X in "$1 + X = 1$", we obtain "$1 + 0 = 1$". If this sentence is an axiom or has already been proved, then we may consider to have proved "$1 + X = 1$ for some X".)

With the last modification we can now rest on our laurels. We may if we wish change the internal design of the machine so that it no longer deigns to print anything but red sentences. When the button is pushed, the machine starts printing axioms, and, by means of the rules of procedure, it goes on to print *theorems* that it can "derive" from the axioms. It can do this in some systematic (say, alphabetic) order.

The millennium is come; the mechanical mathematician is complete. Push one button and sit back. One after another the theorems of elementary arithmetic will appear on the tape. If you wait long enough, sooner or later you will see all theorems pass before your eyes. The machine never talks nonsense, and the machine never tells lies. If you find the machine somewhat boring, possibly repetitious, and much much too slow for your merely human patience, that is not its fault.

Are there contradictions?

We have incorporated into the machine all that we ourselves, its builders, know about elementary arithmetic. The internal workings of the machine *are* elementary arithmetic. The external theory of the machine, its design and its structural properties, are part of another discipline often called *metamathematics* (or, in this case, metaarithmetic). The following question is typical of the ones that can be asked in metamathematics: "Will the machine ever print both a sentence P and its negation $\neg P$?" Should it ever do so, we would probably express our displeasure at this state of affairs by saying that elementary arithmetic is inconsistent. Fortunately this is not so: arithmetic is consistent. The proof of consistency depends on a very sophisticated, definitely non-elementary study of the structure of the "machine" we've been describing. The study is "non-elementary" in several senses of the word. In the most precise technical sense the fact is that the proof that the machine is consistent is not one of the theorems that the machine itself is able to prove. To say it again: the machine will never contradict itself, but it is not able to prove that it won't.

Can everything be proved?

Another interesting metamathematical question is this: "Is the machine complete, in the sense that it either proves or disproves every sentence of elementary arithmetic?" I've already told you the answer to this question, but the point will bear repetition. As the machine now stands, everything it prints it proves. If I am interested in a particular arithmetical sentence, I may write that sentence on a piece of paper and then, after setting the machine into operation, compare each successive output of the machine with my prepared slip. If the slip says P and if, at some stage, the machine also says P, I retire victorious: my P is proved. If the slip says P, and if, at some stage, the machine says $\neg P$, I retire in ignominy: my P is disproved. Isn't there, however, a third possibility? Couldn't it happen that the machine will never print P and neither will it ever print $\neg P$? Couldn't it happen that the machine will never decide the P versus $\neg P$ controversy? It could, and it does, and we have thereby reached the end of the alphabet. G is for Gödel, the brilliant twentieth-century logician. In the early 1930's Gödel proved, by a delicate and ingenious analysis of the arithmetic machine, that there are sentences (many of them) that the machine never decides. His proof is quite explicit: he gives a complete set of directions for writing down an undecidable sentence. The proof that the sentence obtained by following his directions is undecidable depends on a detailed examination of those very directions themselves. There is nothing wrong, there is no paradox, and it all hangs together. The fact that no one has ever bothered actually to write down Gödel's undecidable sentence is, once again, the fault of human impatience and the brevity of human life.

I said when I raised the question of completeness that I had already answered it. Indeed, consider a sentence (written out formally in the dozen formal symbols of arithmetic) that says that arithmetic is consistent. It is not at all clear that the apparently meager formal apparatus of arithmetic is capable of expressing such a sentence; it is one of Gödel's accomplishments to have shown that it is capable of doing so. If we take that for granted, and if we call one such sentence P, then what we know is that P is not provable in elementary arithmetic. What about $\neg P$? Well, clearly, $\neg P$ cannot be provable either. Reason: everything that is provable is true, as we already know from our earlier thoughts on the subject. (This depends on the fact that arithmetic is consistent.) The sentence $\neg P$ is certainly not true. (Recall that $\neg P$ denies the consistency of arithmetic.) Conclusion: neither P nor $\neg P$ is provable. (Note: of the two, P is the one that is true.)

There is more

That is the end of the road, for us, for now. It is by no means the end of the road for mathematical logic. What I've been reporting to you happened in the 1930's, and science has not stood still since then. Gödel himself has contributed several other striking results to our knowledge of formal logic. Many others have taken up the field and opened up unexpected applications and complications. Who, for instance, could have expected formal mathematical logic to turn out to be one of the most important tools in the design of honest-to-goodness circuits-and-printouts electronic computing machines? Mathematical logic is alive and well; much remains to be done; it'll be a long time before anyone can describe mathematical logic from A to Z.

Tiling the Plane with Congruent Pentagons

Doris Schattschneider

Mathematics Magazine
51 (1978), 29–44

Editors' Note: Doris Schattschneider received her PhD from Yale University in 1966. She has written extensively on polyhedra, tiling, symmetry, and the connections between geometry and art, especially the work of M. C. Escher. Her books on Escher are *M.C. Escher Kaleidocycles,* with Wallace Walker (Pomegranate, 1987), *M. C. Escher's Legacy: A Centennial Celebration*, with Michele Emmer, (Springer, 2003), and *M. C. Escher: Visions of Symmetry* (new edition, Harry Abrams, 2004).

From 1981 to 1985 Professor Schattschneider served as editor of *Mathematics Magazine*. In 1993 she received the MAA's Award for Distinguished College or University Teaching of Mathematics. For the article we see here she was awarded the MAA's Allendoerfer Award in 1979.

She is Professor Emerita of Mathematics at Moravian College, where she joined the faculty in 1968 after appointments at Northwestern University and the University of Illinois, Chicago.

The importance of recreational mathematics and the involvement of amateur mathematicians has been dramatically demonstrated recently in connection with the problem of tiling the plane with congruent pentagons. The problem is to describe completely all convex pentagons whose congruent images will tile the plane (without overlaps or gaps). The problem was thought to have been solved by R. B. Kershner, who announced his results in 1968 [18], [19]. In July, 1975, Kershner's article was the main topic of Martin Gardner's column, "Mathematical Games" in *Scientific American*. Inspired by the challenge of the problem, at least two readers attempted their own tilings with pentagons and each discovered pentagons missing from Kershner's list. New interest in the problem has been aroused and both amateur and professional mathematicians are currently working on its solution.

Tilings by convex polygons

A polygon is said **to tile** the plane if its congruent images cover the plane without gaps or overlaps. The pattern formed in this manner is called a **tiling** of the plane, and the congruent polygons are called its tiles. A **vertex of the tiling** is a point at which 3 or more tiles meet. It is well known that any triangle or any quadrilateral can tile the plane. In fact, any single triangle or quadrilateral can be used as a "generating" tile in a tiling of the plane which is **tile-transitive** (isohedral). This simply means that the generating tile can be mapped onto any other tile by an isometry of the tiling. The translations, rotations, reflections, and glide-reflections which map a tiling onto itself make up the **symmetry group** of the tiling. Thus, in terms of group theory, a tile-transitive tiling is one whose symmetry group acts transitively on the tiles.

Also, for an arbitrary triangle or quadrilateral there always exists a tiling which in addition is **edge-to-edge**. This means that for any two tiles, exactly one of the following holds: (i) they have no points in common, (ii) they have exactly one point in common, which is a vertex of each tile (such a point is also a vertex of the tiling), (iii) their intersection is an edge of each tile. Typical tilings of triangles and of quadrilaterals having these properties are shown in [5], [25].

If we ask the natural question, "Do convex polygons of five or more sides tile the plane?", the obvious general answer is "Not always." It is clear that not every convex pentagon tiles the plane—a regular pentagon is a prime example. However, regular

1. In a given tiling, pentagons marked with a dot are oppositely congruent to those which are unmarked, i.e., the plain tiles are 'face up', and the marked tiles are 'face down'.

2. In each tiling there is outlined a minimal block of one or more pentagons which generates the tiling when acted on by the symmetry group of the tiling.

3. Angles A, B, C, D, E of one pentagonal tile are identified in each tiling for use with Tables 1, 2, 3. Sides a, b, c, d, e of that tile correspond to the following labeling:

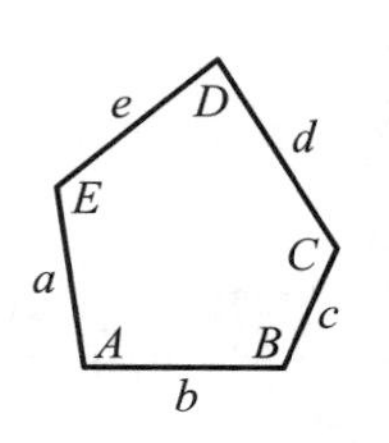

Key to all diagrams

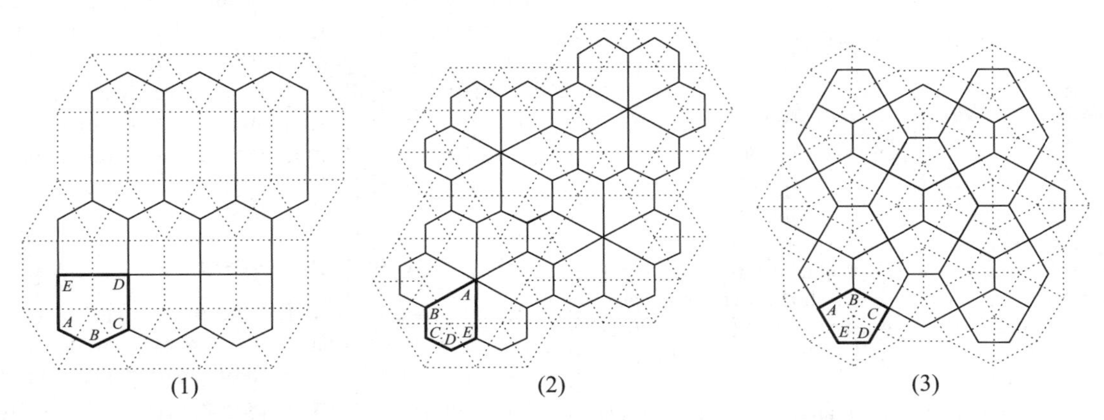

Figure 1. The three pentagonal tilings which are duals of Archimedean tilings. The underlying Archimedean tiling is shown in dotted outline.

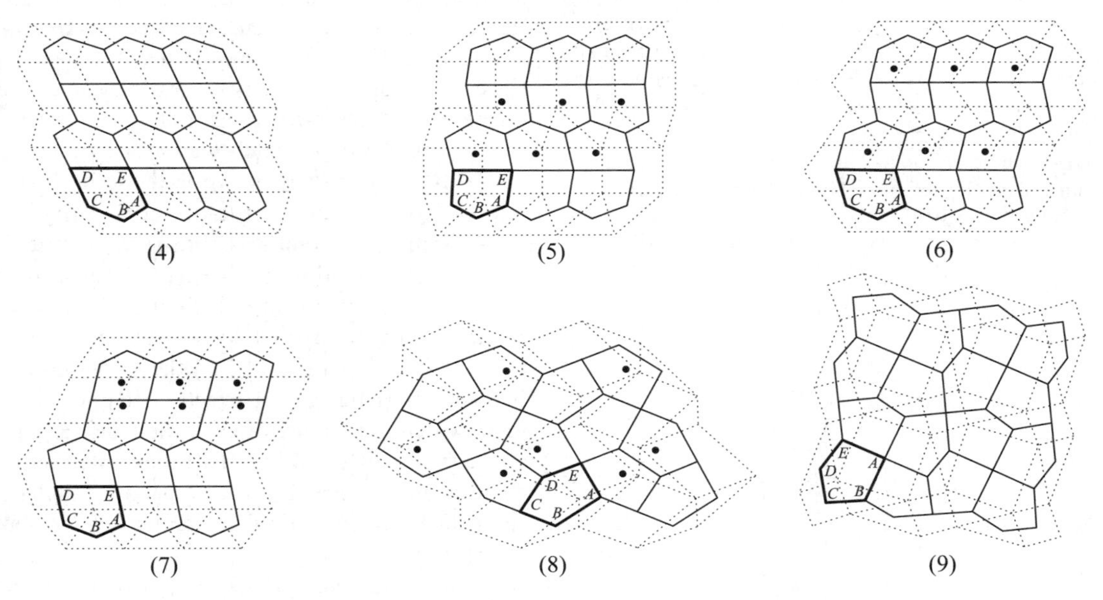

Figure 2. Tile-transitive, edge-to-edge tilings by pentagons, obtained as duals of vertex-transitive tilings (shown in dotted outline). Only in the case of (6) is a non-convex tile necessary in the underlying tiling. Tiling (8) is discussed in [4].

hexagons do tile the plane (how many times have you seen this pattern on a 1930's bathroom floor?), but not all hexagons tile. A complete description of all hexagons which do tile the plane was discovered independently by several mathematicians and is discussed in [2], [5], [14], [18], [19], [29]. It can also be demonstrated that no convex polygon of more than six sides can tile the plane. Thus, the problem of describing all convex pentagons which tile the plane is the only unanswered part of our question. In what follows, we make several observations related to the pentagonal tiling problem and report on the most recent contributions to its solution.

Discovering tilings by pentagons

Three of the oldest known pentagonal tilings are shown in Figure 1. As Martin Gardner observed in [5], they possess "unusual symmetry." This symmetry is no accident, for these three tilings are the duals of the only three Archimedean tilings whose vertices are of valence 5. The underlying Archimedean tilings are shown in dotted outline. Tiling (3) of Figure 1 has special aesthetic appeal. It is said to appear as street paving in Cairo; it is the cover illustration for Coxeter's *Regular Complex Polytopes*, and was a favorite pattern of the Dutch artist, M. C. Escher. Escher's sketchbooks reveal that this tiling is the unobtrusive geometric network which underlies his beautiful "shells and starfish" pattern. He also chose this pentagonal tiling as the bold network of a periodic design which appears as a fragment in his 700 cm. long print "Metamorphosis II."

Tiling (3) can also be obtained in several other ways. Perhaps most obviously it is a grid of pentagons which is formed when two hexagonal tilings are superimposed at right angles to each other. F. Haag noted that this tiling can also be obtained by joining points of tangency in a circle packing of the plane [12]. It can also be obtained by dissecting a square into four congruent quadrilaterals and then joining the dissected squares together [27]. The importance of these observations is that by generalizing these techniques, other pentagonal tilings can be discovered.

The three Archimedean tilings which have as duals the pentagonal tilings in Figure 1 are vertex transitive tilings. An edge-to-edge tiling by polygons is called **vertex transitive** (isogonal) if the symmetry group of the tiling is transitive on the vertices of the tiling. Figure 2 shows six other edge-to-edge pentagonal tilings that arise as duals of vertex-transitive tilings of the plane. Recently, B. Grünbaum and G. C. Shephard showed that the nine tilings of Figures 1 and 2 are the only distinct "types" of pentagonal tilings which are edge-to-edge and tile-transitive. Roughly speaking, two "types" of tilings will differ if they have different symmetry groups or if the relationship of tiles to their adjacent tiles differs. Details are given in [6]. In addition, these mathematicians have classified all vertex-transitive tilings, and their list shows that no such tiling by convex pentagons is possible [7].

If we begin with a tiling by congruent convex hexagons, then pentagonal tilings can arise in two different ways. First, it may be possible to superimpose the hexagonal tiling on itself so as to pro-

duce a tiling by congruent pentagons. Tilings (8) and (9) of Figure 2 can arise in this way. Also, beginning with a hexagonal tiling, it may be possible to dissect each hexagon into two or more congruent pentagons, thus producing a pentagonal tiling. Many tilings of Figures 1, 2, and 3 can be viewed in this manner. Three other examples of such tilings given in Figure 4 also serve to illustrate other properties of pentagonal tilings that can occur. Note that (24) is tile-transitive but not edge-to-edge, tiling (25) is edge-to-edge but not tile-transitive, and tiling (26) is neither edge-to-edge nor tile-transitive.

Finally, experimentation in fitting pieces together or adding or removing lines from other geometric tilings can lead to the discovery of pentagonal tilings. Figure 5 illustrates this with two tilings by a simple "house" shape pentagon.

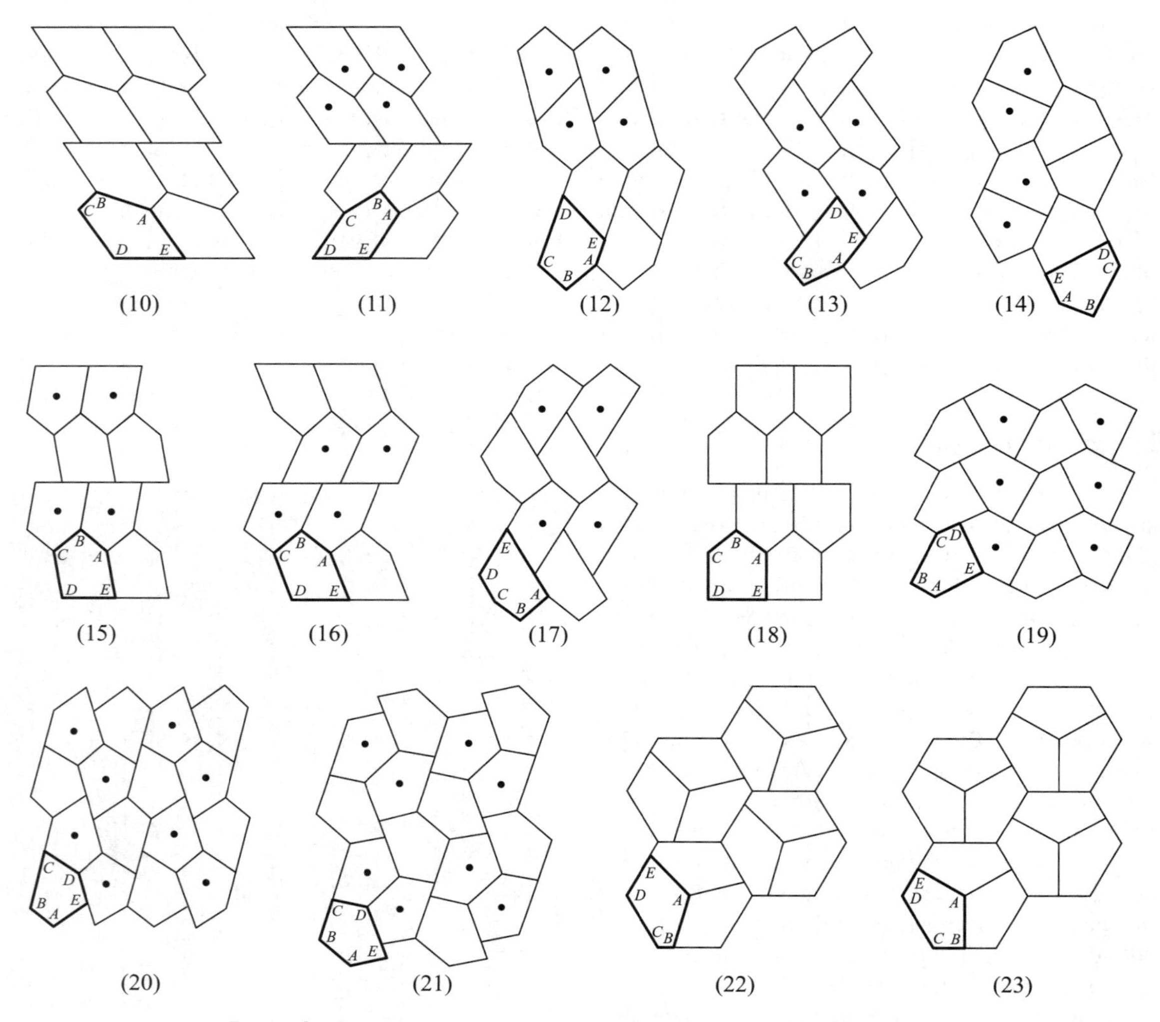

Figure 3. Tile-transitive tilings by pentagons that are not edge-to-edge.

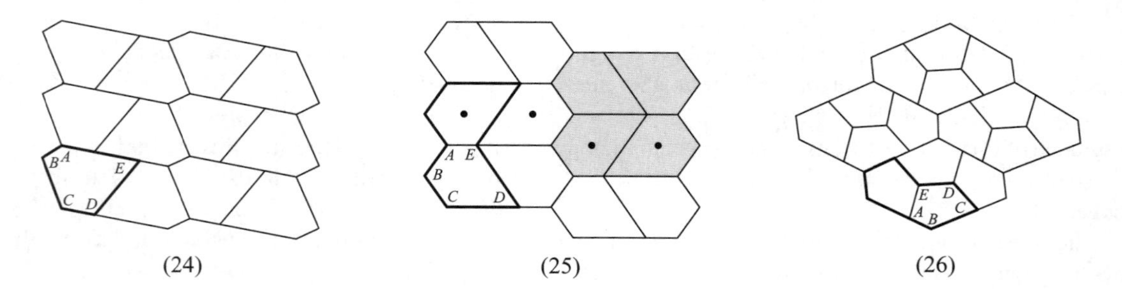

Figure 4. Tilings by pentagons obtained by dissecting hexagonal tilings. Tiling (24) is tile-transitive, but in (25) and (26), it is impossible to map one pentagon in an outlined block onto the other pentagon in that block by a symmetry of the tiling. The shaded portion of (25) shows a "double hexagon" which has been dissected into 4 congruent pentagons.

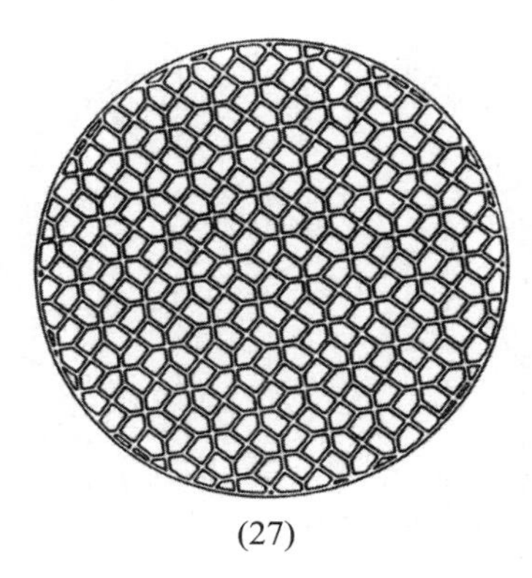
(27)

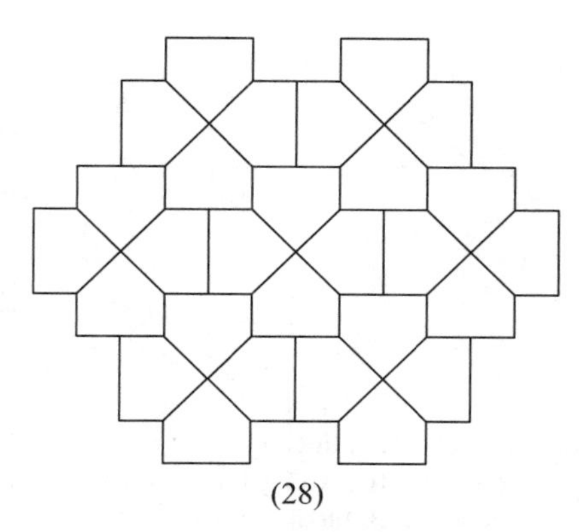
(28)

Figure 5. A "house shape" pentagon tiles in a variety of ways. Tiling (27) is a Chinese lattice design (*Chinese Lattice Designs*, D. S. Dye, Dover, 1974, p. 340); (28) is a familiar geometric pattern of interlocked St. Andrews crosses. Other patterns are (1), (14), and (18).

Methodical attacks

Pentagonal tilings appear as illustrations in several early papers which explore the general problem of classifying plane repeating patterns (especially [11]), but it appears that the first methodical attack on classifying pentagons which tile the plane was done in 1918 by K. Reinhardt in his doctoral dissertation at the University of Frankfurt [29]. He discovered five distinct types of pentagons, each of which tile the plane. More precisely, he stated five different sets of conditions on angles and sides of a pentagon such that each set of conditions is sufficient to ensure that (i) a pentagon fulfilling these conditions exists, and (ii) at least one tiling of the plane by such a pentagon exists. Each of these five sets of conditions defines a *type* of pentagon; pentagons are considered to be of different types only if they do not satisfy the same set of conditions. Many distinct tilings can exist for pentagons of a given type. Reinhardt no doubt hoped that his five types constituted a complete solution to the problem, but he was unable to show that a tiling pentagon was necessarily one of these types.

Each of the five types described by Reinhardt (called types 1–5 in Table 1) can generate a tile-transitive tiling of the plane. His thesis completely settled the problem of describing all convex hexagons which tile the plane—there are just three types, and each of these can generate a tile-transitive tiling. The fact that if a hexagon can tile the plane at all, then that same hexagon can generate a tile-transitive tiling, considerably simplifies the hexagonal tiling problem. Unfortunately, this result is not true for pentagons and this may be the reason that Reinhardt did not pursue the problem further by trying to find other types of pentagons which tile.

In [14], pp. 81–91, Heesch and Kienzle methodically explore the problem of describing types of pentagons which can generate tile-transitive tilings and affirm that Reinhardt's five types are the only convex ones possible. Most recently, B. Grünbaum and G. C. Shephard, using a classification scheme for tilings, found exactly 81 "types" of tile-transitive tilings of the plane by quite general "tiles" [6], and, as a result, have not only confirmed these earlier results, but also have shown that there are exactly twenty-four distinct "types" of tile-transitive tilings by pentagons according to their classification scheme [9]. Nine of these are edge-to-edge (tilings (1)–(9)); the other fifteen are illustrated by tilings (10)–(24). Table 1 summarizes information on these tilings.

In 1968, R. B. Kershner of Johns Hopkins University announced that there are 8 types of pentagons that tile the plane [18], [19]. He devised a method different from Reinhardt's for classifying pentagons that can tile, and this scheme left out any assumptions of tile transitivity for associated tilings. Happily, his search yielded three classes of pentagons not on Reinhardt's list. His three new pentagonal tiles (types 6, 7, and 8 on Table 2) each have an associated tiling which is edge-to-edge and not tile-transitive (Figure 6).

Although Kershner's claim—that these three additional types of pentagons completed the list of pentagons which tile—was later shown false, still his discovery was important. It confirmed that there are pentagons whose associated tilings cannot be tile-transitive, thereby answering a question raised by J. Milnor in [24, p. 499]. Kershner's search for pentagons which tile had been a methodical one; yet still, in his own words, he "made at least 2 errors, one of commission, and one of omission."

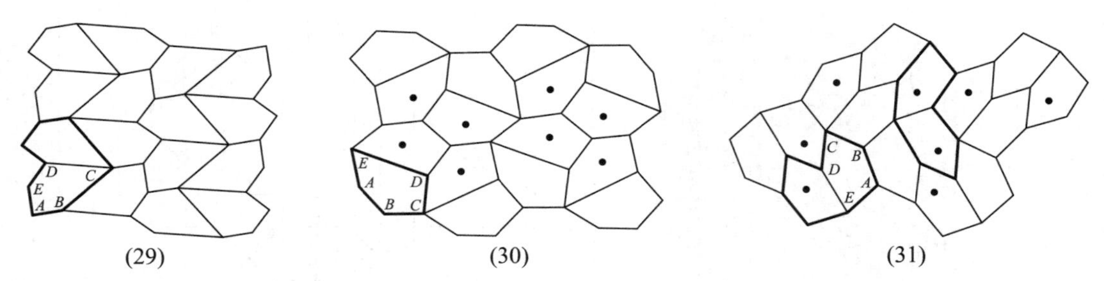

Figure 6. Tilings for each of the three types of pentagons discovered by R. B. Kershner. Each tiling is 2-block transitive, and there exists no tile-transitive tiling for these types of pentagons. Two non-congruent 2-blocks are outlined in tiling (31); each of these generates a block-transitive tiling.

Type, with characterizing conditions	Tile-transitive tilings	Additional conditions necessary for tiling	International notation for symmetry group of tiling (see [31])	Isohedral type (from [6])	Type from [9]	Other tilings shown
1. $D + E = \pi$	(24)		$p2$	$IH4$	$P_5 - 4$	
	(10)	$a = d$	$p2$	$IH4$	$P_5 - 5$	
	(11)	$a = d$	pgg	$IH5$	$P_5 - 8$	
	(4)	$a = d$	$p2$	$IH23$	$P_5 - 18$	
	(5)	$a = d$	pmg	$IH24$	$P_5 - 19$	
	(12)	$b = c$	pgg	$IH5$	$P_5 - 6$	(25); (33) if $a = e,\ b = d$, $D = E = \pi/2$
	(13)	$a + e = d$	pgg	$IH5$	$P_5 - 7$	
	(14)	$a + d = c$	pgg	$IH6$	$P_5 - 12$	
	(15)	$a = d,\ b = c$	pg	$IH2$	$P_5 - 1$	(35) if $A + 2D = 2\pi$, $a = b = c = d$
	(16)	$a = d,\ b = c$	pgg	$IH5$	$P_5 - 9$	
	(6)	$a = d,\ b = c$	cm	$IH22$	$P_5 - 17$	(40), (41) if $D = 80°$ and $a = b = c = d = e$
	(7)	$a = d,\ b = c$	pgg	$IH25$	$P_5 - 20$	
	(17)	$b = c, a = d + e$	pg	$IH2$	$P_5 - 2$	
	(18)	$D = E = \pi/2$, $A = C,\ a = d,\ b = c$	pmg	$IH15$	$P_5 - 14$	
	(1)	$D = E = \pi/2$, $A = C,\ a = d,\ b = c$	cmm	$IH26$	$P_5 - 21$	(27), (28) if $B = \pi/2$
	(19)	$D + B = \pi,\ c = e$, $a = b + d$	pg	$IH3$	$P_5 - 3$	
2. $C + E = \pi$, $a = d$	(20)		pgg	$IH6$	$P_5 - 10$	(26) if $A + C = \pi,\ d = e$
	(21)		pgg	$IH6$	$P_5 - 11$	
	(8)	$c = e$	pgg	$IH27$	$P_5 - 22$	(36) if $D + 2E = 2\pi$, $a = c = d = e$
3. $A = C = D = 2\pi/3$, $a = b$, $d = c + e$	(22)		$p3$	$IH7$	$P_5 - 13$	
	(23)	$B = E = \pi/2$	$p31m$	$IH16$	$P_5 - 15$	
4. $A = C = \pi/2$, $a = b$, $c = d$	(9)		$p4$	$IH28$	$P_5 - 23$	(27) if $E = \pi/2$ and $a = b = e$
	(3)	$D = E$	$p4g$	$IH29$	$P_5 - 24$	
5. $C = 2A = 2\pi/3$, $a = b,\ c = d$	(2)		$p6$	$IH21$	$P_5 - 16$	

Table 1. The five types of pentagons which can generate tile-transitive tilings of the plane. The tiles for a given transitive tiling are not always uniquely of one type (e.g. the tiles of tiling (19) satisfy conditions on angles and sides of both types 1 and 2). In this table we have listed each transitive tiling only once, identifying its tiles by just one type.

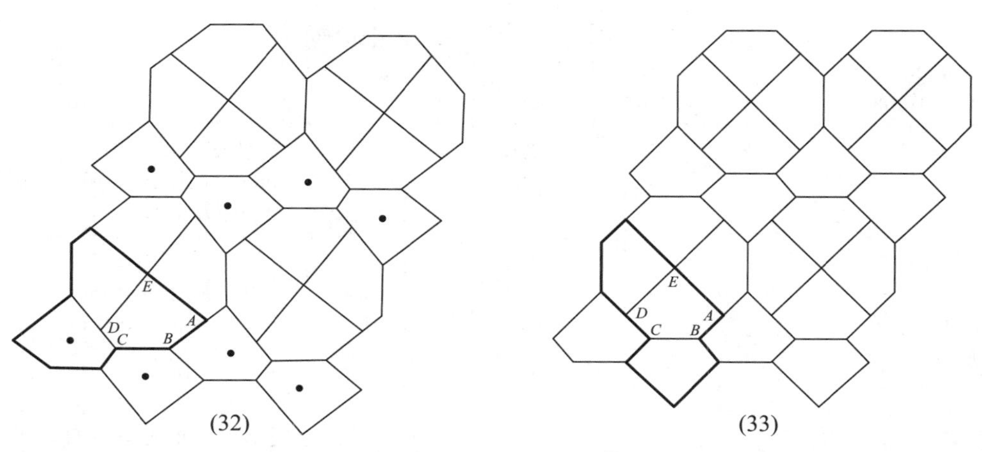

Figure 7. Tilings by type of pentagon discovered by Richard James III. The tilings consist of strips of attached octagons, separated by strips of bow ties. Each octagon contains four pentagons, and each bow tie contains two pentagons, which are in opposite orientation from those in the octagons. The tilings are 3-block transitive.

Contributions by amateurs

The publication of the problem and Kershner's list by Martin Gardner in [5] stimulated amateurs to try to find pentagons that tile. Richard James III, a computer scientist, read the problem and decided not to look at Kershner's list, but see if he could find some pentagonal tilings himself. Familiar with the common tiling by regular octagons and squares, and noting that an octagon is easily dissected into four congruent pentagons by perpendicular lines through its center, he attempted to change the familiar tiling into a pentagonal one. He was successful and sent an example to Gardner [17]. Figure 7 shows two tilings by pentagons of James's type (type 10 on Table 2). James's discovery served to point out a hidden assumption in Kershner's search—he had, in fact, only been looking for pentagons which could tile in an edge-to-edge manner, or in a manner in which every tile was surrounded by six vertices of the tiling (as, for example, in tilings (10) through (24)). James's pentagons are only capable of tiling in a manner which is not tile-transitive and not edge-to-edge. In addition, some pentagons in this tiling are surrounded by 5 vertices of the tiling, while others are surrounded by 7 vertices of the tiling.

Marjorie Rice, a Californian with no mathematical training beyond "the bare minimum they required ... in high school over 35 years ago," also read Gardner's column and began her own methodical attack on the problem. Her approach was to consider the different ways in which the vertices of a single pentagon could "come together" to form a vertex of a tiling by congruent images of that pentagon. These considerations forced conditions on the angles and sides of the pentagon if it was to tile, thus giving either a description of a pentagon which could tile in a prescribed manner, or forcing the conclusion that no pentagon could be constructed which satisfied the conditions.

This essentially combinatorial search yielded over forty different tilings by pentagons and included a tiling by a new type of tile not on Kershner's list. Her discovery (type 9 on Table 2) showed that Kershner's search (which was similar to hers) erroneously eliminated the possibility of this type of edge-to-edge tiling. A later methodical search by Rice considered twelve different classes of pentagons, each class corresponding to a description of which sides of a given pentagon are equal. Possible tilings for each class were sought—and for every class at least one tiling was found. She produced over 58 diagrams of distinct tilings in this effort, most of them non-transitive tilings by tiles of type 1. Even though she missed several of the 24 tile-transitive tilings, her scheme was complete enough to produce a tiling for every one of the pentagons of types 1–10 in Tables 1 and 2. No other new types of tiles were produced in this effort. Figure 8 shows three of Rice's tilings for the class of pentagons having four equal sides, including tiling (34) associated to her type 9. (The pentagons in tilings (30) and (31), Figure 6, also have four equal sides.)

Block transitive tilings

The solution of the hexagonal tiling problem was simplified by a theorem which reduced the problem to one of hexagons capable of producing tile-

Type, with characterizing conditions	Illustrative tile and tiling number	Symmetry group of tiling	Remarks
6. $C + E = \pi$ $A = 2C$ $a = b = e$ $c = d$	(29)	$p2$	2-block transitive. Associated block tiling (29)-B is isohedral type $IH4$.
7. $2B + C = 2\pi$ $2D + A = 2\pi$ $a = b = c = d$	(30)	pgg	2-block transitive. Associated block tiling (30)-B is isohedral type $IH6$.
8. $2A + B = 2\pi$ $2D + C = 2\pi$ $a = b = c = d$	(31)	pgg	2-block transitive. Associated block tilings (31)-B and (31)-B′ are type $IH6$.
9. $2E + B = 2\pi$ $2D + C = 2\pi$ $a = b = c = d$	(34)	pgg	2-block transitive. Associated block tiling (34)-B is type $IH53$.
10. $E = \frac{\pi}{2}$ $A + D = \pi$ $2B - D = \pi$ $2C + D = 2\pi$ $a = e = b + d$	(32) (33) if $D = \frac{\pi}{2}$, $b = d$	$p2$ cmm	D is bounded: $\pi - \tan^{-1}(\frac{4}{3}) < D < \tan^{-1}(\frac{4}{3})$ 3-block transitive. Associated block tilings (32)-B and (33)-B are type $IH4$.
11. $A = \frac{\pi}{2}$ $C + E = \pi$ $2B + C = 2\pi$ $d = e = 2a + c$	(37)	pgg	2-block transitive. Associated block tiling is type $IH6$.
12. $A = \frac{\pi}{2}$ $C + E = \pi$ $2B + C = 2\pi$ $e + c = d = 2a$	(38)	pgg	2-block transitive. Associated block tiling is type $IH6$.
13. $A = C = \frac{\pi}{2}$ $B = E = \pi - \frac{D}{2}$ $c = d$ $2c = e$	(39)	pgg	2-block transitive. Associated block tiling is $IH5$.

Table 2. Eight types of pentagons which tile, but for which no tile-transitive tiling exists.

transitive tilings. Although this theorem is clearly false for the case of pentagons which tile, we can observe that for each pentagon of types 1 through 10, there exists a tiling containing a minimal 'block' of congruent pentagons which has the property that (i) the tiling consists of congruent images of this block and (ii) this block can be mapped onto any other congruent block by an isometry of the tiling. If a minimal such block contains n pentagons, we will say that the tiling is ***n*-block transitive**. Thus a tile-transitive tiling is 1-block transitive. We remark that for $n \geq 2$, a given n-block transitive tiling may have several non-congruent minimal n-blocks. We have outlined two such 2-blocks in tiling (31).

If we remove the interior edges of pentagons in the heavily outlined blocks shown in the tilings in Figures 6, 7, and 8, the resulting transitive block tilings reveal information not immediately apparent from these pentagonal tilings alone. In Figure 9, we can see that each of Kershner's tilings (29), (30),

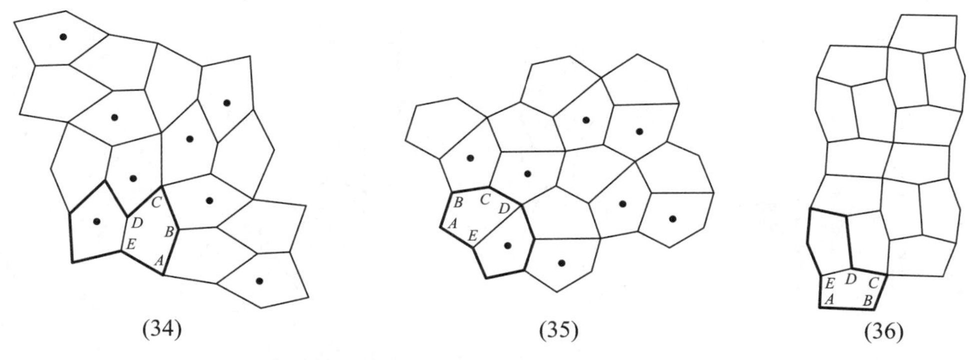

Figure 8. Three tilings discovered by Marjorie Rice, each containing congruent pentagons having four equal sides. The tile in (34) is type 9, and was a new addition to Kershner's list. The tile in (35) is type 1, the tile in (36) is type 2. All tilings are 2-block transitive, and for type 9, no tile-transitive tiling exists.

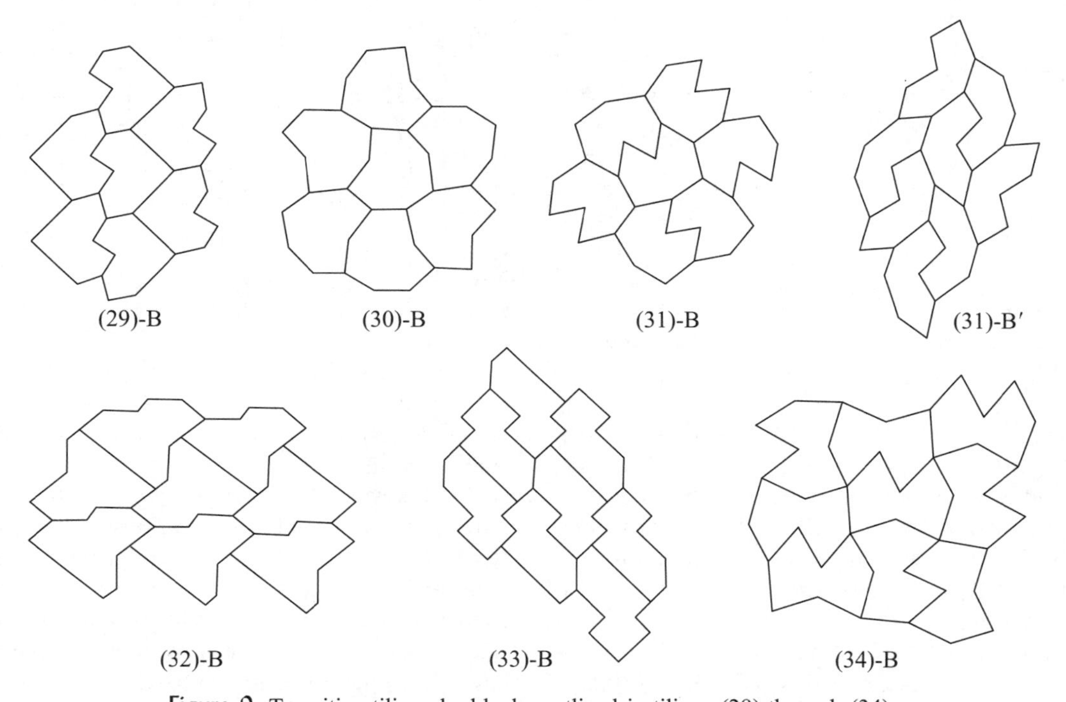

Figure 9. Transitive tilings by blocks outlined in tilings (29) through (34).

(31) produces a block tiling in which each block is surrounded by six vertices of the tiling. Thus, these block tilings are formed by non-convex tiles which are topological hexagons. The Kershner tilings are obtained from these 'hexagonal' tilings by bisecting each 'hexagon' into two congruent pentagons. This method parallels the technique noted earlier of obtaining pentagonal tilings by bisecting hexagonal tilings. The reader can verify that the 2-block transitive tilings (25), (26), (35) are also obtained by bisecting blocks which tile as topological hexagons.

Tilings (32) and (33) of James's pentagons also have associated block tilings of topological hexagons (Figure 9). In this case, however, each 'hexagon' has been dissected into 3 congruent pentagons, an occurrence that has no parallel in any of the pentagonal tilings known prior to James's discovery. It might be appropriate to note here that the symmetry group of the block tiling (33)-B of Figure 9 is $p2$, which is a proper subgroup of the symmetry group of the pentagonal tiling (33). (The names of tiling symmetry groups are given in Tables 1 and 2.) This occurrence is not surprising, since the removal of some edges of the pentagonal tiling can cause loss of some symmetries of the pattern (in this case, reflections).

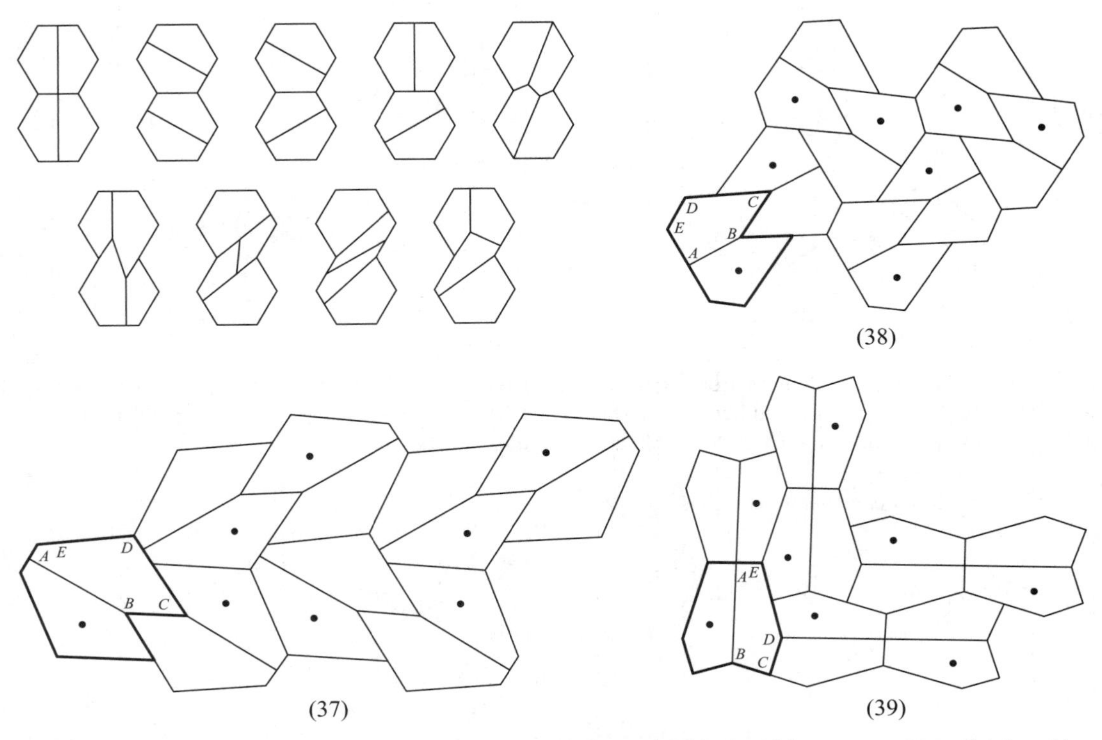

Figure 10. Marjorie Rice's schematic drawings of the "double hexagon" blocks of 4 pentagons. By adjusting sides and angles, many blocks of 4 congruent pentagons which tiled were found. Tilings (37) and (38) are for pentagons of types 11 and 12, respectively. Both are formed by blocks of the type shown at the lower left in the schematic diagram. Tiling (39) is for a type 13 pentagon, and is formed by blocks like those shown at the upper left in the schematic diagram.

Figure 9 shows the surprising fact that Rice's tiling (34) of type 9 pentagons has as its associated block tiling one in which each block is surrounded by just 4 vertices of the tiling. Thus tiling (34) is produced by dissecting a transitive tiling of topological quadrilaterals (Rice's tiling (36) is also obtained from 'quadrilateral' blocks). This dissection of a 'quadrilateral' tiling to produce a pentagonal tiling is most unexpected, since ordinary quadrilateral tiles (convex or non-convex), when bisected, produce either triangles or new quadrilaterals.

It is now easy to speculate that new types of pentagons which tile can be discovered by considering blocks of two or more congruent pentagons and determining if such blocks can tile transitively. A preliminary version of this article prompted Rice to examine a particular family of blocks, with hopes of determining new 2-block transitive pentagonal tilings. She observed that several of the 2-block transitive tilings previously discussed could also be viewed as tilings by blocks of four pentagons where these larger blocks had the outline of two hexagons stuck together. Figure 10 contains a schematic diagram of these "double hexagon" blocks, with their various dissections considered by Rice. Over sixty 2-block transitive tilings of pentagons were discovered in this way, some previously known ((25) and (26), for example), and some new. Best of all, two of the new tilings showed new types of pentagons! Just as this article was going to press, Marjorie Rice discovered yet another new tile as the result of a further search for new 2-block transitive tilings. Figure 10 shows the tilings associated to these new tiles (types 11, 12, and 13 in Table 2).

It is quite likely that still other new types of pentagons which tile can be discovered by considering dissections of transitive block tilings. The enumeration of the 81 types of isohedral tilings in [6] (complete with helpful diagrams) makes this task feasible. Checking possibilities will be an extremely lengthy task, however, with a great deal of built-in repetition. This is assured by the fact that a given pentagonal tiling may have many distinct n-blocks which produce transitive block tilings and the additional fact that a single pentagonal tile may produce many distinct tilings. In order to determine if the list of types is complete, a theorem is needed to put some kind of bound on the possibilities. For the known pentagons

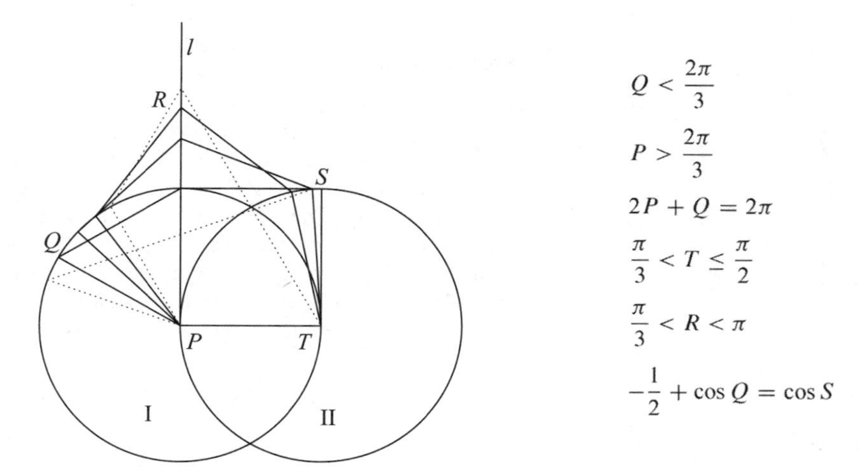

Figure 11. The family of equilateral pentagons for which $2P + Q = 2\pi$ can be envisioned as generated by a flexible equilateral pentagon, hinged at the vertices, with side PT held fixed, while R rides up and down line l (causing vertices Q and S to ride along arcs on circles I and II respectively). The formal construction is as follows:

Choose angle $P > 2\pi/3$. Construct sides $PT = PQ$ by drawing circle I with center at P. Construct line l perpendicular to PT at P. Find vertex R on line l so $QR = QP$ (then $2P + Q = 2\pi$). Draw circle II with center T, radius PT. Find vertex S on circle II so that $RS = ST$ and $PQRST$ is convex.

The conditions on the angles listed above follow easily from this construction.

Three pentagons in this family are known to tile the plane. In addition to the unique type 7 and unique type 8 (which is also type 2) listed in Table 3, there is a unique type 1 pentagon: $P = 150°$, $Q = 60°$, $R = 150°$, $S = 90°$, $T = 90°$.

which tile, there always exists an n-block transitive tiling for $n \le 3$. It is natural to hope that the following theorem is true: A pentagon tiles the plane only if there exists an n-block transitive tiling by that pentagon for $n \le 3$.

Equilateral pentagons which tile

Although regular pentagons cannot tile the plane, a surprising variety of equilateral pentagons do. Of the thirteen types of pentagons which tile it is obvious that types 3, 10, 11, 12, and 13 cannot be equilateral. Using construction techniques, together with familiar trigonometric relations, and the extension of these relations found in [20], we determined all possible equilateral pentagons of known types which tile. Types 1 and 2 provide distinct infinite families of equilateral pentagons. For types 4, 7 and 8, there is a unique equilateral pentagon of each type. Types 5, 6 and 9 cannot be equilateral. In order to investigate the equilateral case for types 7, 8 and 9, we studied the general class of equilateral pentagons $PQRST$ satisfying the condition $2P + Q = 2\pi$. Figure 11 contains information on this class. Table 3 contains a summary of details for all known types of equilateral pentagons which tile.

Since any two congruent equilateral pentagons will match edge-to-edge in any order, the possibilities for tilings are great. In attempting to determine all equilateral pentagons which tile, Marjorie Rice produced many interesting tilings. In addition, Martin Gardner's article inspired George Szekeres and Michael Hirschhorn of the University of New South Wales to conduct a week-long study group of Form 5 high school students on tilings by equilateral pentagons [16]. Both Rice and the Australian class discovered that a particular type 1 pentagon ($A = 140°$, $B = 60°$, $C = 160°$, $D = 80°$, $E = 100°$) could tile in curious ways (Figure 12). Both discovered tiling (40) composed of zigzag bands in which the pentagons can fit together in two distinct ways. The particularly beautiful tiling (41), having only rotational symmetry, was discovered by Hirschhorn. Since there are eleven ways in which angles of this pentagon can be summed to 360°, several other unusual tilings are also possible, including another design having only rotational symmetry. For this reason, Hirschhorn has dubbed it the "versa-tile."

Some questions

The general problem of determining all convex pentagons which can tile the plane remains unsolved. Is our list of 13 types which tile complete? We doubt it.

Type 1. $D + E = \pi$

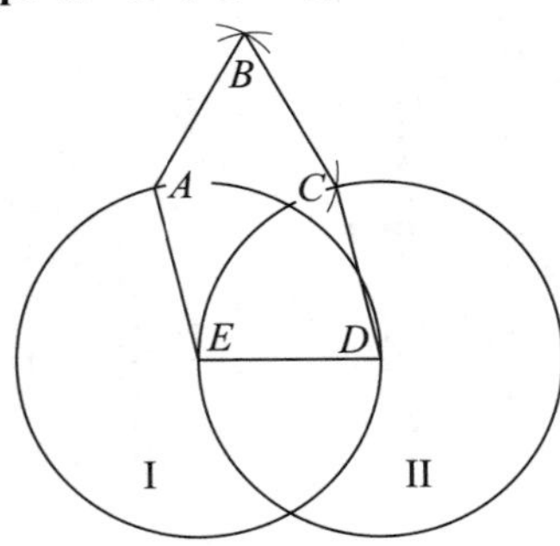

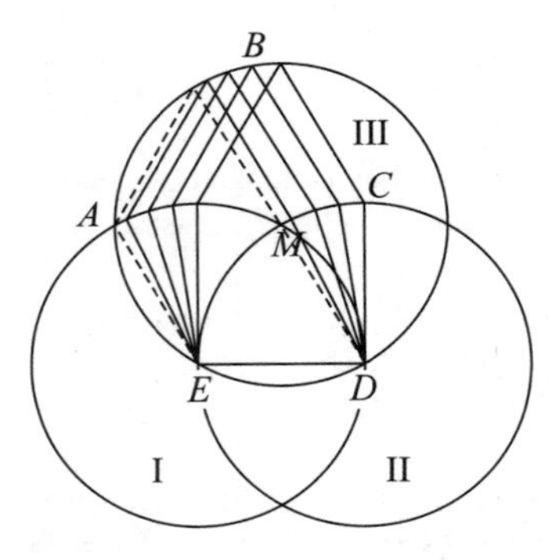

Construction

Choose angle E.

Construct sides $AE = ED$ by drawing circle I with center at E. Construct circle II with center D and radius DE. Find vertex C on circle II so that $DC \| AE$. Construct $CB = AB = AE$. The pentagon $ABCDE$ has the outline of an equilateral triangle atop a rhombus. Draw circle III with radius equal to DE, and with center M, the intersection of circle I and circle II. Then for each pentagon $ABCDE$, vertex A lies on circle I, vertex C lies on circle II, and vertex B lies on circle III (see lower diagram).

The diagram shows representatives of this type for $\frac{\pi}{2} \leq E < \frac{2\pi}{3}$.

Angles

$$\frac{\pi}{3} < E < \frac{2\pi}{3} \qquad A = \frac{4\pi}{3} - E \qquad C = \frac{\pi}{3} + E$$

$$B = \frac{\pi}{3} \qquad D = \pi - E$$

Illustrative tilings

Tile-transitive:
(4), (5), (6), (7), (10),
(11), (12), (15), (16), (24)
For $E = \frac{\pi}{2}$: (1), (18)

Non-transitive:
(25)
For D = 80°: (35), (40), (41)

Type 2. $C + E = \pi$

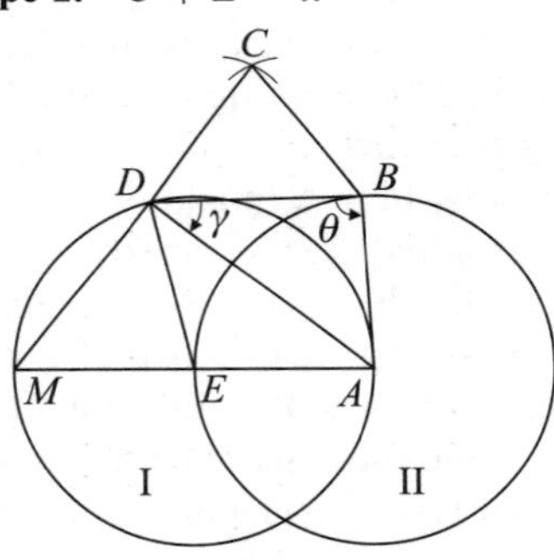

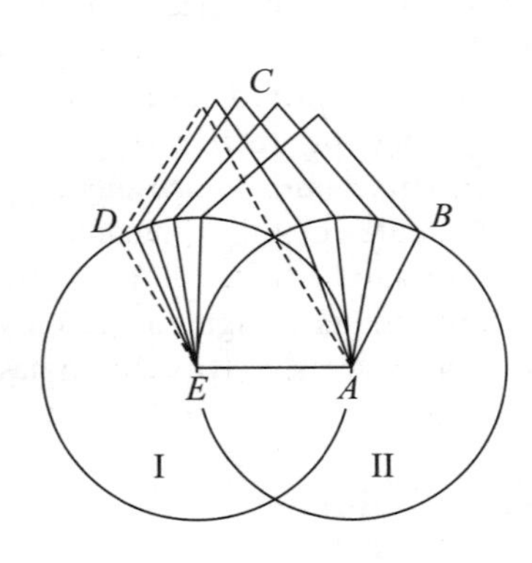

Construction

Choose angle E.

Construct sides $AE = ED$ by drawing circle I with center at E. Extend EA to M on circle I. Construct circle II with center A and radius AE. Find point B on circle II so that $DB = DM$; construct $\triangle DCB \approx \triangle DEM$, (so $C = \pi - E$). Join AB to form pentagon $ABCDE$.

To obtain the description of the angles, draw segment DA, and denote $\gamma = \sphericalangle BDA$ and $\theta = \sphericalangle DBA$. Note $\triangle MDA$ is a right triangle, with $\sphericalangle DMA = E/2$ and $\sphericalangle DAM = C/2$. These facts, together with the law of cosines for $\triangle DBA$ lead to the angle relationships given.

The lower diagram shows representatives of this type for $\frac{\pi}{2} \leq E < \frac{2\pi}{3}$.

Angles

$$\frac{\pi}{3} < E < \frac{2\pi}{3} \qquad \theta = \arccos\left(\frac{1 + 4\cos E}{4\cos E/2}\right)$$

$$C = \pi - E \qquad D = \frac{\pi}{2} + \gamma$$

$$\gamma = \arccos\left(\frac{3}{4\sin E}\right); \qquad B = \frac{E}{2} + \theta$$

$$\frac{\pi}{6} < \gamma \leq \arccos\left(\frac{3}{4}\right) \qquad A = \frac{3\pi}{2} - \frac{E}{2} - \theta - \gamma$$

Illustrative tilings

Tile-transitive:
(8), (20), (21)
Non-transitive: (36)

Special cases:
Types 4 and 8 below.

Table 3. **Equilateral Pentagons Known to Tile.** Types 1–2

Type 4. $A = C = \frac{\pi}{2}$

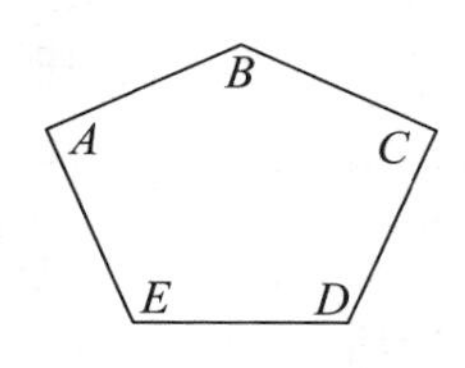

Construction

Follow the construction of type 2 above for chosen angle $E = \pi/2$, then relabel the vertices of the pentagon replacing E by A, D by B, etc., in clockwise order. Angles can be established easily using right triangle relationships.

Angles

$$A = C = \frac{\pi}{2} \qquad B = 2\pi - 2D \sim 131°24'$$

$$D = E = \frac{\pi}{4} + \arccos\frac{\sqrt{2}}{4} \sim 114°18'$$

Illustrative tiling: (3) and tilings listed for type 2 above.

Type 5.
$A = \frac{\pi}{3}, C = \frac{2\pi}{3}$

Impossible. The conditions $A = \pi/3$, $C = 2\pi/3$ imply that the pentagon would have to be of type 2 above. For $A = \pi/3$, the construction yields a limiting quadrilateral of the type 2 family, i.e., an equilateral "pentagon" with $E = \pi$.

Type 6.
$C + E = \pi, A = 2C$

Impossible. The conditions $C + E = \pi$ and $A = 2C$ imply that the pentagon would have to be of type 2 above, with $\pi/3 < C < \pi/2$. Then $\pi/2 < E < 2\pi/3$ implies $-1 < \cos E < -1/2$ and $\cos E/2 > 0$, so $\arccos\theta < 0$. Thus $\theta > \pi/2$, and since $\gamma > \pi/6$, we have $3C/2 = A - C/2 = \pi - \theta - \gamma < \pi/3$, a contradiction to $\pi/3 < C$.

Type 7.
$2B + C = 2\pi, 2D + A = 2\pi$

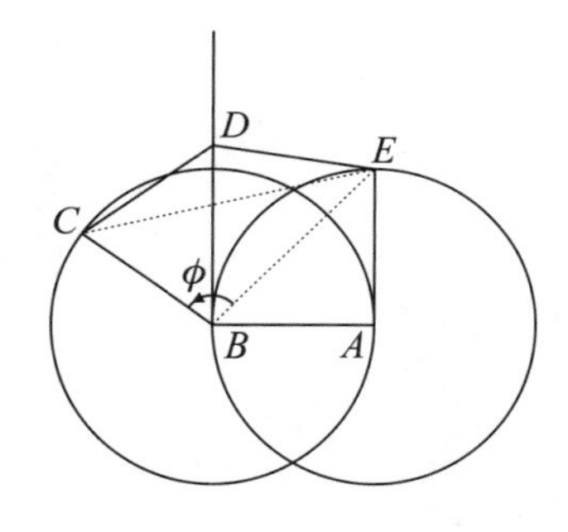

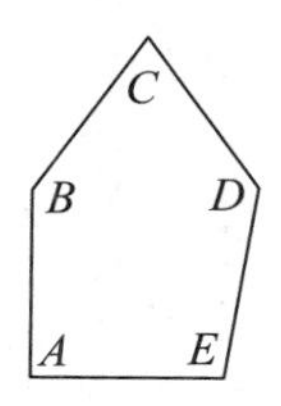

Construction

Let $B = P$ and $C = Q$ in Figure 11; then $D = R$, $E = S$, $A = T$. Adjust vertex D on line l until $2D+A = 2\pi$ (a unique solution). Let $\phi = \sphericalangle CBE$. The angle relations indicated in Figure 11, together with the additional condition $2D + A = 2\pi$, give the following general conditions on angles:

$$B = \frac{\pi}{2} - \frac{A}{2} + \phi \qquad D = \pi - \frac{A}{2}$$

$$C = \pi + A - 2\phi \qquad E = \frac{\pi}{2} - A + \phi$$

The extended law of cosines in [20] gives $\sin\phi = 1 - \cos A$; $\phi = \pi - \arcsin(\sin\phi)$. In addition, the following equations can be obtained, the first from Figure 11, the second by combining law of cosines for $\triangle CDE$ and law of sines for $\triangle CBE$:

$$\cos C - \cos E = 1/2$$

$$16\sin^4 D + 8\sin^2 D + 4\cos D = 5$$

To obtain the angle approximations given, we noted by construction that A was close to 89°. We then computed the angles of the pentagon as A ranged over values within 1° of 89°, and computed the error in the above equations for this range. The angles given produce an error of less than .001 in both equations.

Angles

$$A \sim 89°16' \qquad C \sim 70°55'$$
$$B \sim 144°32'30'' \qquad D \sim 135°22' \qquad E \sim 99°54'30''$$

Illustrative tiling: (30)

Table 3. **Equilateral Pentagons Known to Tile.** Types 4–7

Type 8.

$2A + B = 2\pi$.
$2D + C = 2\pi$

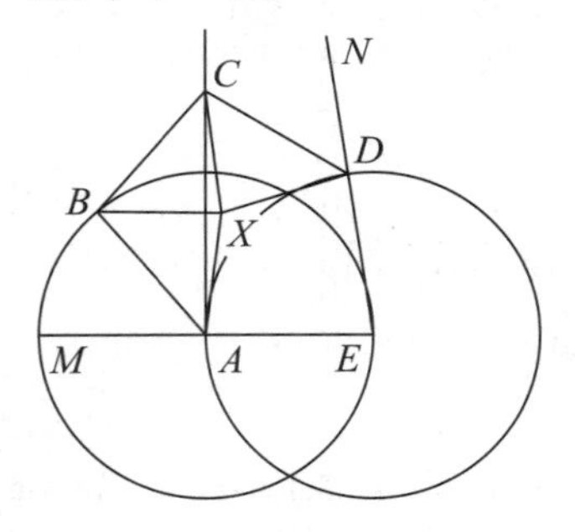

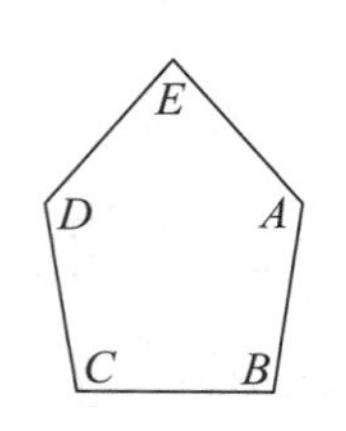

Construction

Let $A = P$ and $B = Q$ in Figure 11; then $C = R$, $D = S$, $E = T$. Adjust vertex C on line l until $2D + C = 2\pi$ (a unique solution). This is the unique pentagon of the family in Figure 11 satisfying $Q = R$, $P = S$. To see this, add construction lines to $ABCDE$ as follows.

Extend EA to M and ED to N. Let X be the intersection of the bisectors of angles B and C; draw AX, BX, CX, DX. Then $\triangle XCD \approxeq \triangle XCB \approxeq \triangle XBA$ so $\sphericalangle XAB = C/2$ and $\sphericalangle XDC = B/2$. Now $\sphericalangle MAB = B/2$, and $\sphericalangle NDC = C/2$, which implies $\sphericalangle XAD = \sphericalangle XDA$. Thus $BX = CX$ and so angles B and C are equal, angles A and D are equal.

The explicit value for $\cos E$ can be obtained using the extended law of cosines in [20].

Angles

$$E = \arccos \frac{\sqrt{13} - 3}{4} \sim 81°18' \qquad A = D = \frac{\pi}{2} + \frac{E}{2} \sim 130°39'$$

$$B = C = \pi - E \sim 98°42'$$

$C + E = \pi$ and $B + E = \pi$ show this is also type 2.

Illustrative tilings

(31) and tilings listed for type 2 above.

Type 9.
$2E + B = 2\pi$,
$2D + C = 2\pi$

Impossible. The angle condition $2D + C = 2\pi$ puts it in the family in Figure 11, with $P = D$, $Q = C$, $E = T$. The condition $2E + B = 2\pi$ implies $E > \pi/2$, which contradicts the condition in Figure 11 that $E \leq \pi/2$.

Table 3. **Equilateral Pentagons Known to Tile.** Types 8–9

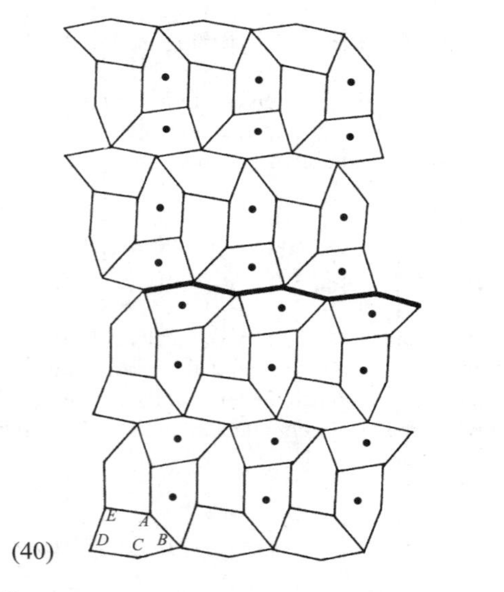

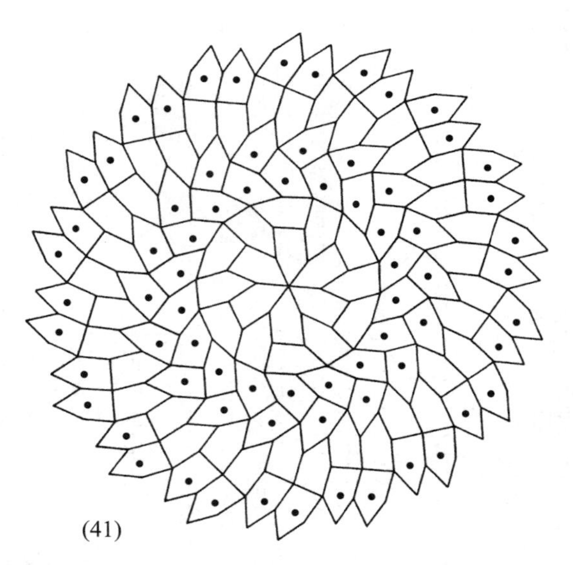

Figure 12. Non-transitive tilings by the equilateral pentagon $A = 140°$, $B = 60°$, $C = 160°$, $D = 80°$, $E = 100°$. Pentagons fill zig-zag horizontal strips in two distinct ways in tiling (40); these strips fit together in innumerable ways to fill the plane. Tiling (41) by Michael Hirschhorn has only 6-fold rotational symmetry.

Even though a complete solution of the problem appears to be difficult (certainly lengthy), perhaps the full answer to some special cases can be obtained. We have noted that the list of types of pentagons which can generate a tile-transitive tiling is complete (types 1 through 5). Are the only tiles capable of edge-to-edge tilings types 1, 2, 4, 5, 6, 7, 8 and 9? Is the list of equilateral pentagons which tile complete? I hope these questions will stimulate further activity on the problem.

References

1. A. D. Bradley, *The Geometry of Repeating Design and Geometry of Design for High Schools*, Teacher's College, Columbia University, 1933.
2. Béla Bollobás, Filling the Plane with Congruent Convex Hexagons without Overlapping, *Annales Universitatis Scientiarum Budapestinensis Sectio Mathematica* (1963) 117–123.
3. Stanley Clemens, Tessellations of Pentagons, *Mathematics Teaching*, No. 67, 18–19.
4. J. A. Dunn, Tessellations with Pentagons, *Mathematical Gazette*, 55, Dec. 1971, 366–369. Follow-up correspondence Vol. 56, pp. 332–335.
5. Martin Gardner, On tessellating the plane with convex polygon tiles, *Scientific American*, July 1975, 112–117. Reprinted with Addendum as "Tiling with Convex Polygons" in *Time Travel and Other Mathematical Bewilderments*, W.H. Freeman and Company, New York, 1988. In CD collection *Martin Gardner's Mathematical Games*, Mathematical Association of America, 2005.
6. B. Grünbaum and G. C. Shephard, The 81 types of isohedral tilings in the plane, *Mathematical Proceedings of the Cambridge Philosophical Society*, 82 (1977), 177–196.
7. ——, The 91 types of isogonal tiling in the plane, *Transactions of the American Mathematical Society*, 242 (1978) 335–353 and 249 (1979) 446.
8. ——, Isotoxal tilings, *Pacific Journal of Mathematics*, 76 (1978) 407–430.
9. ——, Isohedral tilings of the plane by polygons, *Commentarii Math. Helvetici*, 53 (1978) 542–571.
10. F. Haag, Die regelmässigen Planteilungen, *Zeitschrift für Kristallographie*, 49 (1911), 360–369.
11. ——, Die regelmässigen Planteilung und Punktsysteme, *Zeitschrift für Kristallographie*, 58 (1923), 478–489.
12. ——, Die pentagonale Anordnung von sich berührenden Kreisen in der Ebene, *Zeitschrift für Kristallographie*, 61 (1925), 339-340.
13. ——, Die Symmetrieverhältnisse einer regelmässigen Planteilung, *Zeitschrift für mathematik und nature.* unterricht, 59 (1926), 262–263.
14. H. Heesch and O. Kienzle, *Flächenschluss*, Springer-Verlag, Berlin–Göttingen–Heidelberg, 1963.
15. H. Heesch, *Reguläres Parkettierungsproblem*, Westdeutscher Verlag, Köln-Opladen, 1968.
16. M. Hirschhorn, The 1976 Summer Science School. Tessellations with convex equilateral pentagons, *Parabola*, 13, Feb/March 1977, 2–5. More tessellations with convex equilateral pentagons, *Parabola*, 13, May/June 1977, 20–22.
17. Richard James, New pentagonal tiling reported by Martin Gardner in *Scientific American*, December 1975, 117–118.
18. R. B. Kershner, On Paving the Plane, *American Mathematical Monthly*, 75 (1968), 839–844.
19. ——, On Paving the Plane, *APL Technical Digest*, 8 (1969), 4–10.
20. ——, The Law of Sines and Cosines for Polygons, *Mathematics Magazine*, 44, No. 3, May 1971, 150–153.
21. F. Laves, Ebenenteilung in Wirkungsbereiche, *Zeitschrift für Kristallographie*, 76 (1931), 277–284.
22. ——, Ebenenteilung und Koordinationszahl, *Zeitschrift für Kristallographie*, 78 (1931), 208–241.
23. P. A. MacMahon, *New Mathematical Pastimes*, Cambridge University Press, London, 1921.
24. J. Milnor, Hilbert's problem 18: on crystallographic groups, fundamental domains, and on sphere packing, Mathematical Developments Arising from Hilbert Problems, *Proceedings of Symposia in Pure Mathematics*, American Mathematical Society, 1976, 441–506.
25. Phares O'Daffer and Stanley Clemens, *Geometry: An Investigative Approach*, Addison-Wesley, 1976.
26. ——, *Laboratory Investigations in Geometry*, Addison-Wesley, 1976.
27. John Parker, Topics, Tessellations of Pentagons, *Mathematics Teaching*, 70, 1975, p. 34.
28. Marjorie Rice, New pentagonal tilings reported to Martin Gardner and the author. (Unpublished)
29. K. Reinhardt, *Über die Zerlegung der Ebene in Polygone*, Dissertation, Universität Frankfurt, 1918.
30. D. Schattschneider, Pentagonal Tilings of Richard James's Type, (unpublished).
31. ——, The Plane Symmetry Groups: their recognition and notation, *American Mathematical Monthly*, 85 (1978) 439–450.
32. A. V. Shubnikov and V. A. Koptsik, *Symmetry in Science and Art*, Plenum Press, 1974.

Afterword

Thirty years after writing this article, I must report that the problem of characterizing all convex pentagons that tile the plane is still unsolved. However, there has been progress, and the problem has received much more attention.

In 1985, Rolf Stein, at the University of Dortmund, found a fourteenth pentagon tiler that is unique up to similarity. It is determined by these relations: $A = \pi/2, C+E = \pi, 2B+C = 2\pi$; $d = e = 2a = a + c$. The announcement [A7] in *Mathematics Magazine* featured a portion of the tiling on its cover. Stein believed this completed the list of convex pentagons that can tile, but his proof was incomplete. Although no additional convex pentagon tilers have been found, a proof that there are no others is still lacking.

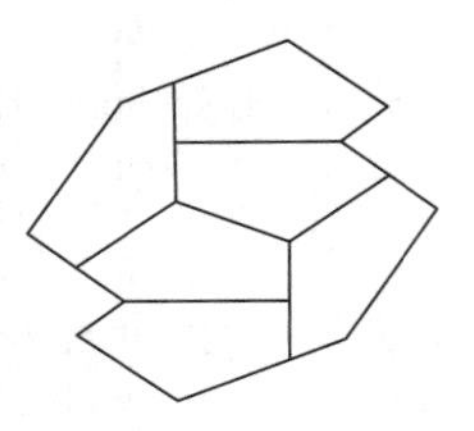

At the end of my article, I asked, "Is the list of equilateral pentagons which tile complete?" The answer is yes, and was first confirmed in 1985 by Michael Hirschhorn and David Hunt [A4], who used a computer to examine all possible angle equations for equilateral pentagon tilers, and determined which equations could be satisfied simultaneously. Recently Olga Bagina gave a shorter proof, using *Mathematica* to aid in some of her calculations [A1].

In response to my article here, Branko Grünbaum suggested that I investigate nonconvex pentagons that tile the plane. Some tilings by nonconvex pentagons had been published in *Mathematics Teaching*, but there was no attempt at characterizing such tiles (see references in [5]). By 1982, I had an enumeration similar to that for convex pentagons, including what I hoped would be a complete list of nonconvex equilateral pentagons that tile. This was presented at some conferences, but never published. In 2002, news of Bagina's result spurred Michael Hirschhorn to resurrect my earlier results and prove that a non-convex equilateral pentagon tiles the plane if and only if two or three angles add to 360°, and tiles "properly" (at least three tiles at every vertex of the tiling) if and only if three angles add to 360° or it is the unique nonconvex equilateral pentagon satisfying the angle conditions of convex type 9 [A5]. This unique pentagon can be seen in [A2]; my article here prompted H. Martin Cundy to write me that he had investigated equilateral pentagons in 1970, and when I sent him this special nonconvex tiler, he posed a problem in the *Mathematical Gazette*.

Ivan Niven contributed a helpful article [A8]; it was widely believed that no convex polygon with seven or more sides could tile the plane, but it had been impossible to find a published reference. Branko Grünbaum and Geoffrey Shephard made an enormous contribution to the understanding of tilings with their tome [A3]; chapter 9 on tilings by polygons provided definitions, characterizations, questions, and extensive references.

Finally, Marjorie Rice, this article's star, continued her interest in tilings (producing many new ones by convex and nonconvex pentagons), studied aperiodic tilings, and made artistic Escher-like tilings based on underlying grids of pentagons [A9], [A11]. A more personal account of her search for tiling pentagons is in [A10]. In 1994, I sent her computer output from a program by Daniel Huson that gave combinatorial possibilities for edge-to-edge 3-isohedral tilings by pentagons. She methodically researched hundreds of cases by hand to determine for which tilings the tiles could be congruent. In 1999, one of her new discoveries was installed in glazed ceramic tile at MAA headquarters in Washington, DC.

References

A1. O. Bagina, Tiling the plane with congruent equilateral convex pentagons, *Journal of Combinatorial Theory, Ser. A* 105 (2004) 221–232.

A2. Solution to problem 67D, posed by H. Martyn Cundy, *The Mathematical Gazette*, 67 (1983) 307–309.

A3. B. Grünbaum and G. C. Shephard, *Tilings and Patterns*, W.H. Freeman, 1987.

A4. M.D. Hirschhorn and D. C. Hunt, Equilateral Convex Pentagons Which Tile the Plane, *Journal of Combinatorial Theory, Ser. A*, 39 (1985) 1–18.

A5. M. D. Hirschhorn, private communication.

A6. M. S. Klamkin and A. Liu, A Note on a Result of Niven on Impossible Tesselations, *American Mathematical Monthly*, 87 (1980) 651–653.

A7. *Mathematics Magazine*, 58, no. 5 (1985) cover and page 308.

A8. I. Niven, Convex Polygons that Cannot Tile the Plane, *American Mathematical Monthly*, 85 (1978) 785–789.

A9. M. Rice, Escher-Like Tilings from Pentagonal Tilings, in *M. C. Escher's Legacy*, Springer-Verlag, 2003.

A10. D. Schattschneider, In Praise of Amateurs, in *The Mathematical Gardner*, ed. David A. Klarner, Prindle, Weber and Schmidt, 1981.

A11. D. Schattschneider and Marjorie Rice, The Incredible Pentagonal Versatile, *Mathematics Teaching*, December 1980, 52–53.

Unstable Polyhedral Structures

Michael Goldberg

Mathematics Magazine
51 (1978), 165–70

Editors' Note: Michael Goldberg was trained as an electrical engineer at the University of Pennsylvania and did graduate work at George Washington University. He then worked for 40 years as a naval engineer. Since 1922 his name has appeared as a problem solver in the *American Mathematical Monthly* (and somewhat later in *Mathematics Magazine*). He wrote extensively on polyhedra, dissection problems, packing problems and linkage mechanisms. A striking paper of his on the isoperimetric problem for polyhedra (a problem still not completely solved) appeared in 1935 in the *Tôhoku Mathematical Journal*, 40 (1935), 226–36. His first article in *Mathematics Magazine* appeared in 1942.

For more problems of the type in this paper see Jack E. Graver's *Counting on Frameworks/Mathematics To Aid the Design of Rigid Structures* (MAA, 2001).

It was shown by Cauchy [1] and Dehn [2] that a convex polyhedron made of rigid plates which are hinged at their edges is a rigid structure. However, if the structure is not convex, but still simply connected, there are several possibilities. It may be any of the following cases:

(a) rigid,
(b) infinitesimally movable (shaky),
(c) two or more stable forms (multi-stable),
(d) a continuously movable linkage.

Shaky polyhedra

The regular icosahedron of twenty triangular faces is convex and rigid. If six pairs of faces with an edge in common are replaced by other pairs of isosceles faces with their edges in common at right angles to the original common edge to make a non-convex icosahedron, we may obtain the orthogonal icosahedron of Jessen [3], shown in Figure 1. Each dihedral angle is $\pi/2$ or $3\pi/2$. This polyhedron can be deformed infinitesimally, making the dihedral angles at all the long edges infinitesimally greater or less, without strain on the faces. Structures of this type are called *shaky* structures. The tensegrity icosa of R. Buckminster Fuller [4] is made by replacing the long edges by rigid columns in compression, while the short edges are replaced by wires in tension. A shaky octahedron was described by Blaschke [5] and is pictured in a paper by Gluck [6]. Other shaky structures will be described in later sections.

Multi-stable structures

A polyhedron can be made by folding its development into a closed surface. Sometimes this can be

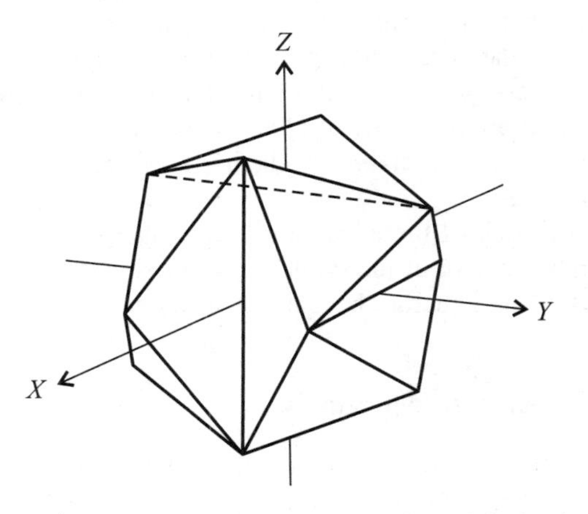

Figure 1. Orthogonal icosahedron (shaky polyhedron)

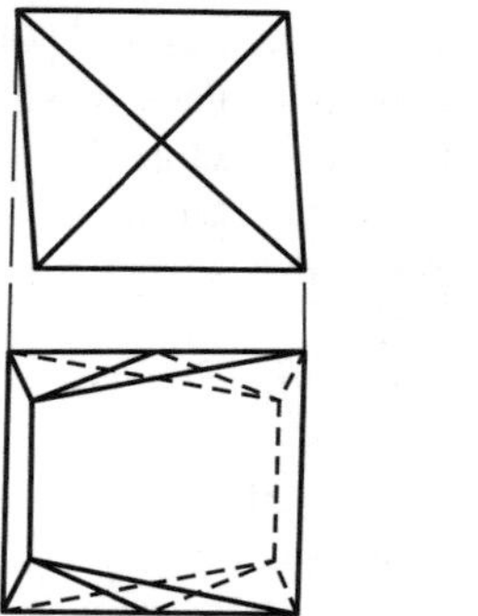
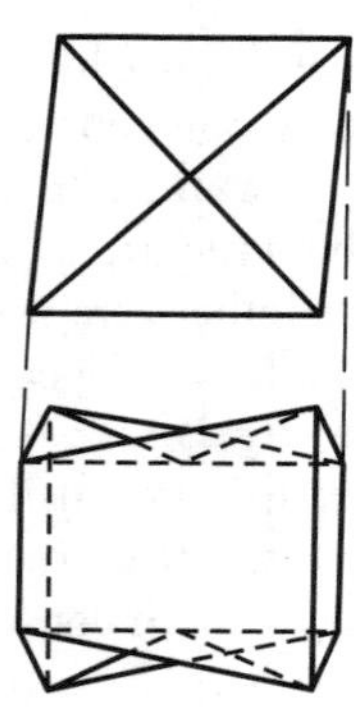
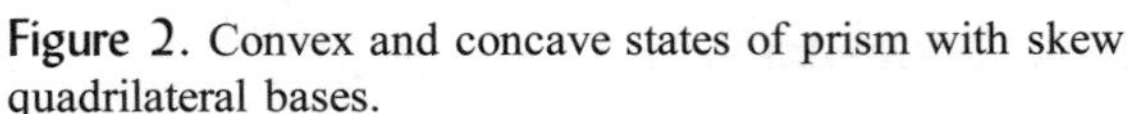
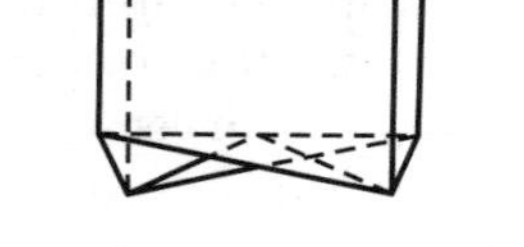
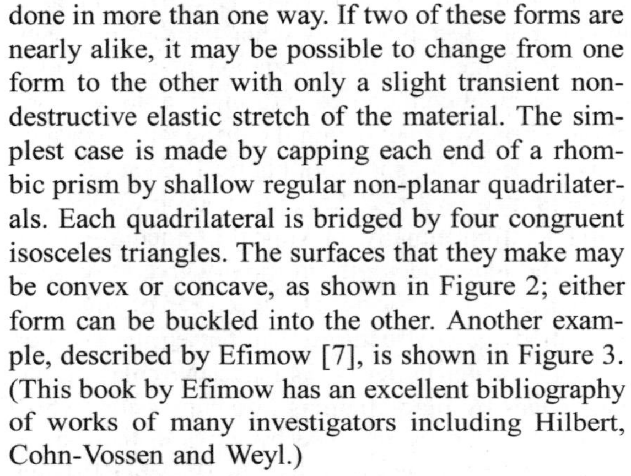

Figure 2. Convex and concave states of prism with skew quadrilateral bases.

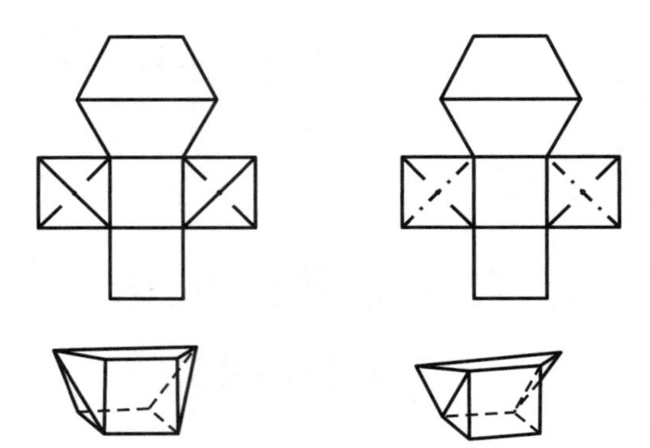

Figure 3. Convex and concave states of polyhedron made from same development (Efimow)

done in more than one way. If two of these forms are nearly alike, it may be possible to change from one form to the other with only a slight transient non-destructive elastic stretch of the material. The simplest case is made by capping each end of a rhombic prism by shallow regular non-planar quadrilaterals. Each quadrilateral is bridged by four congruent isosceles triangles. The surfaces that they make may be convex or concave, as shown in Figure 2; either form can be buckled into the other. Another example, described by Efimow [7], is shown in Figure 3. (This book by Efimow has an excellent bibliography of works of many investigators including Hilbert, Cohn-Vossen and Weyl.)

Another type of example is a non-convex triangular antiprism described by Wunderlich [8]. The chain of six lateral plates, shown in Figure 4, can be closed to make a buckled ribbon surface in either of two ways. By adding two triangles, a complete closed octahedron is produced. Either form can be made into the other while undergoing a slight temporary elastic strain. This structure can be generalized to other antiprisms, both regular and irregular, provided the bases are congruent cyclic polygons. The simplest cases to design and construct are those in which all the faces can be collapsed into a plane. Then the true shapes of all the faces are shown, as in Figure 5.

In each of these examples, there is a special case in which two of the stable positions coincide. In this case, the structure becomes a shaky structure.

The Siamese dipyramids

Some years ago, in his search for simply connected closed structures which have multiple stable positions, the author devised the following Siamese dipyramids. These are made of two regular dipyramids which are combined to form a closed polyhedron, as shown in Figure 6. If one of the dipyramids is compressed, then the other dipyramid increases in height. Consider the case of a Siamese dipyramid made of twenty equilateral triangles of unit edge. If the height of one dipyramid is $2x$, the height of the other dipyramid is $2y$, the radii of one dipyramid are r, and the angles are θ and 2θ as marked, then the variables r, θ, x, y are connected by the relations

$$\sin\theta = l/2r, \qquad x^2 = 1 - r^2,$$

$$\begin{aligned} y &= r\sin 5\theta = r(5\sin\theta - 20\sin^3\theta + 16\sin^5\theta) \\ &= r\sin\theta(5 - 20\sin^2\theta + 16\sin^4\theta) \\ &= (5 - 5/r^2 + 1/r^4)/2. \end{aligned}$$

Hence,

$$2yr^4 = 5r^4 - 5r^2 + 1,$$

or

$$2y(1-x^2)^2 = 5x^4 - 5x^2 + 1.$$

If $2x = 2y$, we obtain the solutions $2x = 2y \approx 0.655$. Other solutions of the equation are $2x \approx 0.142$ and $2y \approx 0.985$, or $2x \approx 0.985$ and $2y \approx 0.142$. Hence, there are three stable configurations that the polyhedron can assume.

If the equilateral triangles are replaced by isosceles triangles whose apex angles are approximately 59°, instead of 60°, then either dipyramid can be completely collapsed, as shown in Figure 7. If the apex angles are approximately 62°, then the three stable positions coincide to make it another shaky icosahedron for which $2x = 2y$.

The foregoing examples are only special cases of a family of dipyramids of $4n$ faces. For $n = 3$, we obtain a shaky dodecahedron of twelve isosceles triangles of apex angle approximately 107°36′.

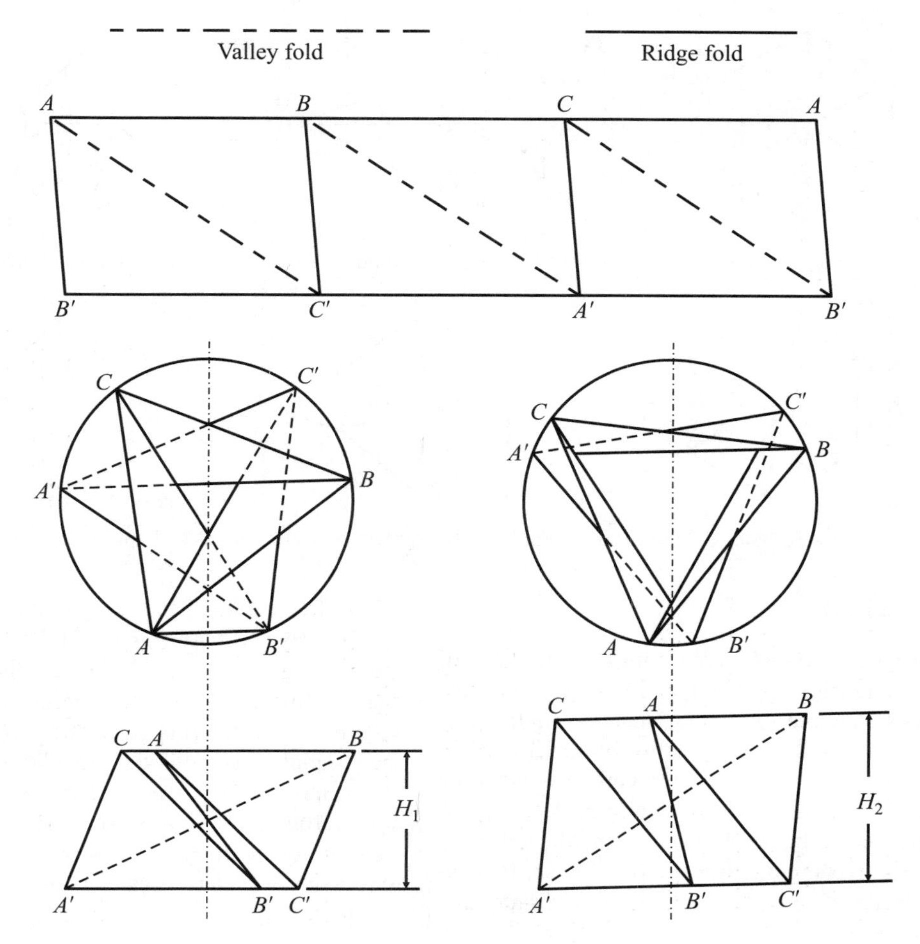

Figure 4. Chain of six congruent triangles and two stable states of Wunderlich triangular antiprism.

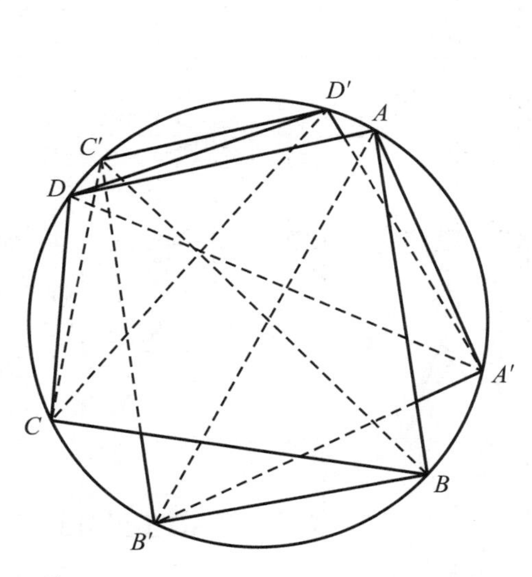

Figure 5. Collapsed state of bistable quadrilateral antiprism.

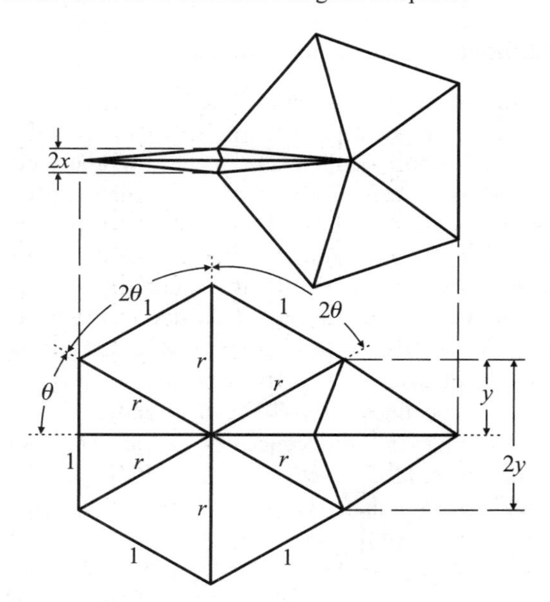

Figure 6. Siamese dipyramid of 20 congruent triangles.

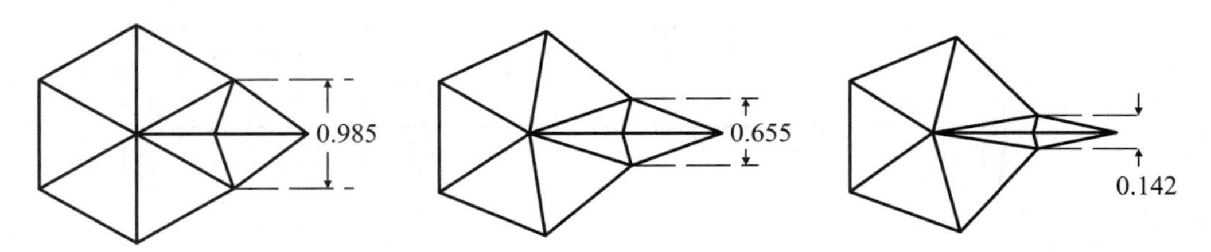

Figure 7. Three stable states of Siamese dipyramid of 20 congruent equilateral triangles.

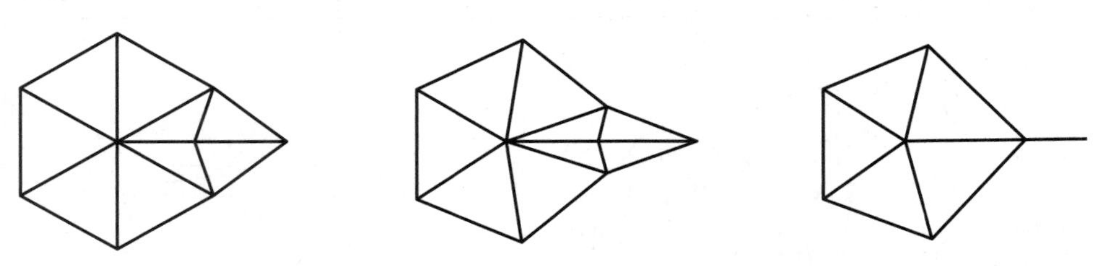

Figure 8. Three stable states of Siamese dipyramid of 20 congruent isosceles triangles of apex angles approximately 59°.

Buckled surfaces

Many examples of polyhedra with several states of stability can be obtained by buckling polyhedral surfaces. For example, two congruent rectangles, that are joined along all four edges to form a closed envelope, can be buckled into a non-convex polyhedron of twelve faces, as shown in Figure 9. A triangular envelope can be buckled into a triangular prism to which three rectangular pyramids are joined. By successive buckling, a large number of stable states can be produced.

The Bricard deformable octahedron

One of the most remarkable plate linkages was discovered by Bricard [9, 10]. A special form of this linkage is doubly collapsible, that is, all the faces of the linkage can be collapsed into a plane in two distinct ways. This form of the linkage can be constructed as follows. Construct two concentric circles of arbitrary radii, as shown in Figure 10a. Choose two arbitrary points A and A' outside of the larger circle. Construct the tangents from A and A' to the circles and determine their intersections B, B', C and C'. The lines BC, $B'C'$, $B'C$ and BC' will be tangent to a third concentric circle. Then the six triangles ABC', ABC, $AB'C$, $A'B'C$, $A'B'C'$, $A'BC'$ taken in that cyclic order, hinged at the common edges, constitute a deformable six-plate linkage. The construction of a workable cardboard model from the net of Figure 11 is not difficult, and it is well worth the effort. The other collapsed form taken by this linkage is shown in Figure 10b.

The free edges of the foregoing six-plate linkage are the edges of two triangles $AB'C'$ and $A'BC$. If these triangles are added as hinged plates to the linkage, a complete closed octahedron is formed. This octahedron is deformable in the same manner that the six-plate linkage is deformable; the added triangles impose no additional restraints. However, this octahedron is not simply connected, since the faces pass through each other. Therefore, a complete working model of material plates is not possible. Part of the missing plates $AB'C'$ and $A'BC$ can be added,

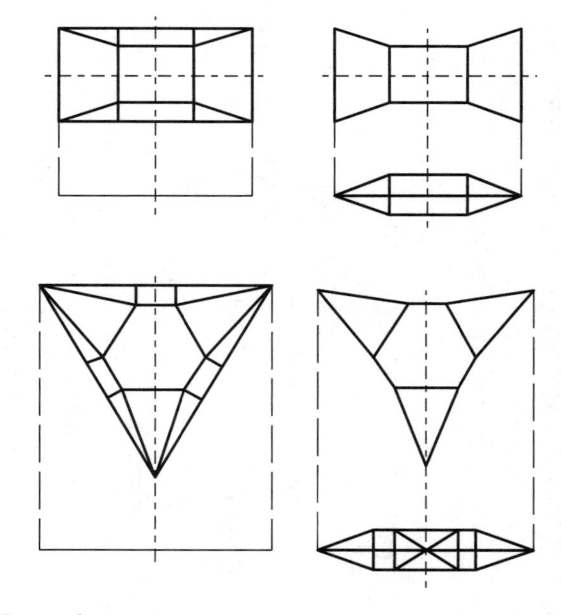

Figure 9. Buckled surfaces made from rectangular and triangular envelopes.

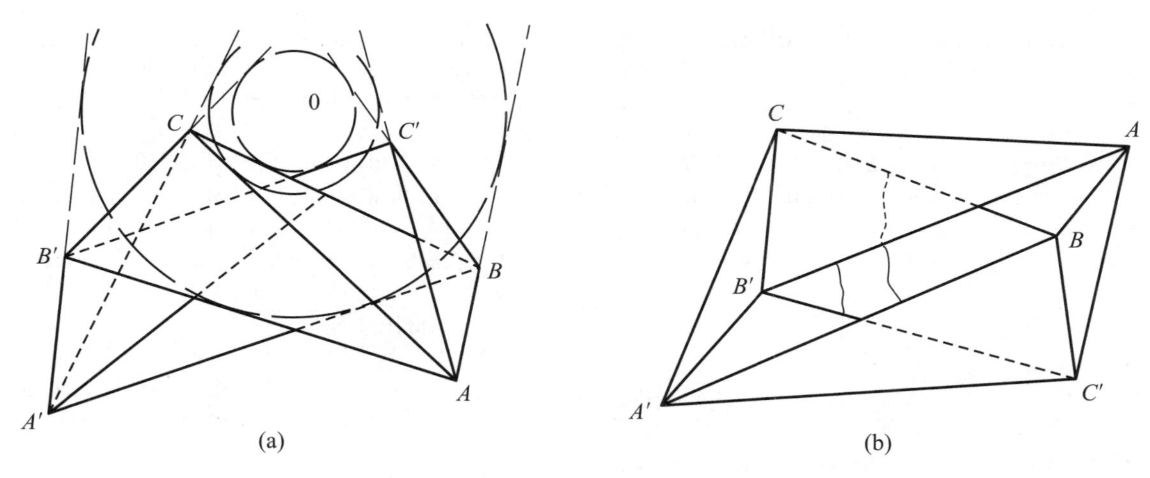

Figure 10. Bricard deformable octahedrons (collapsed)

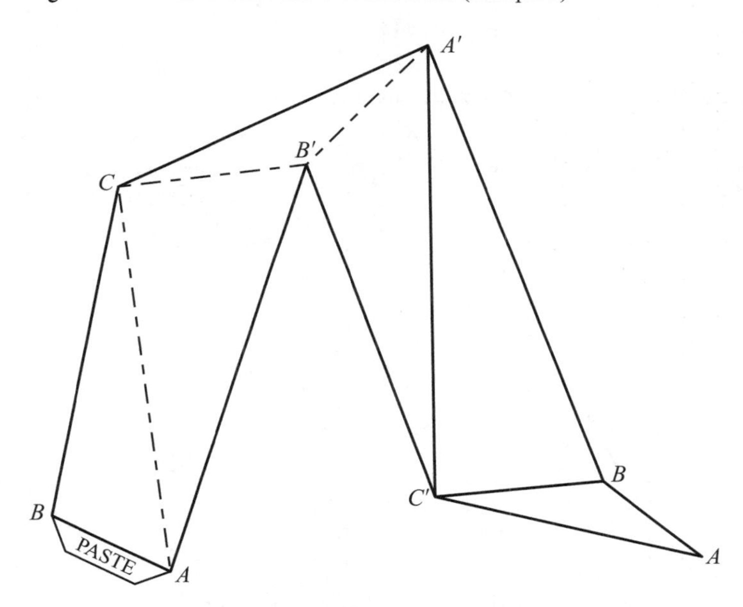

Figure 11. Net for Bricard deformable octahedron.

as shown in Figure 10b, and still permit motion of the linkage.

The closed flexible polyhedron

Since the time of Cauchy, mathematicians have sought a simply connected closed polyhedral structure which is continuously movable as a linkage. Recently Robert Connelly [13] has produced such a linkage by an ingenious modification of the Bricard octahedron. The plates $A'B'C$ and ABC' are retained unchanged. However, on each of the other six plates, a rigid trihedral structure has been built externally, and the original six plates have been eliminated. The resulting linkage has no interference, but it has a very limited motion. The relative motions of the eight links in this linkage are the same as in the original Bricard linkage. By a further modification of two hinges, the mobility has been greatly increased. This modification consists of cutting a notch in a hinge, and then closing the opening by the addition of six small plates. This "crinkle mechanism" is described in detail in a forthcoming paper [14].

Extensions and open questions

Besides the Bricard octahedron, there are other nonsimple polyhedral structures that are deformable.

There are many three-dimensional linkage mechanisms in which each link can be replaced by a tetrahedron or a prism. Then, the linkage becomes a deformable plate linkage. Examples are described and pictured in papers by the author [11, 12]. The possibilities of such three-dimensional linkage mechanisms have not been exhausted. They are being actively investigated by mechanical engineers.

References

1. A. Cauchy, Sur les polygones et les polyèdres, Second Mémoire, *J. École Polytechnique* 9 (1813) 87.
2. M. Dehn, Über die Starkheit konvexer Polyeder, *Math. Ann.* 77 (1916) 466–473.
3. B. Jessen, Orthogonal icosahedron, *Nordisk Mat. Tidskr.* 15 (1967) 90–96.
4. Hugh Kenner, *Bucky, A Guided Tour of Buckminster Fuller* (1973) 236.
5. W. Blaschke, Wackelige Achtfläche, *Math. Z.* 6 (1920) 85–93.
6. Herman Gluck, Almost all simply connected closed surfaces are rigid, *Lecture Notes in Math.* No. 438, Springer-Verlag, Heidelberg, 1975, 225–239.
7. N. W. Efimow, *Flächenverbiegung im Grossen*, Akademie-Verlag, Berlin, 1957, p. 130.
8. W. Wunderlich, Starre, kippende, wackelige und bewegliche Achtfläche, *Elem. Math.* 20 (1965) 25–32.
9. R. Bricard, Mémoire sur la théorie de l'octaèdre articulé, *Liouville J.* (5)3 (1897) 113–148.
10. G. T. Bennett, Deformable octahedra, *Proc. London Math. Soc.* 10 (1912) 309–343.
11. Michael Goldberg, New five-bar and six-bar linkages in three dimensions, *Trans. ASME* 65 (1943) 649-661.
12. Michael Goldberg, Tubular linkages, *J. Mathematical Phys.* 26 (1947) 10–21.
13. Robert Connelly, An immersed polyhedral surface which flexes, *Indiana University Mathematics Journal* 25 (1976) 965–972.
14. Robert Connelly, A counterexample of the rigidity conjecture for polyhedra, *Institut des Hautes Études Scientifiques, Pub. Math.* 47 (1977), 333–338.

Part VI

The 1980s

Leonhard Euler, 1707–1783

J. J. Burckhardt

Mathematics Magazine
56 (1983), 262–73

Editors' Note: Johann Jacob Burckhardt was one of the foremost experts on Euler at the time this article was published in 1983, the bicentennial of the death of Euler. He was one of the three members of the editorial committee that produced a major Euler memoir published that year in Switzerland, *Leonhard Euler 1707–1783, Beiträge zu Leben und Werk, Gedenkband des Kantons Basel-Stadt* (Birkhäuser, 1983).

Burckhardt, a native of Basel (the birthplace of Euler), received his PhD from the University of Zürich and taught there from 1933 to 1977. For 31 years he was managing editor of the *Commentarii Mathematici Helvetici* and was president of the Swiss Mathematical Society 1954–56. Primarily a mathematical historian, he has also published research in the field of crystallography. In 2003 a tribute to him was published in *Elemente der Mathematik*, to honor him on his 100th birthday, July 13th of that year.

Here he gives a quick sketch of the life and enormous accomplishments of that greatest of Swiss mathematicians and one of the greatest scientists of all time. The November 1983 issue of the *Magazine* is devoted to Euler and his work, with this article on Euler's life, George Andrews on the pentagonal number theorem, Harold M. Edwards on quadratic reciprocity, Paul Erdős and coauthor Underwood Dudley on problems in number theory, and Morris Kline on series, among others. The issue was nicely illustrated with portraits, pictures of medals, title pages from Euler's own books, and a picture of a well-known Swiss banknote with Euler's portrait.

Born in 1707, Leonhard Euler grew up in the town of Riehen, near Basel, Switzerland. Encouraged by his father, Paulus, a minister, young Leonhard received very early instruction from Johann I Bernoulli, who immediately recognized Euler's talents. Euler completed his work at the University of Basel at age 15, and at age 19 won a prize in the competition organized by the Academy of Sciences in Paris. His paper discussed the optimal arrangement of masts on sailing ships (*Meditationes super problemate nautico...*). In 1727 Euler attempted unsuccessfully to obtain a professorship of physics in Basel by submitting a dissertation on sound (*Dissertatio physica de sono*); however, this failure, in retrospect, was fortunate. Encouraged by Nicholas and Daniel, sons of his teacher Johann Bernoulli, he went to the St. Petersburg Academy in Russia, a field of action that could accommodate his genius and energy.

In St. Petersburg Euler was met by compatriots Jacob Hermann and Daniel Bernoulli and soon befriended the diplomat and amateur mathematician Christian Goldbach. During the years 1727–1741 spent there, Euler wrote over 100 scientific papers and his fundamental work on mechanics. In 1741, at the invitation of Fredrick the Great, he went to the Akademie in Berlin. During his 25 years in Berlin, his incredible mathematical productivity continued. He created, among other works, the calculus of variations, wrote the *Introductio in analysin infinitorum*, and translated and rewrote the treatise on artillery by Benjamin Robins.

Disputes with the Court led Euler in 1766 to accept a very favorable invitation by Katherine II to return to St. Petersburg. There he was received in a princely manner, and he spent the rest of his life in St. Petersburg. Although totally blind, he wrote, with the help of his students, the famous *Algebra* and over 400 scientific papers; he left many unpublished manuscripts.

In recent decades, numerous important materials concerning Euler have been discovered in the archives in the Academy of Sciences of the USSR. It would seem that there is probably little chance of now discovering an unknown manuscript or something important about his life. Euler himself acknowledged the advantageous circumstances he found at the Academy. Judith Kh. Kopelevic notes, "Euler's tombstone, erected by the Academy; his bust in the building of the Presidium of the Academy; the two-centuries-long efforts of the Academy to care for his enormous heritage and publish it—all these show clearly that Euler's encounter

INTRODUCTIO
IN ANALYSIN
INFINITORUM.
AUCTORE
LEONHARDO EULERO,
Professore Regio BEROLINENSI, & *Academiæ Imperialis Scientiarum* PETROPOLITANÆ *Socio.*

TOMUS PRIMUS

LAUSANNÆ,
Apud MARCUM-MICHAELEM BOUSQUET & Socios.
MDCCXLVIII.

with the Petersburg Academy of Sciences was a happy one for both sides."

The legacy of Euler's writings

Euler's productivity is astonishing in its range of content and in the sheer volume of written pages. He wrote landmark books on the subjects of mathematical analysis, analytic and differential geometry, the calculus of variations, mechanics, and algebra. He published over 760 research papers, many of which won awards in competitions, and at his death left hundreds of unpublished works; even today there remain unpublished over 3,000 pages in notebooks. In view of this prodigious collection of written material, it is not surprising that soon after Euler's death the task of surveying and publishing his works encountered extraordinary difficulty.

N. I. Fuss made efforts to publish more writings of the master, but only his son P.-H. Fuss succeeded (with the help of C. G. J. Jacobi) to generate interest among others, including Ostrogradskii. An enterprise in this direction was undertaken in Belgium (1838–1839), but failed after the publication of the fifth volume. In 1844, the Petersburg Academy decided on publication of the manuscripts, but this was not carried out. However, in 1849 the *Commentationes arithmeticae collectae*, edited by P.-H. and N. Fuss, were published; this contains, among others, the important manuscript *Tractatus de doctrina numerorum*.

The centennial of Euler's death in 1883 rekindled interest in Euler's works and in 1896 the most valuable preliminary to any complete publication appeared—the *Index operum Leonhardi Euleri* by J. G. Hagen. As the bicentennial of Euler's birth neared, new life was infused into the project, which was thoroughly discussed by the academies of Petersburg and Berlin in 1903. Although the project was abandoned at this time, the celebrations of the bicentennial of Euler's birth provided the needed impetus for the publication of the *Opera omnia*. The untiring efforts of Ferdinand Rudio led to the decision by the Schweizerische Naturforschende Gesellschaft [Swiss Academy of Sciences] in 1909 to undertake the publication, based on the list of Euler's writings prepared by Gustaf Eneström (1852–1923). He lists 866 papers and books published by then. The financial side appeared assured through gifts and subscriptions. But the first World War led to unforeseen difficulties. We are indebted to Andreas Speiser for his efforts, which made it possible to continue the publication, and who overcame financial and publication difficulties so that at the start of World War II about one half of the project was completed. After the war, Speiser, succeeded by Walter Habicht, completed the series 1 (29 volumes), 2 (31 volumes) and 3 (12 volumes) of the *Opera omnia* except for a few volumes.

In 1947–1948 the manuscripts which had been lent by the St. Petersburg Academy to the Swiss Academy of Sciences were returned to the archives of the Academy of Sciences of the USSR in Leningrad. Their systematic study was started under the supervision of the Academician V. I. Smirnov, with the goal of publishing a fourth series of the *Opera omnia*. As a first result, there appeared in 1965 a new edition of the correspondence between Euler and Goldbach, edited by A. P. Juškevič and E. Winter. In 1967, the Swiss Academy of Sciences and the Academia Nauk of the USSR formed an International Committee, to which was entrusted the publication of Euler's correspondence in a series 4A, and a critical publication of the remaining manuscripts in a series 4B.

To mark the passage of 200 years since Euler's death, a memorial volume has been produced by the Canton of Basel, *Leonhard Euler 1707–1783, Beiträge zu Leben und Werk*, edited by J. J. Burckhardt, E. A. Fellmann, and W. Habicht (Birkhäuser Verlag, Basel and Boston). From a contemporary point of view, this volume presents the insights of outstanding scientists on various aspects of Euler's achievements and their influence on later works. The complete list of essays and their authors appears at the end of this article. The memorial volume ends with a list, compiled by J. J. Burckhardt, of over 700 papers which are devoted to the work of Euler. It should be stressed that this is certainly an incomplete list, and it is hoped that it will lead to many additional listings which will then be published in an appropriate form. It is hoped that papers little known till now will receive the attention they deserve, and that this effort will lead to an improvement in the collaboration of scientists of all countries.

In the present article, we give a brief overview of the work of Euler. In order to include information from recently discovered work as well as the observations and insights of modern scholars, we draw freely from material found in the memorial volume.

Number Theory

Euler had a passionate lifelong interest in the theory of numbers. Approximately one-sixth of his published work in pure mathematics is in this area; the same is true of the manuscripts left unpublished at his death. Although he had an active correspondence with Goldbach, he complained about the lack of response on the part of other contemporary mathematicians such as Huygens, Clairaut, and Daniel Bernoulli, who considered number theory investigations a waste of time, and were even unaware of Fermat's theorem. (Forty years passed before Euler's investigations into Goldbach's problem were followed up by Lagrange.)

André Weil has commented that if one were to distinguish between "theoretical" and "experimental" researchers, as is done for physicists, then Euler's constant preoccupation with number theory would place him among the former. But in view of his insistence on the "inductive" method of discovery of arithmetic truths, carrying out a wealth of numerical calculations for special cases before tackling the general question, one could equally well call him an "experimental" genius.

At the beginning of the eighteenth century—50 years after Fermat's death—the number theoretical work of Fermat was practically forgotten. In a letter dated December 1, 1727, Christian Goldbach brought to Euler's attention Fermat's assertion that numbers of the form

$$2^{2^{p-1}} + 1, \; p \text{ prime},$$

(i.e., $3, 5, 17, 257, \ldots$) are also prime; this led Euler to a study of Fermat's works. His investigations included Fermat's theorem and its generalizations, representations of numbers as sums of squares of polygonal numbers, and elementary quadratic forms.

In the decade between 1740 and 1750, Euler created the basis of a new theory which, until this day, has not essentially changed its character. The question which motivated this work was posed by Naudé on September 12, 1740, who asked Euler the number of ways in which a given integer can be represented as a sum of integers. For this problem, the "partitio numerorum," as well as for related problems, Euler found solutions by associating with a number-theoretic function its generating function, which can be investigated by analytical methods. Euler clearly understood the importance of his discovery. Although he had not found the proof of several central theorems of his theory, he incorporated the basic ideas and a few elementary but remarkable special results in his fundamental text in analysis, *Introductio in analysin infinitorum*. V. Scharlau comments, "Even today it is hard to imagine a more convincing and interesting introduction to this theory."

Euler used this theory in attempting to find a formula for prime numbers, where he considered the function $\sigma(n)$, the sum of all divisors of n. He obtained the formula

$$\sigma(p^k) = \frac{p^{k+1}-1}{p-1}, \quad \text{for } p \text{ prime}$$

from which the computation of $\sigma(n)$ follows. Euler also formulated the recursion rule for $\sigma(n)$,

$$\sigma(n) = \sigma(n-1)+\sigma(n-2)-\sigma(n-5)-\sigma(n-7)+\cdots$$

and observed its similarity to the one for $p(n)$, the number of partitions of n. In 1750, Euler brought these investigations to a conclusion by formulating the identity

$$\sum_{i=1}^{\infty}(1-x^i) = 1 + \sum_{m=1}^{\infty}(-1)^m(x^{\frac{1}{2}(3m^2-m)} + x^{\frac{1}{2}(3m^2+m)})$$

which is a cornerstone for all his related results.

Another interesting application of generating functions can be found in Euler's various investigations of "population dynamics," which probably originated in the years 1750–1755. Scharlau writes:

> From today's point of view it is possibly not surprising that Euler found no additional results on generating functions; indeed it took many decades—almost a century—after the end of his activity before his achievements were substantially surpassed. It is remarkable how little attention was given to Euler's ideas by the mathematicians of the 18th and 19th centuries.... There are very few mathematical theories whose character has changed so little since Euler's time as the theory of generating functions and the partitions of numbers.

Among the unpublished fragments of Euler's work (a total of about 3,000 pages, mainly bound in numbered notebooks) are over 1,000 pages which are devoted to number theory, mostly from the years 1736–1744 and 1767–1783. Euler's technique of investigation emerges clearly from these. After lengthy efforts which at times span many years, he reaches his results based on observations, tables, and empirically established facts.

G. P. Matvievskaja and E. P. Ozigova, who have perused these fragments, note that "the handwritten materials widen our views of Euler's activity in the field of number theory. The same holds for other directions of his research. The manuscripts enable us to recognize the sources of his mathematical discoveries." A few examples serve to illustrate these points. On page 18 in notebook N 131 is the problem of deciding whether a given integer is prime. The same notebook contains an entry about the origin of the zeta function, as well as the first mention of the theorem of four squares, to which Euler returns in notebook N 132 (1740–1744). A particularly interesting entry in notebook N 134 (1752–1755) contains Euler's formulation, a hundred years before Bertrand, of the "Bertrand postulate," that there is at least one prime between any integer n and $2n$.

Analysis

Euler was occupied throughout his life with the concept of function; the treatises he produced in analysis were fundamental to the development of the modern foundations of analysis. As early as 1727 Euler had written a fifteen-page manuscript *Calculus differentialis*; it's interesting to compare this fledgling work with his later treatise *Institutiones calculi differentialis* (1755). Here Euler explains the calculus of finite differences of finite increments and considers calculations with infinitely small quantities. D. Laugwitz, one of the contributors to the modern development of analysis through the adjoining of an infinity symbol Ω, remarks that anyone who reads this work, or Euler's *Introductio in analysin infinitorum* (1748), must be struck by the confidence with which Euler utilizes the calculus of both infinitely large and infinitely small magnitudes. Laugwitz indicates that it is possible to formulate Euler's ideas in the modern setting of nonstandard analysis; hence Euler receives a belated justification of his unorthodox techniques.

The richness and diversity of Euler's work in analysis can be seen by a brief summary of the book *Introductio in analysin infinitorum*. The first chapter discusses the definition of "function" which originated with Johann Bernoulli. In the second, Euler formulates the "fundamental theorem of algebra" and sketches a proof; he presents results on real and complex solutions of algebraic equations, a topic resumed in chapter 12 which deals with the decomposition of rational functions into partial fractions. The third chapter contains the so-called "Euler substitution," and the important replacement of a nonexplicit functional dependence by a parametric rep-

EVOLUTIONE FACTORUM ORTIS. 223

ubi alii numeri non occurrunt, niſi qui ex his duobus 2 & 3 per multiplicationem originem trahunt; ſeu qui alios Diviſores præter 2 & 3 non habent. CAP. XV.

273. Si igitur pro α, β, γ, δ, &c., unitas per ſingulos omnes numeros primos ſcribatur, ac ponatur

$$P = \frac{1}{(1-\frac{1}{2})(1-\frac{1}{3})(1-\frac{1}{5})(1-\frac{1}{7})(1-\frac{1}{11})(1-\frac{1}{13})\ \&c.},$$

fiet

$$P = 1 + \frac{1}{2} + \frac{1}{3} + \frac{1}{4} + \frac{1}{5} + \frac{1}{6} + \frac{1}{7} + \frac{1}{8} + \frac{1}{9} + \&c.,$$

ubi omnes numeri tam primi, quam qui ex primis per multiplicationem naſcuntur, occurrunt. Cum autem omnes numeri vel ſint ipſi primi, vel ex primis per multiplicationem oriundi, manifeſtum eſt, hic omnes omnino numeros integros in denominatoribus adeſſe debere.

274. Idem evenit, ſi numerorum primorum Poteſtates quæcunque accipiantur: ſi enim ponatur

$$P = \frac{1}{(1-\frac{1}{2^n})(1-\frac{1}{3^n})(1-\frac{1}{5^n})(1-\frac{1}{7^n})(1-\frac{1}{11^n})\&c.},$$

fiet

$$P = 1 + \frac{1}{2^n} + \frac{1}{3^n} + \frac{1}{4^n} + \frac{1}{5^n} + \frac{1}{6^n} + \frac{1}{7^n} + \frac{1}{8^n} + \&c.,$$

ubi omnes numeri naturales nullo excepto occurrunt. Quod ſi autem in Factoribus ubique ſignum + ſtatuatur, ut ſit

$$P = \frac{1}{(1+\frac{1}{2^n})(1+\frac{1}{3^n})(1+\frac{1}{5^n})(1+\frac{1}{7^n})(1+\frac{1}{11^n})\ \&c.},$$

erit

Euleri *Introduct. in Anal. infin. parv.* Ff $P =$

The Euler product-sum formula from chapter 15 of *Introductio in analysin infinitorum*

resentation. Particularly remarkable is Euler's strict theory of logarithms, and the consideration of the exponential function in chapter 6. Euler asserts that the logarithms of rationals are either rational or transcendental, a fact which was proved only two hundred years later. Weakly convergent series are considered in chapter 7, as well as the question of convergence of series and the relation between a function and its representation outside the circle of convergence. Subsequent chapters deal with transcendental functions and their representation as series or products. The starting point of Bernhard Riemann's investigation of the distribution of primes is in chapter 15, in the formula

$$\sum_n \frac{1}{n^x} = \prod_p \left(\frac{1}{1-1/p^x}\right)$$

in which the summation extends over all positive integers and the product over all primes (see p. 80). In chapter 16 Euler turns to the new topic—rife with

DE PARTITIONE NUMERORUM. 253

CAP. XVI.

CAPUT XVI.

De Partitione numerorum.

297. PRopoſita ſit iſta expreſſio

$$(1 + x^{\alpha}z)(1 + x^{\beta}z)(1 + x^{\gamma}z)(1 + x^{\delta}z)(1 + x^{\varepsilon}z)\ \&c.,$$

quæ cujuſmodi induat formam, ſi per multiplicationem evolvatur, inquiramus. Ponamus prodire

$$1 + Pz + Qz^2 + Rz^3 + Sz^4 + \&c.,$$

atque manifeſtum eſt P fore ſummam Poteſtatum $x^{\alpha} + x^{\beta} + x^{\gamma} + x^{\delta} + x^{\varepsilon} + \&c.$. Deinde Q eſt ſumma Factorum ex binis Poteſtatibus diverſis, ſeu Q erit aggregatum plurium Poteſtatum ipſius x, quarum Exponentes ſunt ſummæ duorum terminorum diverſorum hujus Seriei

$$\alpha,\ \beta,\ \gamma,\ \delta,\ \varepsilon,\ \zeta,\ \eta,\ \&c.$$

Simili modo R erit aggregatum Poteſtatum ipſius x, quarum Exponentes ſunt ſummæ trium terminorum diverſorum. Atque S erit aggregatum Poteſtatum ipſius x, quarum Exponentes ſunt ſummæ quatuor terminorum diverſorum ejuſdem Seriei, $\alpha, \beta, \gamma, \delta, \varepsilon$, &c., & ita porro.

298. Singulæ hæ Poteſtates ipſius x, quæ in valoribus literarum P, Q, R, S, &c., inſunt, unitatem pro coëfficiente habebunt, ſi quidem earum Exponentes unico modo ex

Ii 3 $\alpha, \beta,$

Chapter 16 title page from *Introductio in analysin infinitorum*

algebraic ideas—of *Partitione numerorum*, the additive decomposition of natural numbers (see p. 81). The developments of power series into infinite series found here were continued only by Ramanujan, Hardy and Littlewood. The expressions found here were later called theta functions, and used by Jacobi in the general theory of elliptic functions. The last chapter, 17, deals with the numerical solution of algebraic equations, following Daniel Bernoulli.

A. O. Gelfond, whose essay in the memorial volume contains a deep analysis of the contents of *Introductio* ..., interprets Euler's ideas in modern terms and stresses the great relevance of this work, even to this day.

Euler's interest in the theory of vibrating strings is legendary. In 1747 d'Alembert formulated the theory and the corresponding partial differential equation; this prompted Euler in 1750 to develop a solution, although restricted to the case in which the vibrations satisfy certain conditions. Euler's friend Daniel Bernoulli contributed (about 1753) two remarkable articles, and presented the solution in the form of a trigonometric series. The problem is fittingly illuminated by Euler's question "what is the law of the vibrating string if it starts with an arbitrary shape" and d'Alembert's answer "in several cases it is not possible to solve the problem, which transcends the resources of the analysis available at this time."

Euler has sometimes been criticized for seeming to ignore the concept of convergence in his freewheeling calculations. Yet in 1740, Euler gave an incomplete formulation of the criterion of convergence that later received Cauchy's name. Euler's last paper was completed in 1783, the year of his death; it contained the germ of the concept of uniform convergence. His example was utilized by Abel in 1826.

After surveying the rich contributions to analysis made in Euler's time, Pierre Dugac declares, "Euler and d'Alembert were the instigators of the most important work on the foundations of analysis in the nineteenth century."

LETTRES
A UNE PRINCESSE
D'ALLEMAGNE
SUR DIVERS SUJETS
de
PHYSIQUE & de PHILOSOPHIE

TOME PREMIER

A SAINT PETERSBOURG
de l'Imprimerie de l'Academie Impériale des Sciences
M DCC LX VIII.

"Applied" Mathematics (Physics)

Euler's investigations and formulations of basic theory in the areas of optics, electricity and magnetism, mechanics, hydrodynamics and hydraulics are among the most fundamental contributions to the development of physics as we know it today. Euler's views on physics had an immediate influence on the study of physics in Russia; this grew out of his close relationship with the contemporary and most influential Russian scientist, M. V. Lomonosov, his several Russian students, and the publication of a translation (by S. J. Rumovskii) of his very popular "Letters to a German Princess." The "Letters ...," which had originated as lessons to the princess of Anhalt-Dessau, niece of the King of Prussia, during Euler's years in Berlin, served as the first encyclopedia of physics in Russia. A. T. Grigor'jan and V. S. Kirsanov observed that the physicist N. M. Speranskii, a noted statesman and author of a physics book (1797), used to read to his students sections from Euler's "Letters ...."

B. L. van der Waerden, in discussing Euler's justification of the principles of mechanics, has asked, "What did Euler mean by saying that in the computation of the total moment of all forces, the inner forces can be neglected because 'les forces internes se détruisent mutuellement'?" He points out that in order to answer that question it is important to know Euler's concept of solids, fluids, and gases. Are they true continua, or aggregates of small particles? The answer can be found in Euler's letters #69 and #70 to a German princess. He does not consider water, wool and air as true continua, but assumes that they consist of separate particles. However, in hydrodynamics, Euler treats liquids and gases as if they were continua. Euler is well aware that this is only an approximation.

A study of the published works of Daniel and Johann Bernoulli, as well as Euler's unpublished works (in particular, Euler's thick notebook from 1725–1727), by G. K. Mikhailov, gives some new and surprising insights into Euler's contributions to the development of theoretical hydraulics. Mikhailov states:

> It is generally known that the creation of the foundations of modern hydrodynamics of ideal fluids is one of the fruits of Euler's scientific activity. Less well known is his role in the development of theoretical hydraulics, that is, as usually understood, the hydrodynamic theory of fluid motion under a one-dimensional flow

model. Traditionally—and with good reason—it is assumed that the foundations of hydraulics were developed by Daniel and Johann Bernoulli in their works published between 1729 and 1743. In fact, during the second quarter of the eighteenth century Euler did not publish even a single paper on the elements of hydraulics. The central theme of most of the recent historical-critical studies on the state of hydraulics in that period is the determination of the respective contributions of Daniel and of Johann Bernoulli. But Euler stood, all this time, just beyond the curtain of the stage on which the action was taking place, although almost no contemporary was aware of that.

Euler's work on the theory of ships culminated in the publication of *Scientia navalis seu tractatus de construendis ac dirigendis navibus*, published in 1749. Walter Habicht notes the fundamental importance of this treatise:

> Following the *Mechanica sive motus scientia analytice exposita* which appeared in 1736, it [the *Scientia navalis* ...] is the second milestone in the development of rational mechanics, and to this day has lost none of its importance. The principles of hydrostatics are presented here, for the first time, in complete clarity; based on them is a scientific foundation of the theory of shipbuilding. In fact, the topics treated here permit insights into all the related developments in mechanics during the eighteenth century.

Although Euler's intense interest in the science of optics appeared before he was 30 and remained with him almost to his death, there is still no monographic evaluation of his contributions to the wide field of physical and geometrical optics. Part of Euler's work is best described by Habicht:

> In the second half of his life, from 1750 on and throughout the sixties, Leonhard Euler worked intensively on problems in geometric optics. His goal was to improve in several ways optical instruments, in particular, telescopes and microscopes. Besides the determination of the enlargement, the light intensity and the field of view, he was primarily interested in the deviations from the point-by-point imaging of objects (caused by the diffraction of light passing through a system of lenses), and also in the even less tractable deviations which arise from the spherical shape of the lenses. To these problems Euler devoted a long series of papers, mainly published by the Berlin academy. He admitted that the computational solution of these problems is very hard. As was his custom, he collected his results in a grandly conceived textbook, the *Dioptricæ* (1769–1771). This book deals with the determination of the path of a ray of light through a system of diffracting spherical surfaces, all of which have their centers on a line, the optical axis of the system. In a first approximation, Euler obtains the familiar formulae of elementary optics. In a second approximation he takes into account the spherical and chromatic aberrations. After passing through a diffracting surface, a pencil of rays issuing from a point on the optical axis is spread out in an interval on the optical axis; this is the so-called "longitudinal aberration." Euler uses the expression "espace de diffusion." If the light passes through several diffracting surfaces, the "espace de diffusion" is determined using a principle of superposition.
>
> Euler had great expectations for his theory, and believed that using his recipes, the optical instruments could be brought to "the highest degree of perfection." Unfortunately, the practical realization of his systems of lenses did not yield the hoped-for success. He searched for the causes of failure in the poor quality of the lenses on the one hand, and also in basic errors in the laws of diffraction which were determined experimentally in a manner completely unsatisfactory from a theoretical point of view. Because of the failure of his predictions, Euler's *Dioptrica* is often underrated.

Habicht notes that Euler's theory can be modified to obtain the general imaging theories developed in the nineteenth century. The crucial gap in Euler's treatment consists in neglecting those aberrations which are caused by the distance of the object and its images from the optical axis; with modification it is possible to determine the spherical aberration errors of the third order directly from Euler's formulas.

A responsible evaluation of Euler's contributions to optics will be possible only after Euler's unpublished letters and manuscripts are edited and made generally accessible. E. A. Fellmann provides an example of Euler's method which helps to place Euler's contribution in a historic context. The problem of diffraction in the atmosphere is one which was first seriously considered by Euler:

He began by deriving a very general differential equation; naturally, it turned out not to be integrable—it would have been a miracle had that not happened. Then he searched for conditions which make a solution possible, and finally he solved the problem in several cases under practically plausible assumptions.

Euler frequently expressed the opinion that the phenomena in optics, electricity and magnetism are closely related (as states of the ether), and that therefore they should receive simultaneous and equal treatment. This prophetic dream of Euler concerning the unity of physics could only be realized after the construction of bridges (experimental as well as theoretical) which were missing in Euler's time. These were later built by Faraday, W. Weber and Maxwell.

Euler was deeply influenced by the work of scientists who preceded him as well as by the work of his contemporaries. This is perhaps best illustrated by his role in the development of potential theory. He acknowledges the influence of the work of Leibniz, the Bernoullis, and Jacob Hermann, whose work he had studied in his days in Basel to 1727. In the decade 1730–1740, the contemporaries Euler, Clairaut and Fontaine all were active in developing the main ideas that would lead to potential theory: the geometry of curves, the calculus of variations, and the study of mechanics. By 1752 Euler's work on fluid mechanics *Principia motus fluidorum* was complete. A summary of his contributions to potential theory is given by Jim Cross:

> He helped, with Fontaine and Clairaut, to develop a logical, well-founded calculus of several variables in a clear notation; he transformed, with Daniel Bernoulli and Clairaut, the Galileo-Leibniz energy equation for a particle falling under gravity, into a general principle applicable to continuous bodies and general forces (the principle of least action with Daniel Bernoulli and Maupertuis forms part of this); and he founded, after the attempts of the Bernoullis, d'Alembert, and especially Clairaut, the modern theory of fluid mechanics on complete differentials for forces and velocities. His work was fruitful: the theories of Lagrange grew from his writings on extremization, fluids and sound, and mechanics; the work of Laplace followed.

Astronomy

Research by Nina I. Nevskaja based on newly available original documents justifies calling Euler a professional astronomer—and even an observer and experimental scientist. Five hundred books and manuscripts from the private library of Joseph Nicholas Delisle have recently come to light and from these one finds that this scientist found Euler a suitable collaborator and valued his knowledge in spherical trigonometry, analysis and probability.

It was a surprise when the records of observations of the Petersburg observatory during its first 21 years—which were presumed lost—were discovered in 1977 in the Leningrad branch of the archives of the Academy of Sciences of the USSR. For almost ten years, Euler was among those who were regularly taking measurements twice daily. Based on these observations, Delisle and Euler computed the instant of true noon, and the noon correction. Euler's entries were so detailed and numerous that it is possible to deduce from them how he gradually mastered the methods of astronomical observations. Utilizing the insights he obtained, Euler found a simple method of computing tables for the meridional equation of the sun; he presented it in the paper *Methodus computandi aequationem meridiei* (1735).

Euler was fascinated by sunspots; his notes from this period contain enthusiastic comments on his observations. The computation of the trajectories of the sunspots by Delisle's method can be considered the beginning of celestial mechanics. The archives also disclose that Euler helped Delisle by working out analytical methods for the determination of the paths of comets.

A little-noted field of Euler's activities, the theory of motion of celestial bodies, is documented by Otto Volk. Euler's first paper, based on generally formulated differential equations of mechanics, is entitled *Recherches sur le mouvement des corps célestes en général* (1747). Using the tables of planets computed by Thomas Street from the pure Keplerian motion of planets around the sun, Euler discusses in Sections 1 to 17 the observed irregularities. In Section 18 he formulates the differential equations of mechanics, and obtains the solution

$$r = a(1 + e\cos v) = \frac{a(1-e^2)}{1 - e\cos\phi}$$

in which r is the radius, v is the eccentric anomaly and s is the true anomaly, while e and a are constants. This is a regularization of the so-called inverse problem of Newton. Later, Euler obtains a trigonometric series for ϕ; such Fourier series are the basis of his computation of perturbations. This is the topic treated in detail in the prize proposal to the Paris Academy, *Recherches sur la question des inégalités du mouvement de Saturne et de Jupiter, sujet proposé pour le prix de l'année 1748*. In it Euler uses, for the first time, Newton's laws of gravitation to compute the mutual perturbations of planets.

In his paper *Considerationes de motu corporum coelestium* (1764), Euler is the first to begin considering the three-body problem, under certain restrictions. Euler notes the intractability of the problem:

> There is no doubt that Kepler discovered the laws according to which celestial bodies move in their paths, and that Newton proved them—to the greatest advantage of astronomy. But this does not mean that the astronomical theory is at the highest level of perfection. We are able to deal completely with Newton's inverse-square law for two bodies. But if a third body is involved, so that each attracts both other bodies, all the arts of analysis are insufficient .... Since the solution of the general problem of three bodies appears to be beyond the human powers of the author, he tried to solve the restricted problem in which the mass of the third body is negligible compared to the other two. Possibly, starting from special cases, the road to the solution of the general problem may be found. But even in the case of the restricted problem the solution encounters difficulties so great that the author has to admit to have spent much effort in vain attempts at solution.

Euler's investigation of the three-body problem was noted only at a later date; the linear solutions to the equation of the fifth degree were (and sometimes still are) called "Lagrange's solutions," without any mention of Euler. But Euler achieved fame through his theory of perturbations, presented in *Nouvelle méthode de déterminer les dérangemens dans le mouvement des corps célestes, causé par leur action mutuelle*. By iteration he determined, for the first time, the perturbations of the elements of the elliptical paths, and then applied this method to determine the motion of three mutually attracting bodies.

Correspondence

The circle of contemporary scholars who were influenced by and in turn, influenced, Euler's investigations was as wide as one could imagine in the eighteenth century. His voluminous correspondence testifies to the fruitful interaction between scientists through queries, conjectures, critical comments, and praise. Some of the correspondence has been published previously in collected works; a standard reference is the collection *Correspondance Mathématique et Physique*, edited by N. Fuss and published in 1843 by the Imperial Academy of Science, St. Petersburg. New discoveries and more complete information have produced recently published collections. The publication in 1965 of the correspondence between Euler and Christian Goldbach has been mentioned earlier.

It is significant that the first volume, Al, published in the fourth series of Euler's *Opera omnia*, contains a complete list of all existing letters to and from Euler (about 3,000), together with a summary of their contents. Volume A5 of this series (1980), edited by A. P. Juškevič and R. Taton, contains Euler's correspondence with A. C. Clairaut, J. d'Alembert, and J. L. Lagrange.

The correspondence between Euler and Lagrange from 1754 to 1775 gives valuable testimony to the development of personal relations between two of the most important scientists of that time. The letter exchange begins with a letter from the 18-year-old Lagrange, who lived in Turin, containing a query in which he mentions the analogy in the development of the binomial $(a+b)^m$ and the differential $d^m(xy)$. Mathematically isolated, Lagrange expresses his admiration for Euler's work, particularly in mechanics. Especially significant is the second letter to Euler (1755). In it Lagrange announces, without details, his new methods in the calculus of variations; Euler at once notes the advantage of these methods over the ones in his *Methodus inveniendi lineas curvas maximi minimive proprietate gaudentes* (1744), and heartily congratulates Lagrange. In 1756 Lagrange develops the differential calculus for several variables and investigates, for the first time, minimal surfaces. After an interruption of three years, Lagrange continues the correspondence by sending his work *La nature et la propagation du son*, and we find interesting discussions on the problem of vibrating strings, which had been carried on since 1749 between d'Alembert, Euler and Daniel Bernoulli.

After a lengthy pause, Euler resumes the correspondence. The first letter (1765) concerns the discussion with d'Alembert on vibrating strings, and the librations of the moon. In a second, Euler tells Lagrange that he has been granted permission by Friedrich II to return to Petersburg, and is attempting to have Lagrange come there. In later correspondence, the emphasis is on questions in the theory of numbers and in algebra. Pell's equation $x^2-ay^2=b$, and in particular $p^2-13q^2=101$, are discussed. Other topics deal with arithmetic, questions concerning developable surfaces, and the motion of the moon.

In 1770 Lagrange writes of his plan to publish Euler's *Algebra* in French, and to add to it an appendix; the published book is mailed on July 13, 1773. The last of Euler's letters, dated March 23, 1775, is remarkable by the exceptionally warm congratulations for Lagrange's work, especially about elliptic integrals. It may be conjectured that this was not the end of the correspondence, but unfortunately no additional letters have survived.

Postscript

This overview of Euler's life and work touches only a small part of the wealth of material to be found in the scholarly essays in the Basel memorial volume. In addition to careful and detailed analysis of many of Euler's scientific and mathematical achievements, these chapters contain new information on all aspects of Euler's private and academic life, his family, his philosophical and religious views, and the fabric of his life and work at the St. Petersburg Academy. In view of the overwhelming volume and diversity of Euler's work, it may never be possible to produce a comprehensive scientific biography of his genius. It is to be hoped that these newest contributions to the study of his life and work will provide impetus for further study and publication of many of the yet unpublished papers which are the unknown legacy of this mathematical giant.

The author and the editor express deep appreciation to Branko Grünbaum, who translated from the German the author's original manuscript. Doris Schattschneider took the trouble to shorten this manuscript from 38 to 22 pages.

Reference

Leonhard Euler 1707–1783, Beiträge zu Leben und Werk, Gedenkband des Kantons Basel-Stadt, edited by J. J. Burckhardt, E. A. Fellmann, and W. Habicht, Birkhäuser Verlag, Basel, 1983.

Table of Contents

Emil A. Fellmann (Basel, CH) Leonhard Euler-Ein Essay über Leben und Werk

Aleksander O. Gelfond (1906-1968) (Moskau, UdSSR) Uber einige charakteristische Züge in den Ideen L. Eulers auf dem Gebiet der mathematischen Analysis und in seiner "Einführung in die Analysis des Unendlichen"

André Weil (Princeton, USA) L'œuvre arithmétique d'Euler

Winfried Scharlau (Münster, BRD) Eulers Beiträge zur *partitio numerorum* und zur Theorie der erzeugenden Funktionen

Galina P. Matvievskaja/Helena P. Ožigova (Taškent-Leningrad UdSSR) Eulers Manuskripte zur Zahlentheorie

Adolf P. Juškevič (Moskau, UdSSR) L. Euler's unpublished manuscript *Calculus Differentialis*

Pierre Dugac (Paris, F) Euler, d'Alembert et les fondements de l'analyse

Detlef Laugwitz (Darmstadt, BRD) Die Nichtstandard-Analysis: Eine Wiederaufnahme der Ideen und Methoden von Leibniz und Euler

Isaac J. Schoenberg (Madison, USA) Euler's contribution to cardinal spline interpolation: The exponential Euler splines

David Speiser (Louvain, Belgien) Eulers Schriften zur Optik, zur Elektrizität und zum Magnetismus

Gleb K. Mikhailov (Moskau, UdSSR) Leonhard Euler und die Entwicklung der theoretischen Hydraulik im zweiten Viertel des 18.Jahrhunderts

Walter Habicht (Basel, CH) Einige grundlegende Themen in Leonhard Eulers Schiffstheorie

Bartel L. van der Waerden (Zürich, CH) Eulers Herleitung des Drehimpulssatzes

Walter Habicht (Basel, CH) Betrachtungen zu Eulers Dioptrik

Emil A. Fellmann (Basel, CH) Leonhard Eulers Stellung in der Geschichte der Optik

Jim Cross (Melbourne, Australien) Euler's contributions to Potential Theory 1730–1755

Otto Volk (Würzburg, BRD) Eulers Beiträge zur Theorie der Bewegungen der Himmelskörper

Nina I. Nevskaja (Leningrad, UdSSR) Leonhard Euler und die Astronomie

Judith Kh. Kopelevič (Leningrad, UdSSR) Leonhard Euler und die Petersburger Akademie

Ašot T. Grigor'jan/V. S. Kirsanov (Moskau, UdSSR) Euler's Physics in Russia

Ivor Grattan-Guinness (Barnet, GB) Euler's Mathematics in French Science, 1795–1815

René Taton (Paris, F) Les relations d'Euler et de Lagrange

Pierre Speziali (Genève, CH) Léonard Euler et Gabriel Cramer

Roger Jaquel (Mulhouse, F) Leonard Euler, son fils Jean-Albrecht et leur ami Jean III Bernoulli

Wolfgang Breidert (Karlsruhe, BRD) Leonhard Euler und die Philosophie

Michael Raith (Riehen bei Basel, CH) Der Vater Paulus Euler. Beiträge zum Verständnis der geistigen Herkunft Leonhard Eulers

René Bernoulli (Basel, CH) Leonhard Eulers Augenkrankheiten

Kurt-Reinhard Biermann (Berlin, DDR) Aus der Vorgeschichte der Euler-Werkausgabe

Johann Jakob Burckhardt (Zürich, CH) Die Eulerkommission der Schweizerischen Naturforschenden Gesellschaft.-Ein Beitrag zur Editionsgeschichte

Johann Jakob Burckhardt (Zürich, CH) Euleriana-Verzeichnis des Schrifttums über Leonhard Euler

Love Affairs and Differential Equations

Steven H. Strogatz

Mathematics Magazine
61 (1988), 35

Editors' Note: Professor Strogatz teaches in the Department of Theoretical and Applied Mechanics at Cornell University. His undergraduate education was at Princeton, with graduate work at Cambridge and Harvard. The present article was written at Harvard while he was on a postdoctoral NSF appointment. After reading this piece we are not surprised to learn that while teaching subsquently at MIT Strogatz won their highest teaching award, the E. M. Baker Award for Excellence in Undergraduate Teaching. He also won a Presidential Young Investigator Award from the National Science Foundation. His research is in nonlinear dynamics and chaos applied to physics, engineering, and biology.

The purpose of this note is to suggest an unusual approach to the teaching of some standard material about systems of coupled ordinary differential equations. The approach relates the mathematics to a topic that is already on the minds of many college students: the time-evolution of a love affair between two people. Students seem to enjoy the material, taking an active role in the construction, solution, and interpretation of the equations.

The essence of the idea is contained in the following example.

Juliet is in love with Romeo, but in our version of this story, Romeo is a fickle lover. The more Juliet loves him, the more he begins to dislike her. But when she loses interest, his feelings for her warm up. She, on the other hand, tends to echo him: her love grows when he loves her, and turns to hate when he hates her.

A simple model for their ill-fated romance is

$$\frac{dr}{dt} = -aj, \qquad \frac{dj}{dt} = br,$$

where

$$r(t) = \text{Romeo's love/hate for Juliet at time } t$$
$$j(t) = \text{Juliet's love/hate for Romeo at time } t.$$

Positive values of r, j signify love, negative values signify hate. The parameters a, b are positive, to be consistent with the story.

The sad outcome of their affair is, of course, a neverending cycle of love and hate; their governing equations are those of a simple harmonic oscillator. At least they manage to achieve simultaneous love one-quarter of the time.

As one possible variation, the instructor may wish to discuss the more general second-order linear system

$$\frac{dr}{dt} = a_{11}r + a_{12}j$$
$$\frac{dj}{dt} = a_{21}r + a_{22}j,$$

where the parameters a_{ik} $(i, k = 1, 2)$ may be either positive or negative. A choice of sign specifies the romantic style. As named by one of my students, the choice $a_{11}, a_{12} > 0$ characterizes an "eager beaver"—someone both excited by his partner's love for him and further spurred on by his own affectionate feelings for her. It is entertaining to name the other three possible styles, and also to contemplate the romantic forecast for the various pairings. For instance, can a cautious lover ($a_{11} < 0$, $a_{12} > 0$) find true love with an eager-beaver?

Additional complications may be introduced in the name of realism or mathematical interest. Nonlinear terms could be included to prevent the possibilities of unbounded passion or disdain. Poets have long suggested that the equations should be nonautonomous ("In the spring, a young man's fancy lightly turns to thoughts of love"—Tennyson). Finally, the term "many-body problem" takes on new meaning in this context.

The Evolution of Group Theory

Israel Kleiner

Mathematics Magazine
59 (1986), 195–213

Editors' Note: Israel Kleiner received his PhD in ring theory under the direction of J. Lambek at McGill University and has been Professor of Mathematics at York University in Ontario since 1965. He became interested in the history of mathematics and has written a long series of articles for MAA and other journals on historical topics. For the one included here he was awarded the Allendoerfer Award in 1987, and then he went on to win another Allendoerfer Award for "Rigor and proof in mathematics: an historical perspective" in 1992; a Lester R. Ford Award (with N. Movshovitz-Hadar) in 1995 for "The role of paradoxes in the evolution of mathematics," which appeared in the *Monthly*; and a Pólya Award in 1990 for "Evolution of the function concept: a brief survey" in the *College Mathematics Journal.*

This article gives a brief sketch of the evolution of group theory. It derives from a firm conviction that the history of mathematics can be a useful and important integrating component in the teaching of mathematics. This is not the place to elaborate on the role of history in teaching, other than perhaps to give one relevant quotation:

> Although the study of the history of mathematics has an intrinsic appeal of its own, its chief raison d'être is surely the illumination of mathematics itself. For example the gradual unfolding of the integral concept from the volume computations of Archimedes to the intuitive integrals of Newton and Leibniz and finally to the definitions of Cauchy, Riemann and Lebesgue—cannot fail to promote a more mature appreciation of modern theories of integration.
>
> — C. H. Edwards [11]

The presentation in one article of the evolution of so vast a subject as group theory necessitated severe selectivity and brevity. It also required omission of the broader contexts in which group theory evolved, such as wider currents in abstract algebra, and in mathematics as a whole. (We will note *some* of these interconnections shortly.) We trust that enough of the essence and main lines of development in the evolution of group theory have been retained to provide a useful beginning from which the reader can branch out in various directions. For this the list of references will prove useful.

The reader will find in this article an outline of the origins of the main concepts, results, and theories discussed in a beginning course on group theory. These include, for example, the concepts of (abstract) group, normal subgroup, quotient group, simple group, free group, isomorphism, homomorphism, automorphism, composition series, direct product; the theorems of J. L. Lagrange, A.-L. Cauchy, A. Cayley, C. Jordan–O. Hölder; the theories of permutation groups and of abelian groups. At the same time we have tried to balance the technical aspects with background information and interpretation. Our survey of the evolution of group theory will be given in several stages, as follows:

1. Sources of group theory.
2. Development of "specialized" theories of groups.
3. Emergence of abstraction in group theory.
4. Consolidation of the abstract group *concept*; dawn of abstract group *theory*.
5. Divergence of developments in group theory.

Before dealing with each stage in turn, we wish to mention the context within mathematics as a whole, and within algebra in particular, in which group theory developed. Although our "story" concerning the evolution of group theory begins in 1770 and extends to the 20th century, the major developments occurred in the 19th century. Some of the general mathematical features of that century which had a bearing on the evolution of group theory are: (a) an increased concern for rigor; (b) the emergence of abstraction; (c) the rebirth of the axiomatic method; (d) the view of mathematics as a human activity, possible without reference to, or motivation from, physical situations. Each of these items deserves extensive elaboration, but this would go beyond the objectives (and size) of this paper.

Up to about the end of the 18th century, algebra consisted (in large part) of the study of solutions of polynomial equations. In the 20th century, algebra became a study of abstract, axiomatic systems. The transition from the so-called classical algebra of polynomial equations to the so-called modern algebra of axiomatic systems occurred in the 19th century. In addition to group theory, there emerged the structures of commutative rings, fields, noncommutative rings, and vector spaces. These developed alongside, and sometimes in conjunction with, group theory. Thus Galois theory involved both groups and fields; algebraic number theory contained elements of group theory in addition to commutative ring theory and field theory; group representation theory was a mix of group theory, noncommutative algebra, and linear algebra.

1 Sources of group theory

There are four major sources in the evolution of group theory. They are (with the names of the originators and dates of origin):

(a) Classical algebra (J. L. Lagrange, 1770)
(b) Number theory (C. F. Gauss, 1801)
(c) Geometry (F. Klein, 1874)
(d) Analysis (S. Lie, 1874; H. Poincaré and F. Klein, 1876)

We deal with each in turn.

1.1 Classical Algebra (J. L. Lagrange, 1770)

The major problems in algebra at the time (1770) that Lagrange wrote his fundamental memoir "Reflexions sur la resolution algebrique des equations" concerned polynomial equations. There were "theoretical" questions dealing with the existence and nature of the roots (e.g., Does every equation have a root? How many roots are there? Are they real, complex, positive, negative?), and "practical" questions dealing with methods for finding the roots. In the latter instance there were exact methods and approximate methods. In what follows we mention exact methods.

The Babylonians knew how to solve quadratic equations (essentially by the method of completing the square) around 1600 B.C. Algebraic methods for solving the cubic and the quartic were given around 1540. One of the major problems for the next two centuries was the algebraic solution of the quintic. This is the task Lagrange set for himself in his paper of 1770.

In his paper Lagrange first analyzes the various known methods (devised by F. Viète, R. Descartes, L. Euler, and E. Bézout) for solving cubic and quartic equations. He shows that the common feature of these methods is the reduction of such equations to auxiliary equations—the so-called resolvent equations. The latter are one degree lower than the original equations. Next Lagrange attempts a similar analysis of polynomial equations of arbitrary degree n. With each such equation he associates a "resolvent equation" as follows: let $f(x)$ be the original equation, with roots $x_1, x_2, \ldots, x_n$. Pick a rational function $R(x_1, x_2, \ldots, x_n)$ of the roots and coefficients of $f(x)$. (Lagrange describes methods for doing so.) Consider the different values which $R(x_1, x_2, \ldots, x_n)$ assumes under all the $n!$ permutations of the roots $x_1, x_2, \ldots, x_n$ of $f(x)$. If these are denoted by $u_1, y_2, \ldots, y_k$ then the resolvent equation is given by $g(x) = (x-y_1)\cdot(x-y_2)\cdots(x-y_k)$. (Lagrange shows that k divides $n!$—the source of what we call Lagrange's theorem in group theory.) For example, if $f(x)$ is a quartic with roots x_1, x_2, x_3, x_4, then $R(x_1, x_2, x_3, x_4)$ may be taken to be $x_1x_2 + x_3x_4$, and this function assumes three distinct values under the 24 permutations of x_1, x_2, x_3, x_4. Thus the resolvent equation of a quartic is a cubic. However, in carrying over this analysis to the quintic, he finds that the resolvent equation is of degree six!

Although Lagrange did not succeed in resolving the problem of the algebraic solvability of the quintic, his work was a milestone. It was the first time that an association was made between the solutions of a polynomial equation and the permutations of its roots. In fact, the study of the permutations of the

Evariste Galois — stamp issued by France in 1984

roots of an equation was a cornerstone of Lagrange's general theory of algebraic equations. This, he speculated, formed "the true principles for the solution of equations." (He was, of course, vindicated in this by E. Galois.) Although Lagrange speaks of permutations without considering a "calculus" of permutations (e.g., there is no consideration of their composition or closure), it can be said that the germ of the group concept (as a group of permutations) is present in his work. For details see [12], [16], [19], [25], [33].

1.2 Number Theory (C. F. Gauss, 1801)

In the *Disquisitiones Arithmeticæ* of 1801 Gauss summarized and unified much of the number theory that preceded him. The work also suggested new directions which kept mathematicians occupied for the entire century. As for its impact on group theory, the *Disquisitiones* may be said to have initiated the theory of finite abelian groups. In fact, Gauss established many of the significant properties of these groups without using any of the terminology of group theory. The groups appear in four different guises: the additive group of integers modulo m, the multiplicative group of integers relatively prime to m, modulo m, the group of equivalence classes of binary quadratic forms, and the group of nth roots of unity. And though these examples appear in number-theoretic contexts, it is as abelian groups that Gauss treats them, using what are clear prototypes of modern algebraic proofs.

For example, considering the nonzero integers modulo p (p a prime), Gauss shows that they are all powers of a single element; i.e., that the group Z_p^* of such integers is cyclic. Moreover, he determines the number of generators of this group (he shows that it is equal to $\phi(p-1)$, where ϕ is Euler's (ϕ-function). Given any element of Z_p^* he defines the order of the element (without using the terminology) and shows that the order of an element is a divisor of $p-1$. He then uses this result to prove P. Fermat's "little theorem," namely, that $a^{p-1} \equiv \text{mod } p$ if p does not divide a, thus employing group-theoretic ideas to prove number-theoretic results. Next he shows that if t is a positive integer which divides $p-1$, then there exists an element in Z_p^* whose order is t—essentially the converse of Lagrange's theorem for cyclic groups.

Concerning the nth roots of 1 (which he considers in connection with the cyclotomic equation), he shows that they too form a cyclic group. In connection with this group he raises and answers many of the same questions he raised and answered in the case of Z_p^*.

The problem of representing integers by binary quadratic forms goes back to Fermat in the early 17th century. (Recall his theorem that every prime of the form $4n+1$ can be represented as a sum of two squares x^2+y^2.) Gauss devotes a large part of the *Disquisitiones* to an exhaustive study of binary quadratic forms and the representation of integers by such forms. (A **binary quadratic form** is an expression of the form $ax^2+bxy+cy^2$, with a, b, c integers.) He defines a composition on such forms, and remarks that if K and K^1 are two such forms one may denote their composition by $K+K^1$. He then shows that this composition is associative and commutative, that there exists an identity, and that each form has an inverse, thus verifying all the properties of an abelian group.

Despite these remarkable insights one should not infer that Gauss had the concept of an abstract group, or even of a finite abelian group. Although the arguments in the *Disquisitiones* are quite general, each of the various types of "groups" he considers is dealt with separately—there is no unifying group-theoretic method which he applies to all cases. For details see [5], [9], [25], [30], [33].

1.3 Geometry (F. Klein, 1872)

We are referring here to Klein's famous and influential (but see [18]) lecture entitled "A Comparative Review of Recent Researches in Geometry," which he delivered in 1872 on the occasion of his admission to the faculty of the University of Erlangen. The aim of this so-called Erlangen Program was the classification of geometry as the study of invariants

Felix Klein

under various groups of transformations. Here there appear groups such as the projective group, the group of rigid motions, the group of similarities, the hyperbolic group, the elliptic groups, as well as the geometries associated with them. (The affine group was not mentioned by Klein.) Now for some background leading to Klein's Erlangen Program.

The 19th century witnessed an explosive growth in geometry, both in scope and in depth. New geometries emerged: projective geometry, noneuclidean geometries, differential geometry, algebraic geometry, n-dimensional geometry, and Grassmann's geometry of extension. Various geometric methods competed for supremacy: the synthetic versus the analytic, the metric versus the projective. At mid-century, a major problem had arisen, namely, the classification of the relations and inner connections among the different geometries and geometric methods. This gave rise to the study of "geometric relations," focusing on the study of properties of figures invariant under transformations. Soon the focus shifted to a study of the transformations themselves. Thus the study of the geometric relations of figures became the study of the associated transformations. Various types of transformations (e.g., collineations, circular transformations, inversive transformations, affinities) became the objects of specialized studies. Subsequently, the logical connections among transformations were investigated, and this led to the problem of classifying transformations and eventually to Klein's group-theoretic synthesis of geometry.

Klein's use of groups in geometry was the final stage in bringing order to geometry. An intermediate stage was the founding of the first major theory of classification in geometry, beginning in the 1850's, the Cayley-Sylvester Invariant Theory. Here the objective was to study invariants of "forms" under transformations of their variables. This theory of classification, the precursor of Klein's Erlangen Program, can be said to be *implicitly* group-theoretic. Klein's use of groups in geometry was, of course, explicit. (For a thorough analysis of implicit group-theoretic thinking in geometry leading to Klein's Erlangen Program, see [33].) In the next section (2.3) we will note the significance of Klein's Erlangen Program (and his other works) for the evolution of group theory. Since the Program originated a hundred years after Lagrange's work and eighty years after Gauss' work, its importance for group theory can best be appreciated *after* a discussion of the evolution of group theory beginning with the works of Lagrange and Gauss and ending with the period around 1870.

1.4 Analysis (S. Lie, 1874; H. Poincaré and F. Klein, 1876)

In 1874 Lie introduced his general theory of (continuous) transformation groups—essentially what we call Lie groups today. Such a group is represented by the transformations

$$x_i' = f_i(x_1, x_2, \dots, x_n, a_1, a_2, \dots, a_n),$$

$i = 1, 2, \dots, n$, where the f_i are analytic functions in the x_i and a_i (the a_i are parameters, with both x_i and a_i real or complex). For example, the transformations given by

$$x' = \frac{ax+b}{cx+d}, \quad \text{where } a, b, c, d, \text{ are real numbers and } ad - be \neq 0,$$

define a continuous transformation group.

Lie thought of himself as the successor of N. H. Abel and Galois, doing for differential equations what they had done for algebraic equations. His work was inspired by the observation that almost all the differential equations which had been integrated by the older methods remain invariant under continuous groups that can be easily constructed. He was

Sophus Lie

Henri Poincaré

then led to consider, in general, differential equations that remain invariant under a given continuous group and to investigate the possible simplifications in these equations which result from the known properties of the given group (cf. Galois theory). Although Lie did not succeed in the actual formulation of a "Galois theory of differential equations," his work was fundamental in the subsequent formulation of such a theory by E. Picard (1883/1887) and E. Vessiot (1892).

Poincaré and Klein began their work on "automorphic functions" and the groups associated with them around 1876. Automorphic functions (which are generalizations of the circular, hyperbolic, elliptic, and other functions of elementary analysis) are functions of a complex variable z, analytic in some domain D, which are invariant under the group of transformations

$$z' = \frac{az+b}{cz+d}, \quad (a, b, c, d \text{ real or complex and } ad - bc \neq 0)$$

or under some subgroup of this group. Moreover, the group in question must be "discontinuous" (i.e., any compact domain contains only finitely many transforms of any point). Examples of such groups are the modular group (in which a, b, c, d are integers and $ad - bc = 1$), which is associated with the elliptic modular functions, and Fuchsian groups (in which a, b, c, d are real and $ad - bc = 1$) associated with the Fuchsian automorphic functions. As in the case of Klein's Erlangen Program, we will explore the consequences of these works for group theory in section 2.3.

2 Development of "specialized" theories of groups

In §1 we outlined four major sources in the evolution of group theory. The first source—classical algebra—led to the theory of permutation groups; the second source—number theory—led to the theory of abelian groups; the third and fourth sources—geometry and analysis—led to the theory of transformation groups. We will now outline some developments within these specialized theories.

2.1 Permutation Groups

As noted earlier, Lagrange's work of 1770 initiated the study of permutations in connection with the study of the solution of equations. It was probably the first clear instance of implicit group-theoretic thinking in mathematics. It led directly to the works of P. Ruffini, Abel, and Galois during the first third of the 19th century, and to the concept of a permutation group.

Ruffini and Abel proved the unsolvability of the quintic by building upon the ideas of Lagrange concerning resolvents. Lagrange showed that a necessary condition for the solvability of the general polynomial equation of degree n is the existence of a resolvent of degree less than n. Ruffini and Abel showed that such resolvents do not exist for $n > 4$. In the process they developed a considerable amount of permutation theory. (See [1], [9], [19], [23], [24], [25], [30], [33] for details.) It was Galois, however, who made the fundamental conceptual advances, and who is considered by many as the founder of (permutation) group theory.

Galois' aim went well beyond finding a method for solvability of equations. He was concerned with gaining insight into general principles, dissatisfied as he was with the methods of his predecessors: "From the beginning of this century," he wrote, "computational procedures have become so complicated that any progress by those means has become impossible" [19, p. 92].

Galois recognized the separation between "Galois theory" (i.e., the correspondence between fields and groups) and its application to the solution of equations, for he wrote that he was presenting "the general principles and just one application" of the theory [19, p. 42]. "Many of the early commentators on Galois theory failed to recognize this distinction, and this led to an emphasis on applications at the expense of the theory" (Kiernan, [19]).

Galois was the first to use the term "group" in a technical sense—to him it signified a collection of permutations closed under multiplication: "if one has in the same group the substitutions S and T one is certain to have the substitution ST" [33, p. 111]. He recognized that the most important properties of an algebraic equation were reflected in certain properties of a group uniquely associated with the equation— "the group of the equation." To describe these properties he invented the fundamental notion of normal subgroup and used it to great effect. While the issue of resolvent equations preoccupied Lagrange, Ruffini, and Abel, Galois' basic idea was to bypass them, for the construction of a resolvent required great skill and was not based on a clear methodology. Galois noted instead that the existence of a resolvent was equivalent to the existence of a normal subgroup of prime index in the group of the equation. This insight shifted consideraition from the resolvent equation to the group of the equation and its subgroups.

Galois defines the group of an equation as follows [19, p. 80]:

> Let an equation be given, whose m roots are $a, b, c, \ldots$ There will always be a group of permutations of the letters $a, b, c, \ldots$ which has the following property: 1) that every function of the roots, invariant under the substitutions of that group, is rationally known [i.e., is a rational function of the coefficients and any adjoined quantities]. 2) conversely, that every function of the roots, which can be expressed rationally, is invariant under these substitutions.

The definition says essentially that the group of the equation consists of those permutations of the roots of the equation which leave invariant all relations among the roots over the field of coefficients of the equation-basically the definition we would give today. Of course the definition does not guarantee the existence of such a group, and so Galois proceeds to demonstrate it. Galois next investigates how the group changes when new elements are adjoined to the "ground field" F. His treatment is amazingly close to the standard treatment of this matter in a modern algebra text.

Galois' work was slow in being understood and assimilated. In fact, while it was done around 1830, it was published posthumously in 1846, by J. Liouville. Beyond his technical accomplishments, Galois "challenged the development of mathematics in two ways. He discovered, but left unproved, theorems which called for proofs based on new, sophisticated concepts and calculations. Also, the task of filling the gaps in his work necessitated a fundamental clarification of his methods and their group theoretical essence" (Wussing, [33]). For details see [12], [19], [23], [25], [29], [31], [33].

The other major contributor to permutation theory in the first half of the 19th century was Cauchy. In several major papers in 1815 and 1844 Cauchy inaugurated the theory of permutation groups as an autonomous subject. (Before Cauchy, permutations were not an object of independent study but rather a useful device for the investigation of solutions of polynomial equations.) Although Cauchy was well aware of the work of Lagrange and Ruffini (Galois' work was not yet published at the time), Wussing suggests that Cauchy "was definitely not inspired directly by the contemporary group-theoretic formulation of the solution of algebraic equations" [33].

In these works Cauchy gives the first systematic development of the subject of permutation groups. In the 1815 papers Cauchy uses no special name for sets of permutations closed under multiplication. However, he recognizes their importance and gives a name to the number of elements in such a closed set, calling it "diviseur indicatif." In the 1844 paper he defines the concept of a group of permutations generated by certain elements [22, p. 65].

> Given one or more substitutions involving some or all of the elements $x, y, z, \ldots$ I call the products of these substitutions, by themselves or by any other, in any order, derived substitutions. The given substitutions, together with the derived ones, form what I call a system of conjugate substitutions.

In these works, which were very influential, Cauchy makes several lasting additions to the terminology, notation, and results of permutation theory. For example, he introduces the permutation notation $\begin{pmatrix} x & y & z \\ x & z & y \end{pmatrix}$ in use today, as well as the cyclic notation for permutations; defines the product of permutations, the degree of a permutation, cyclic permutation, transposition; recognizes the identity permutation as a permutation; discusses what we would call today the direct product of two groups; and deals with the alternating groups extensively. Here is a sample of some of the results he proves.

(i) Every even permutation is a product of 3-cycles.

(ii) If p (prime) is a divisor of the order of a group, then there exists a subgroup of order p. (This is known today as "Cauchy's theorem," though it was stated without proof by Galois.)

(iii) Determined all subgroups of S_3, S_4, S_5, S_6 (making an error in S_6.)

(iv) All permutations which commute with a given one form a group (the centralizer of an element).

It should be noted that all these results were given and proved in the context of permutation groups. For details see [6], [8], [23], [24], [25], [33].

The crowning achievement of these two lines of development—a symphony on the grand themes of Galois and Cauchy—was Jordan's important and influential *Traité des substitutions et des équations algébriques* of 1870. Although the author states in the preface that "the aim of the work is to develop Galois' method and to make it a proper field of study, by showing with what facility it can solve all principal problems of the theory of equations," it is in fact group theory per se—not as an offshoot of the theory of solvability of equations—which forms the central object of study.

Camille Jordan

The striving for a mathematical synthesis based on key ideas is a striking characteristic of Jordan's work as well as that of a number of other mathematicians of the period (e.g., F. Klein). The concept of a (permutation) group seemed to Jordan to provide such a key idea. His approach enabled him to give a unified presentation of results due to Galois, Cauchy, and others. His application of the group concept to the theory of equations, algebraic geometry, transcendental functions, and theoretical mechanics was also part of the unifying and synthesizing theme. "In his book Jordan wandered through all of algebraic geometry, number theory, and function theory in search of interesting permutation groups" (Klein, [20]). In fact, the aim was a survey of all of mathematics by areas in which the theory of permutation groups had been applied or seemed likely to be applicable. "The work represents... a review of the whole of contemporary mathematics from the standpoint of the occurrence of group-theoretic thinking in permutation-theoretic form" (Wussing, [33]).

The *Traité* embodied the substance of most of Jordan's publications on groups up to that time (he wrote over 30 articles on groups during the period 1860–1880) and directed attention to a large number of difficult problems, introducing many fundamental concepts. For example, Jordan makes explicit the notions of isomorphism and homomorphism for (substitution) groups, introduces the term "solvable group" for the first time in a technical sense, introduces the concept of a composition series, and proves part of the Jordan-Hölder theorem, namely, that the indices in two composition series are the same (the concept of a quotient group was not explicitly recognized at this time); and he undertakes a very thorough study of transitivity and primitivity for permutation groups, obtaining results most of which have not since been superseded. Jordan also gives a proof that A_n is simple for $n > 4$.

An important part of the treatise is devoted to a study of the "linear group" and some of its subgroups. In modern terms these constitute the so-called classical groups, namely, the general linear group, the unimodular group, the orthogonal group, and the symplectic group. Jordan considers these groups only over finite fields, and proves their simplicity in certain cases. It should be noted, however, that he considers these groups as permutation groups rather than groups of matrices or linear transformations (see [29], [33]).

Jordan's *Traité* is a landmark in the evolution of group theory. His permutation-theoretic point of view, however, was soon to be overtaken by the conception of a group as a group of transformations (see 2.3 below). "The *Traité* marks a pause in the evolution and application of the permutation-theoretic group concept. It was an expression of Jordan's deep desire to effect a conceptual synthesis of the mathematics of his time. That he tried to achieve such a synthesis by relying on the concept of a permutation group, which the very next phase of mathematical development would show to have been unduly restricted, makes for both the glory and the limitations of the *Traité*..." (Wussing, [33]). For details see [9], [13], [19], [20], [22], [24], [29], [33].

2.2 Abelian Groups

As noted earlier, the main source for abelian group theory was number theory, beginning with Gauss' *Disquisitiones Arithmeticæ*. In contrast to permutation theory, group-theoretic modes of thought in number theory remained implicit until about the last third of the 19th century. Until that time no explicit use of the term "group" was made, and there was no link to the contemporary, flourishing theory of permutation groups. We now give a sample of some implicit group-theoretic work in number theory, especially in algebraic number theory.

Algebraic number theory arose in connection with Fermat's conjecture concerning the equation $x^n + y^n = z^n$, Gauss' theory of binary quadratic forms, and higher reciprocity laws. Algebraic number fields and their arithmetical properties were the main objects of study. In 1846 G. L. Dirichlet studied the units in an algebraic number field and established that (in our terminology) the group of these units is a direct product of a finite cyclic group and a free abelian group of finite rank. At about the same time E. Kummer introduced his "ideal numbers," defined an equivalence relation on them, and derived, for cyclotomic fields, certain special properties of the number of equivalence classes (the so-called class number of a cyclotomic field; in our terminology, the order of the ideal class group of the cyclotomic field). Dirichlet had earlier made similar studies of *quadratic* fields.

In 1869 E. Schering, a former student of Gauss, investigated the structure of Gauss' (group of) equivalence classes of binary quadratic forms. He found certain fundamental classes from which all classes of forms could be obtained by composition. In group-theoretic terms, Schering found a basis for the abelian group of equivalence classes of binary quadratic forms.

L. Kronecker generalized Kummer's work on cyclotomic fields to arbitrary algebraic number fields. In a paper in 1870 on algebraic number theory, entitled "Auseinandersetzung einiger Eigenschaften der Klassenzahl idealer complexer Zahlen," he began by taking a very abstract point of view: he considered a finite set of arbitrary "elements," and defined an abstract operation on them which satisfied certain laws-laws which we may take nowadays as axioms for a finite abelian group:

> Let $\theta', \theta'', \theta''', \ldots$ be finitely many elements such that with any two of them we can associate a third by means of a definite procedure. Thus, if f denotes the procedure and θ', θ" are two (possibly equal) elements, then there exists a θ''' equal to $f(\theta', \theta'')$. Furthermore, $f(\theta', \theta'') = f(\theta'', \theta')$, $f(\theta', f(\theta'', \theta''')) = f(f(\theta', \theta''), \theta''')$ and if θ" is different from θ''' then $f(\theta', \theta'')$ is different from $f(\theta', \theta''')$. Once this is assumed we can replace the operation $f(\theta', \theta'')$ by multiplication $\theta' \cdot \theta''$ provided that instead of equality we employ equivalence. Thus using the usual equivalence symbol "$\sim$" we define the equivalence $\theta' \cdot \theta$" $\sim \theta'''$ by means of the equation $f(\theta', \theta'') = \theta'''$.

Kronecker aimed at working out the laws of combination of "magnitudes," in the process giving an implicit definition of a finite abelian group. From the above abstract considerations Kronecker deduces the following consequences:

(i) If θ is any "element" of the set under discussion, then $\theta^k = 1$ for some positive integer k. If k is the smallest such then θ is said to "belong to k." If θ belongs to k and $\theta^m = 1$ then k divides m.

(ii) If an element θ belongs to k, then every divisor of k has an element belonging to it.

(iii) If θ and θ' belong to k and k' respectively, and k and k' are relatively prime, then $\theta\theta'$ belongs to kk'.

(iv) There exists a "fundamental system" of elements $\theta_1, \theta_2, \theta_3, \ldots$ such that the expression $\theta_1^{h_1}\theta_2^{h_2}\theta_3^{h_3}\cdots$ $(h_i = 1, 2, 3, \ldots, n_i)$ represents each element of the given set of elements just once. The numbers $n_1, n_2, n_3, \ldots$ to which, respectively, $\theta_1, \theta_2, \theta_3, \ldots$ belong, are such that each is divisible by its successor; the product $n_1 n_2 n_3 \cdots$ is equal to the totality of elements of the set.

The above can, of course, be interpreted as well known results on finite abelian groups; in particular (iv) can be taken as the basis theorem for such groups. Once Kronecker establishes this general framework, he applies it to the special cases of equivalence classes of binary quadratic forms and to ideal classes. He notes that when applying (iv) to the former one obtains Schering's result.

Although Kronecker did not relate his implicit definition of a finite abelian group to the (by that time) well established concept of a permutation group, of which he was well aware, he clearly recognized the advantages of the abstract point of view which he adopted:

> The very simple principles... are applied not only in the context indicated but also frequently, elsewhere—even in the elementary parts of number theory. This shows, and it is otherwise easy to see, that these principles belong to a more general and more abstract realm of ideas. It is therefore proper to free their development from all inessential restrictions, thus making it unnecessary to repeat the same argument when applying it in different cases.... Also, when stated with all admissible generality, the presentation gains in simplicity and, since only the truly essential features are thrown into relief, in transparency.

Georg Frobenius

The above lines of development were capped in 1879 by an important paper of G. Frobenius and L. Stickelberger entitled "On groups of commuting elements." Although Frobenius and Stickelberger built on Kronecker's work, they used the concept of an abelian group explicitly and, moreover, made the important advance of recognizing that the abstract group concept embraces congruences and Gauss' composition of forms as well as the substitution groups of Galois. (They also mention, in footnotes, groups of infinite order, namely groups of units of number fields and the group of all roots of unity.) One of their main results is a proof of the basis theorem for finite abelian groups, including a proof of the uniqueness of decomposition. It is interesting to compare their explicit, "modern," formulation of the theorem to that of Kronecker ((iv) above):

> A group that is not irreducible [indecomposable] can be decomposed into purely irreducible factors. As a rule, such a decomposition can be accomplished in many ways. However, regardless of the way in which it is carried out, the number of irreducible factors is always the same and the factors in the two decompositions

can be so paired off that the corresponding factors have the same order [33, p. 235].

They go on to identify the "irreducible factors" as cyclic groups of prime power orders. They then apply their results to groups of integers modulo m, binary quadratic forms, and ideal classes in algebraic number fields.

The paper by Frobenius and Stickelberger is "a remarkable piece of work, building up an independent theory of finite abelian groups on its own foundation in a way close to modern views" (Fuchs, [30]). For details on this section (2.2), see [5], [9], [24[, [30], [33].

2.3 Transformation Groups

As in number theory, so in geometry and analysis, group-theoretic ideas remained implicit until the last third of the 19th century. Moreover, Klein's (and Lie's) explicit use of groups in geometry influenced conceptually rather than technically the evolution of group theory, for it signified a genuine shift in the development of that theory from a preoccupation with permutation groups to the study of groups of transformations. (That is not to imply, of course, that permutation groups were no longer studied.) This transition was also notable in that it pointed to a turn from finite groups to infinite groups.

Klein noted the connection of his work with permutation groups but also realized the departure he was making. He stated that what Galois theory and his own program have in common is the investigation of "groups of changes," but added that "to be sure, the objects the changes apply to are different: there [Galois theory] one deals with a finite number of discrete elements, whereas here one deals with an infinite number of elements of a continuous manifold" [33, p. 191]. To continue the analogy, Klein notes that just as there is a theory of permutation groups, "we insist on a *theory of transformations*, a study of groups generated by transformations of a given type" [33, p. 191].

Klein shunned the abstract point of view in group theory, and even his technical definition of a (transformation) group is deficient: "Now let there be given a sequence of transformations $A, B, C, \ldots$. If this sequence has the property that the composite of any two of its transformations yields a transformation that again belongs to the sequence, then the latter will be called a group of transformations" [33, p. 185]. His work, however, broadened considerably the conception of a group and its applicability in other fields of mathematics. Klein did much to promote the view that group-theoretic ideas are fundamental in mathematics: "Group theory appears as a distinct discipline throughout the whole of modern mathematics. It permeates the most varied areas as an ordering and classifying principle" [33, p. 228].

There was another context in which groups were associated with geometry, namely, "motion-geometry;" i.e., the use of motions or transformations of geometric objects as group elements. Already in 1856 W. R. Hamilton considered (implicitly) "groups" of the regular solids. Jordan, in 1868, dealt with the classification of all subgroups of the group of motions of Euclidean 3-space. And Klein in his *Lectures on the Icosahedron* of 1884 "solved" the quintic equation by means of the symmetry group of the icosahedron. He thus discovered a deep connection between the groups of rotations of the regular solids, polynomial equations, and complex function theory. (In these *Lectures* there also appears the "Klein 4-group".)

Already in the late 1860's Klein and Lie had undertaken, jointly, "to investigate geometric or analytic objects that are transformed into themselves by *groups of changes*." (This is Klein's retrospective description, in 1894, of their program.) While Klein concentrated on discrete groups, Lie studied continuous transformation groups. Lie realized that the theory of continuous transformation groups was a very powerful tool in geometry and differential equations and he set himself the task of "determining all groups of... [continuous] transformations" [33, p. 214]. He achieved his objective by the early 1880's with the classification of these groups (see [33] for details). A classification of discontinuous transformation groups was obtained by Poincaré and Klein a few years earlier.

Beyond the technical accomplishments in the areas of discontinuous and continuous transformation groups (extensive theories developed in both areas and both are still nowadays active fields of research), what is important for us in the founding of these theories is that

(i) They provided a major extension of the scope of the concept of a group—from permutation groups and abelian groups to transformation groups;

(ii) They provided important examples of infinite groups—previously the only objects of study were finite groups;

(iii) They greatly extended the range of applications of the group concept to include number theory,

> the theory of algebraic equations, geometry, the theory of differential equations (both ordinary and partial), and function theory (automorphic functions, complex functions).

All this occurred prior to the emergence of the abstract group concept. In fact, these developments were instrumental in the emergence of the concept of an abstract group, which we describe next. For further details on this section (2.3), see [5], [7], [9], [17], [18], [20], [24], [29], [33].

3 Emergence of abstraction in group theory

The abstract point of view in group theory emerged slowly. It took over one hundred years from the time of Lagrange's implicit group-theoretic work of 1770 for the abstract group concept to evolve. E. T. Bell discerns several stages in this process of evolution towards abstraction and axiomatization:

> The entire development required about a century. Its progress is typical of the evolution of any major mathematical discipline of the recent period; first, the discovery of isolated phenomena, then the recognition of certain features common to all, next the search for further instances, their detailed calculation and classification; then the emergence of general principles making further calculations, unless needed for some definite application, superfluous; and last, the formulation of postulates crystallizing in abstract form the structure of the system investigated [2].

Although somewhat oversimplified (as all such generalizations tend to be), this is nevertheless a useful framework. Indeed, in the case of group theory, first came the "isolated phenomena"—e.g., permutations, binary quadratic forms, roots of unity; then the recognition of "common features" —the concept of a finite group, encompassing both permutation groups and finite abelian groups (cf. the paper of Frobenius and Stickelberger cited in section 2.2); next the search for "other instances"—in our case transformation groups (see section 2.3); and finally the formulation of "postulates"—in this case the postulates of a group, encompassing both the finite and infinite cases. We now consider when and how the intermediate and final stages of abstraction occurred.

In 1854 Cayley, in a paper entitled "On the theory of groups, as depending on the symbolic equation $\theta^n = 1$," gave the first abstract definition of a finite group. (In 1858 R. Dedekind, in lectures on Galois theory at Göttingen, gave another.) Here is Cayley's definition:

Arthur Cayley

> A set of symbols $1, \alpha, \beta, \ldots$ all of them different, and such that the product of any two of them (no matter in what order), or the product of any one of them into itself, belongs to the set, is said to be a *group*.

Cayley goes on to say that

> These symbols are not in general convertible [commutative], but are associative,

and

> it follows that if the entire group is multiplied by any one of the symbols, either as further or nearer factor [i.e., on the left or on the right], the effect is simply to reproduce the group.

Cayley then presents several examples of groups, such as the quaternions (under addition), invertible matrices (under multiplication), permutations, Gauss' quadratic forms, and groups arising in elliptic function theory. Next he shows that every abstract group is (in our terminology) isomorphic to a permutation group, a result now known as "Cayley's theorem." He seems to have been well aware of the concept of isomorphic groups, although he does not define it explicitly. He introduces, however, the multiplication table of a (finite) group and asserts that an abstract group is determined by its multiplication table. He then goes on to determine all the groups of orders four and six, showing there are two of each by displaying multiplication tables. Moreoever, he notes that the cyclic group of order n "is in every respect analogous to the system of the roots of the ordinary equation $x^n - 1 = 0$," and that there exists only one group of a given prime order.

Cayley's orientation towards an abstract view of groups-a remarkable accomplishment at this time of the evolution of group theory-was due, at least in part, to his contact with the abstract work of G. Boole. The concern with the abstract foundations of mathematics was characteristic of the circles around Boole, Cayley, and Sylvester already in the 1840's. Cayley's achievement was, however, only a personal triumph. His abstract definition of a group attracted no attention at the time, even though Cayley was already well known. The mathematical community was apparently not ready for such abstraction: permutation groups were the only groups under serious investigation, and more generally, the formal approach to mathematics was still in its infancy. As M. Kline put it in his inimitable way [21]: "Premature abstraction falls on deaf ears, whether they belong to mathematicians or to students." For details see [22], [23], [24], [25], [29], [33].

It was only a quarter of a century later that the abstract group concept began to take hold. And it was Cayley again who in four short papers on group theory written in 1878 returned to the abstract point of view he adopted in 1854. Here he stated the general problem of finding all groups of a given order and showed that any (finite) group is isomorphic to a group of permutations. But, as he remarked, this "does not in any wise show that the best or easiest mode of treating the general problem is thus to regard it as a problem of substitutions; and it seems clear that the better course is to consider the general problem in itself, and to deduce from it the theory of groups of substitutions" [22, p. 141]. These papers of Cayley, unlike those of 1854, inspired a number of fundamental group-theoretic works.

Another mathematician who advanced the abstract point of view in group theory (and more generally in algebra) was H. Weber. It is of interest to see his "modern" definition of an abstract (finite) group given in a paper of 1882 on quadratic forms [23, p. 113]:

> A system G of h arbitrary elements $\theta_1, \theta_2, \ldots, \theta_h$ is called a group of degree h if it satisfies the following conditions:
> I. By some rule which is designated as composition or multiplication, from any two elements of the same system one derives a new element of the same system. In symbols 090 = 0t.
> II. It is always true that $(\theta_r\theta_s)\theta_t = \theta_r(\theta_s\theta_t) = \theta_r\theta_s\theta_t$.
> III. From $\theta\theta_r = \theta\theta_s$ or from $\theta_r\theta = \theta_s\theta$ it follows that $\theta_r = \theta_s$.

Heinrich Weber

Weber's and other definitions of abstract groups given at the time applied to finite groups only. They thus encompassed the two theories of permutation groups and (finite) abelian groups, which derived from the two sources of classical algebra (polynomial equations) and number theory, respectively. Infinite groups, which derived from the theories of (discontinuous and continuous) transformation groups, were not subsumed under those definitions. It was W. von Dyck who, in an important and influential paper in 1882 entitled "Group-theoretic studies," consciously included and combined, for the first time, all of the major historical roots of abstract group theory—the algebraic, number theoretic, geometric, and analytic. In von Dyck's own words:

> The aim of the following investigations is to continue the study of the properties of a group in its abstract formulation. In particular, this will pose the question of the extent to which these properties have an invariant character present in all the different realizations of the group, and the question of what leads to the exact determination of their essential group-theoretic content.

Von Dyck's definition of an abstract group, which included both the finite and infinite cases, was given in terms of generators (he calls them "operations") and defining relations (the definition is somewhat long-see [7, pp. 5, 6]). He stresses that "in this way all... isomorphic groups are included in a *single* group," and that "the *essence* of a group is no longer expressed by a particular presentation form of its operations but rather by their mutual relations." He then goes on to construct the free group on n generators, and shows (essentially, without using the terminology) that every finitely generated group is a quotient group of a free group of finite rank. What is important from the point of view of postulates for group theory is that von Dyck was the first to require explicitly the existence of an inverse in his definition of a group: "We require for our considerations that a group which contains the operation T_k must also contain its inverse T_k^{-1}." In a second paper (in 1883) von Dyck applied his abstract development of group theory to permutation groups, finite rotation groups (symmetries of polyhedra), number theoretic groups, and transformation groups.

Although various postulates for groups appeared in the mathematical literature for the next twenty years, the abstract point of view in group theory was not universally applauded. In particular, Klein, one of the major contributors to the development of group theory, thought that the "abstract formulation is excellent for the working out of proofs but it does not help one find new ideas and methods," adding that "in general, the disadvantage of the [abstract] method is that it fails to encourage thought" [33, p. 228].

Despite Klein's reservations, the mathematical community was at this time (early 1880's) receptive to the abstract formulations (cf. the response to Cayley's definition of 1854). The major reasons for this receptivity were:

(i) There were now several major "concrete" theories of groups-permutation groups, abelian groups, discontinuous transformation groups (the finite and infinite cases), and continuous transformation groups, and this warranted abstracting their essential features.

(ii) Groups came to play a central role in diverse fields of mathematics, such as different parts of algebra, geometry, number theory and several areas of analysis, and the abstract view of groups was thought to clarify what was essential for such applications and to offer opportunities for further applications.

(iii) The formal approach, aided by the penetration into mathematics of set theory and mathematical logic, became prevalent in other fields of mathematics, for example, various areas of geometry and analysis.

In the next section we will follow, very briefly, the evolution of that abstract point of view in group theory.

4 Consolidation of the abstract group *concept*; dawn of abstract group *theory*

The abstract group concept spread rapidly during the 1880's and 1890's, although there still appeared a great many papers in the areas of permutation and transformation groups. The abstract viewpoint was manifested in two ways:

(a) Concepts and results introduced and proved in the setting of "concrete" groups were now reformulated and reproved in an abstract setting;

(b) Studies originating in, and based on, an abstract setting began to appear.

An interesting example of the former case is a reproving by Frobenius, in an abstract setting, of Sylow's theorem, which was proved by Sylow in 1872 for permutation groups. This was done in 1887, in a paper entitled "Neuer Beweis Sylowschen Satzes." Although Frobenius admits that the fact that every finite group can be represented by a group of permutations proves that Sylow's theorem must hold for all finite groups, he nevertheless wishes to establish the theorem abstractly: "Since the symmetric group, which is introduced in all these proofs, is totally alien to the context of Sylow's theorem, I have tried to find a new derivation of it.... " (For a case study of the evolution of abstraction in group theory in connection with Sylow's theorem see [28] and [32].)

Hölder was an important contributor to abstract group theory, and was responsible for introducing a number of group-theoretic concepts abstractly. For example, in 1889 he introduced the abstract notion of a quotient group (the "quotient group" was first seen as the Galois group of the "auxiliary equation", later as a homomorphic image and only in Hölder's time as a group of cosets), and "completed" the proof of the Jordan-Hölder theorem, namely, that the quotient groups in a composition series are invariant up to isomorphism (see section 2(a) for Jordan's contribution). In 1893, in a paper on groups of order

p^3, pq^2, pqr, and p^4, he introduced abstractly the concept of an automorphism of a group. Hölder was also the first to study simple groups abstractly. (Previously they were considered in concrete cases-as permutation groups, transformation groups, and so on.) As he says [29, p. 338]. "It would be of the greatest interest if a survey of all simple groups with a finite number of operations could be known." (By "operations" Hölder meant elements.) He then goes on to determine the simple groups of order up to 200.

Other typical examples of studies in an abstract setting are the papers by Dedekind and G. A. Miller in 1897/1898 on Hamiltonian groups—i.e., non-abelian groups in which all subgroups are normal. They (independently) characterize such groups abstractly, and introduce in the process the notions of the commutator of two elements and the commutator subgroup (Jordan had previously introduced the notion of commutator of two permutations).

The theory of group characters and the representation theory for finite groups (created at the end of the 19th century by Frobenius and Burnside/Frobenius/Molien, respectively) also belong to the area of abstract group theory, as they were used to prove important results about abstract groups. See [17] for details.

Although the abstract group *concept* was well established by the end of the 19th century, "this was not accompanied by a general acceptance of the associated method of presentation in papers, textbooks, monographs, and lectures. Group-theoretic monographs based on the abstract group concept did not appear until the beginning of the 20th century. Their appearance marked the birth of abstract group *theory*" (Wussing, [33]).

The earliest monograph devoted entirely to abstract group theory was the book by J. A. de Séguier of 1904 entitled *Elements of the Theory of Abstract Groups* [27]. At the very beginning of the book there is a set-theoretic introduction based on the work of Cantor: "De Séguier may have been the first algebraist to take note of Cantor's discovery of uncountable cardinalities" (B. Chandler and W. Magnus, [7]). Next is the introduction of the concept of a semigroup with two-sided cancellation law and a proof that a finite semigroup is a group. There is also a proof, by means of counterexamples, of the independence of the group postulates. De Séguier's book also includes a discussion of isomorphisms, homomorphisms, automorphisms, decomposition of groups into direct products, the Jordan-Hölder theorem, the first isomorphism theorem, abelian groups including the basis theorem, Hamiltonian groups, and finally, the theory of p-groups. All this is done in the abstract, with "concrete" groups relegated to an appendix. "The style of de Seguier is in sharp contrast to that of Dyck. There are no intuitive considerations... and there is a tendency to be as abstract and as general as possible..." (Chandler and Magnus, [7]).

De Séguier's book was devoted largely to finite groups. The first abstract monograph on group theory which dealt with groups in general, relegating finite groups to special chapters, was O. Schmidt's *Abstract Theory of Groups* of 1916 [26]. Schmidt, founder of the Russian school of group theory, devotes the first four chapters of his book to group properties common to finite and infinite groups. Discussion of finite groups is postponed to chapter 5, there being ten chapters in all. See [7], [10], [33].

5 Divergence of developments in group theory

Group theory evolved from several different sources, giving rise to various concrete theories. These theories developed independently, some for over one hundred years (beginning in 1770) before they converged (early 1880's) within the abstract group concept. Abstract group theory emerged and was consolidated in the next thirty to forty years. At the end of that period (around 1920) one can discern the divergence of group theory into several distinct "theories." Here is the barest indication of *some* of these advances and new directions in group theory, beginning in the 1920's (with contributors and approximate dates):

(a) *Finite group theory.* The major problem here, already formulated by Cayley (1870's) and studied by Jordan and Hölder, was to find all finite groups of a given order. The problem proved too difficult and mathematicians turned to special cases (suggested especially by Galois theory): to find all simple or all solvable groups (cf. the Feit-Thompson theorem of 1963, and the classification of all finite simple groups in 1981). See [14], [15], [30].

(b) *Extensions of certain results from finite group theory to infinite groups with finiteness conditions*; e.g., O. J. Schmidt's proof, in 1928, of the Remak-Krull-Schmidt theorem. See [5].

(c) *Group presentations* (*Combinatorial Group Theory*), begun by von Dyck in 1882, and continued in the 20th century by M. Dehn, H. Tietze, J. Nielsen, E. Artin, O. Schreier, et al. For a full account, see [7].

(d) *Infinite abelian group theory* (H. Prifer, R. Baer, H. Ulm et al.—1920's to 1930's). See [30].

(e) *Schreier's theory of group extensions* (1926), leading later to the cohomology of groups.

(f) *Algebraic groups* (A. Borel, C. Chevalley et al.—1940's).

(g) *Topological groups*, including the extension of group representation theory to continuous groups (Schreier, E. Cartan. L. Pontrjagin, I. Gelfand, J. von Neumann et al.—1920's and 1930's). See [4].

Figure 1 gives a diagrammatic sketch of the evolution of group theory as outlined in the various sections and as summarized at the beginning of this section.

References

We give references here to *secondary* sources. Extensive references to *primary* sources, including works referred to in this article, may be found in [25] and [33].

1. R. G. Ayoub, Paolo Ruffini's contributions to the quintic, *Arch. Hist. Ex. Sc.*, 23 (1980) 253–277.
2. E. T. Bell, *The Development of Mathematics*, McGraw Hill, 1945.
3. G. Birkhoff, Current trends in algebra, *Amer. Math. Monthly*, 80 (1973) 760–782 and 81 (1974) 746.
4. ——, The rise of modem algebra to 1936, in *Men and Institutions in American Mathematics*, eds. D. Tarwater, J. T. White and J. D. Miller, Texas Tech. Press, 1976, pp. 41–63.
5. N. Bourbaki, *Eléments d'Histoire des Mathématiques*, Hermann, 1969.
6. J. E. Burns, The foundation period in the history of group theory, *Amer. Math. Monthly*, 20 (1913) 141–148.
7. B. Chandler and W. Magnus, *The History of Combinatorial Group Theory: A Case Study in the History of Ideas*, Springer-Verlag, 1982.
8. A. Dahan, Les travaux de Cauchy sur les substitutions. Etude de son approche du concept de groupe, *Arch. Hist. Ex. Sc.*, 23 (1980) 279–319.
9. J. Dieudonné (ed.), *Abrégé d'Histoire des Mathématiques, 1700–1900*, 2 vols., Hermann, 1978.

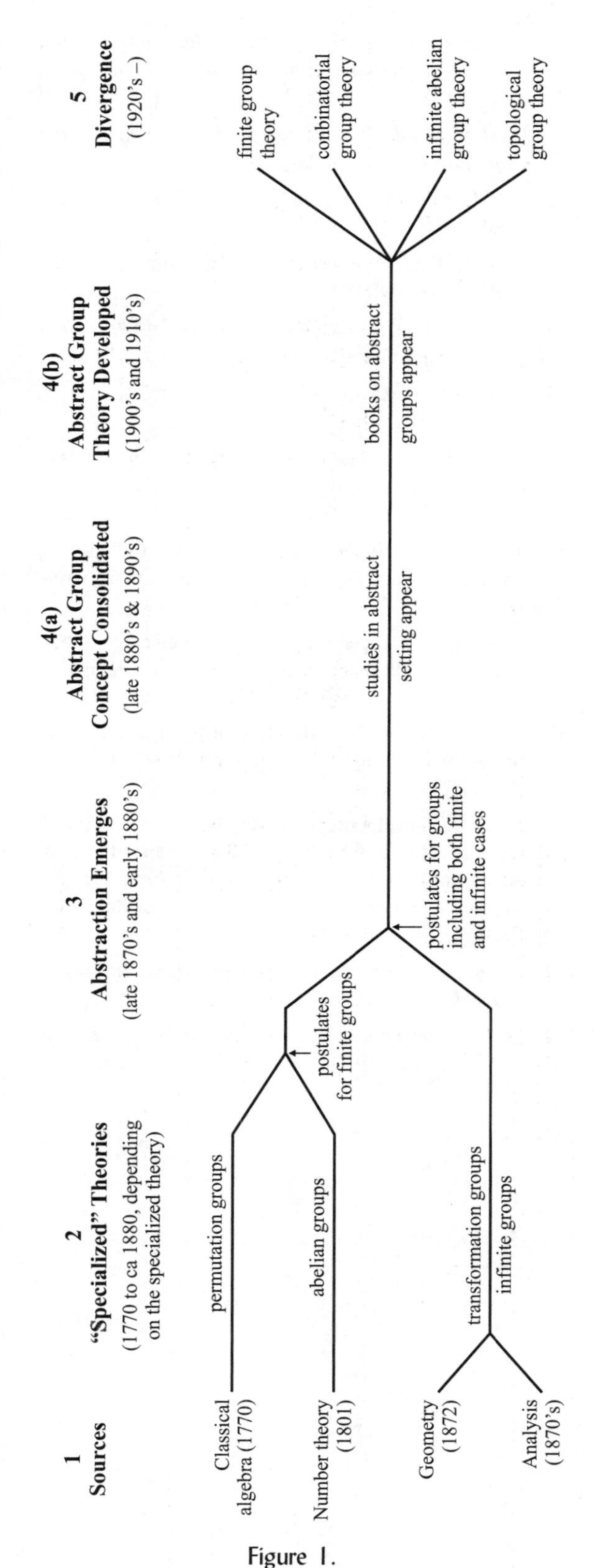

Figure 1.

10. P. Dubreil, L'algèbre, en France, de 1900 a 1935, *Cahiers du seminaire d'histoire des mathématiques*, 3 (1981) 69–81.

11. C. H. Edwards, *The Historical Development of the Calculus*, Springer-Verlag, 1979.

12. H. M. Edwards, *Galois Theory*, Springer-Verlag, 1984.

13. J. A. Gallian, The search for finite simple groups, *Math. Mag.*, 49 (1976) 163–179.

14. D. Gorenstein, *Finite Simple Groups: An Introduction to Their Classification*, Plenum Press, 1982.

15. ——, *The Classification of Finite Simple Groups*, Plenum Press, 1983.

16. R. R. Hamburg, The theory of equations in the 18th century: The work of Joseph Lagrange, *Arch. Hist. Ex. Sc.*, 16 (1976/77) 17–36.

17. T. Hawkins, Hypercomplex numbers, Lie groups, and the creation of group representation theory, *Arch. Hist. Ex. Sc.*, 8 (1971/72) 243–287.

18. ——, The Erlanger Programme of Felix Klein: Reflections on its place in the history of mathematics, *Hist. Math.*, 11 (1984) 442–470.

19. B. M. Kiernan, The development of Galois theory from Lagrange to Artin, *Arch. Hist. Ex. Sc.*, 8 (1971/72) 40–154.

20. F. Klein, Development of Mathematics in the 19th Century (transl. from the 1928 German ed. by M. Ackerman), in *Lie Groups: History, Frontiers and Applications*, vol. IX, ed. R. Hermann, Math. Sci. Press, 1979, pp. 1–361.

21. M. Kline, *Mathematical Thought from Ancient to Modern Times*, Oxford Univ. Press, 1972.

22. D. R. Lichtenberg, *The Emergence of Structure in Algebra*, Doctoral Dissertation, Univ. of Wisconsin, 1966.

23. U. Merzbach, *Development of Moder Algebraic Concepts from Leibniz to Dedekind*, Doctoral Dissertation, Harvard Univ., 1964.

24. G. A. Miller, *History of the theory of groups*, Collected Works, 3 vols., pp. 427–467, pp. 1–18, and pp. 1–15, Univ. of Illinois Press, 1935, 1938, and 1946.

25. L. Novy, *Origins of Modern Algebra*, Noordhoff, 1973.

26. O. J. Schmidt, *Abstract Theory of Groups*, W. H. Freeman & Co., 1966. (Translation by F. Holling and J. B. Roberts of the 1916 Russian edition.)

27. J.-A. de Séguier, *Théorie des Groupes Finis. Eléments de la Théorie des Groupes Abstraits*, Gauthier Villars, Paris, 1904.

28. L. A. Shemetkov, Two directions in the development of the theory of non-simple finite groups, *Russ. Math. Surv.*, 30 (1975) 185–206.

29. R. Silvestri, Simple groups of finite order in the nineteenth century, *Arch. Hist. Ex. Sc.*, 20 (1979) 313–356.

30. J. Tarwater, J. T. White, C. Hall, and M. E. Moore (eds.), *American Mathematical Heritage: Algebra and Applied Mathematics*, Texas Tech. Press, 1981. Has articles by Feit, Fuchs, and MacLane on the history of finite groups, abelian groups, and abstract algebra, respectively.

31. B. L. Van der Waerden, Die Algebra seit Galois, *Jahresbericht der Deutsch. Math. Ver.*, 68 (1966) 155–165.

32. W. C. Waterhouse, The early proofs of Sylow's theorem, *Arch. Hist. Ex. Sc.*, 21 (1979/80) 279–290.

33. H. Wussing, *The Genesis of the Abstract Group Concept*, MIT Press, 1984. (Translation by A. Shenitzer of the 1969 German edition.)

Design of an Oscillating Sprinkler

Bart Braden

Mathematics Magazine
58 (1985), 29–38

Editors' Note: Charles Bart Braden took his PhD at the University of Oregon, having written a dissertation on Lie algebras under the direction of Charles Curtis. He is now Professor Emeritus of Mathematics and Computer Science at Northern Kentucky University. Between 1994 and 1998 he was editor of the MAA's *College Mathematics Journal*. He is also the author of *Discovering Calculus with Mathematica* (Wiley, 1992).

For this paper he won the Allendoerfer Award in 1986 and two years later he won the Allendoerfer Award again for a paper entitled "Pólya's Geometric Picture of Complex Contour Integrals."

The common oscillating lawn sprinkler has a hollow curved sprinkler arm, with a row of holes on top, which rocks slowly back and forth around a horizontal axis. Water issues from the holes in a family of streams, forming a curtain of water that sweeps back and forth to cover an approximately rectangular region of lawn. Can such a sprinkler be designed to spread water uniformly on a level lawn?

We break the analysis into three parts:

1. How should the sprinkler arm be curved so that streams issuing from evenly spaced holes along the curved arm will be evenly spaced when they strike the ground?
2. How should the rocking motion of the sprinkler arm be controlled so that each stream will deposit water uniformly along its path?
3. How can the power of the water passing through the sprinkler be used to drive the sprinkler arm in the desired motion?

The first two questions provide interesting applications of elementary differential equations. The third, an excursion into mechanical engineering, leads to an interesting family of plane curves which we've called curves of constant diameter. A serendipitous bonus is the surprisingly simple classification of these curves.

The following result, proved in most calculus textbooks, will play a fundamental role in our discussion.

Lemma. *Ignoring air resistance, a projectile shot upward from the ground with speed v at an angle θ from the vertical, will come down at a distance $(v^2/g)\sin 2\theta$. (Here g is the acceleration due to gravity.)*

Note that $\theta = \pi/4$ gives the maximum projectile range, since then $\sin 2\theta = 1$. Textbooks usually express the projectile range in terms of the 'angle of elevation,' $\pi/2-\theta$; but since $\sin 2(\pi/2-\theta) = \sin 2\theta$, the range formula is unaffected when the zenith angle is used instead.

The sprinkler arm curve

In Figure 1, a (half) sprinkler arm is shown in a vertical plane, which we take to be the xy plane throughout this section. Let L be the length of the arc from the center of the sprinkler arm to the outermost hole, and let $x = x(s)$, $y = y(s)$ be parametric equations for the curve, using the arc length s,

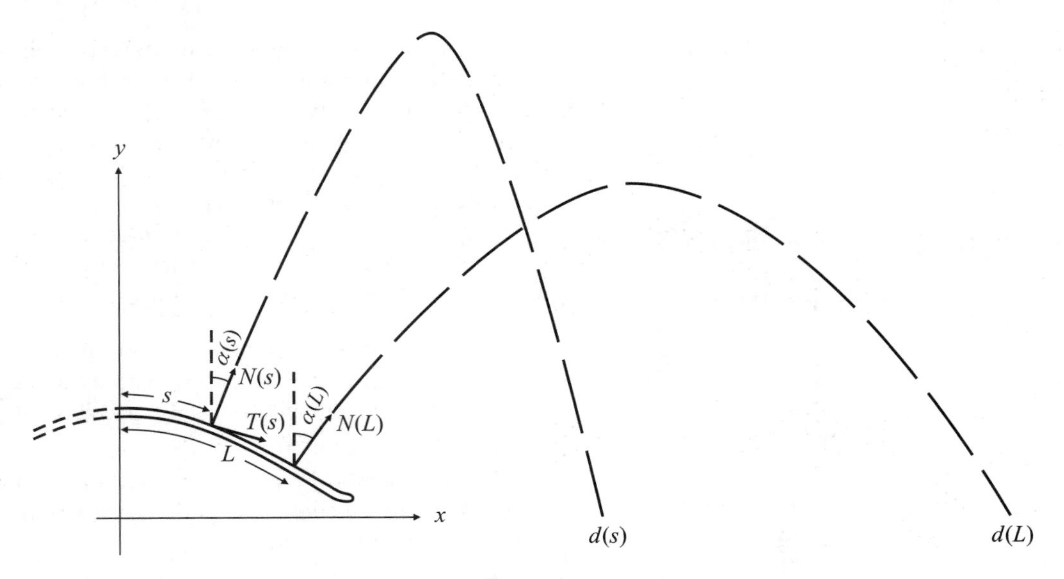

Figure 1.

$0 \le s \le L$, as parameter. Let $\alpha(s)$ denote the angle between the vertical and the outward normal to the arc at the point $(x(s), y(s))$.

We'll see that the functions $x(s)$ and $y(s)$ which define the curve are completely determined (once L, $\alpha(L)$ and $y(0)$ have been chosen) by the requirement that streams passing through evenly-spaced holes on the sprinkler arm should be uniformly spaced when they strike the ground.

If there were a hole at the point $(x(s), y(s))$ on the sprinkler arm, the direction vector of the stream issuing from this hole would be $N(s) = \langle \sin\alpha(s), \cos\alpha(s)\rangle$, and this stream would reach the ground at a distance

$$d(s) = \frac{v^2}{g}\sin^2\alpha(s).$$

The condition that evenly-spaced holes along the arm produce streams which are evenly spaced when they reach the ground is that $d(s)$ be proportional to s:

$$\frac{d(s)}{d(L)} = \frac{s}{L},$$

or equivalently,

$$\sin 2\alpha(s) = \frac{s}{L}\sin 2\alpha(L). \tag{1}$$

(We have made the assumption that the dimensions of the sprinkler are small in comparison to the dimensions of the area watered. This simplifies the calculations, and the errors introduced are not significant.)

The unit tangent vector to the sprinkler arm curve at $(x(s), y(s))$ is $T(s) = \langle x'(s), y'(s)\rangle$, so the unit outward normal vector (obtained by rotating $T(s)$ counterclockwise by $\pi/2$) is $N(s) = \langle -y'(s), x'(s)\rangle$. Comparing this with our earlier expression for $N(s)$, we have

$$x'(s) = \cos\alpha(s), \qquad y'(s) = -\sin\alpha(s).$$

Since $\sin 2\alpha(s) = 2\sin\alpha(s)\cos\alpha(s)$, equation (1) for the sprinkler arm curve becomes

$$-2x'(s)y'(s) = \frac{s}{L}\sin 2\alpha(L).$$

The value of $\alpha(L)$, the angle between the vertical and the outermost stream as it leaves the sprinkler, is a parameter under the designer's control; once it is chosen, the value of $\sin 2\alpha(L)$ is determined—call it k, where $0 < k \le 1$. Our equation then becomes

$$x'(s)y'(s) = \frac{-k}{2L}s. \tag{2}$$

Since $N(s)$ is a unit vector, also

$$x'(s)^2 + y'(s)^2 = 1. \tag{3}$$

Fortunately, the pair of nonlinear differential equations (2) and (3) for $x(s)$ and $y(s)$ simplifies algebraically:

$$x'^2 + \left(\frac{-ks}{2Lx'}\right)^2 = 1,$$

or

$$x'^4 - x'^2 + \frac{k^2s^2}{4L^2} = 0$$

Solving by the quadratic formula, we find that

$$x'(s)^2 = \frac{1 \pm \sqrt{1 - \left(\frac{ks}{L}\right)^2}}{2}.$$

Since the sprinkler arm is horizontal at its midpoint, the unit tangent vector $T(0) = \langle x'(0), y'(0) \rangle$ is $(1, 0)$. Thus $x'(0) = 1$, which means we must use the $+$ sign in the quadratic formula. Substituting

$$x'(s) = \frac{1}{\sqrt{2}}\left[1 + \sqrt{1 - \left(\frac{ks}{L}\right)^2}\right]^{1/2}$$

in equation (2) gives

$$y'(s) = \frac{-1}{\sqrt{2}}\left[1 - \sqrt{1 - \left(\frac{ks}{L}\right)^2}\right]^{1/2}.$$

Since $x(0) = 0$ and $y(0)$ is arbitrary, we conclude that

$$x(s) = \frac{1}{\sqrt{2}}\int_0^s \left[1 + \sqrt{1 - \left(\frac{kt}{L}\right)^2}\right]^{1/2} dt$$

and

$$y(s) = y(0) - \frac{1}{\sqrt{2}}\int_0^s \left[1 - \sqrt{1 - \left(\frac{kt}{L}\right)^2}\right]^{1/2} dt.$$

These integrals can be evaluated in closed form, using the identity (kindly supplied by a reviewer)

$$\frac{1 \pm \sqrt{1 - \left(\frac{kt}{L}\right)^2}}{2} = \left[\frac{\sqrt{1 + \frac{kt}{L}}}{2} \pm \frac{\sqrt{1 - \frac{kt}{L}}}{2}\right]^2,$$

with the result

$$x(s) = \frac{L}{3K}\left[\left(1 + \frac{ks}{L}\right)^{3/2} - \left(1 - \frac{ks}{L}\right)^{3/2}\right],$$

$$y(s) = y(0) - \frac{2L}{3k}\left[\left(1 + \frac{ks}{L}\right)^{3/2} + \left(1 - \frac{ks}{L}\right)^{3/2}\right].$$

This sprinkler arm curve is drawn in Figure 2. Note that the curve is determined by the requirement that the streams be evenly spaced along the ground when the plane of the sprinkler arm is vertical. Later we indicate what happens when this plane makes an angle ϕ with the vertical.

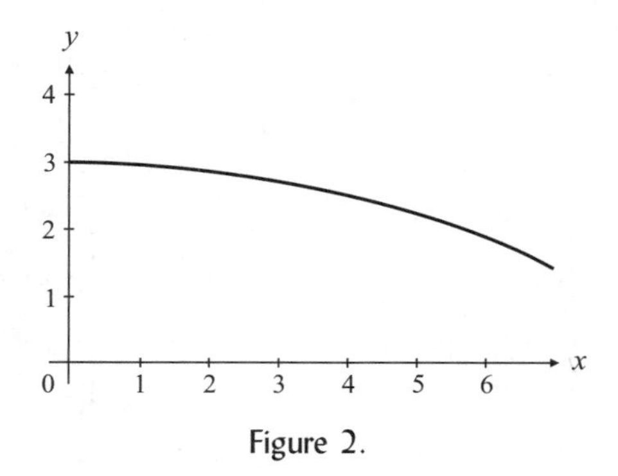

Figure 2.

The rocking motion of the sprinkler arm

We wish the sprinkler arm to oscillate in such a way that each stream will deposit water uniformly along its path, or what is the same, the speed of the point of impact of the stream with the ground should be constant on each pass of the sprinkler. As it turns out, this condition cannot be satisfied by all the streams simultaneously, so we shall concentrate our attention on the central stream.

Henceforth, let's choose a coordinate system in space, as indicated in Figure 3, with the z-axis vertical and the axis of rotation of the sprinkler arm the y-axis, with the center of the arm on the positive z-axis. When the plane of the sprinkler arm makes an angle ϕ with the vertical, the central stream will reach the ground on the x-axis, at $x = (v^2/g)\sin 2\phi$). The oscillation of the sprinkler arm is described by the function $\phi(t)$, and the corresponding speed of the central stream over the lawn is the derivative $x'(t) = (2v^2/g)\cos 2\phi(t)\phi'(t)$. Setting $x'(t) = Kv^2/g$, a conveniently labelled constant, we see that uniform coverage by the central stream is equivalent to the requirement that $\phi(t)$ be a solution of the (separable) differential equation

$$2\cos 2\phi(t)\phi'(t) = K. \tag{4}$$

Integration of (4) gives the solution

$$\sin 2\phi(t) = Kt + c. \tag{5}$$

The parameters K and c have no apparent significance, so we next try to find an expression for the angular variation $\phi(t)$ in terms of two other constants which are easily interpreted. Suppose the sprinkler arm rocks back and forth in the range $-\phi_0 \leq \phi(t) \leq \phi_0$, where the maximum tilt, ϕ_0, is a design parameter in the range $0 < \phi_0 < \pi/4$. Let the time required for the sprinkler arm to rotate between the vertical and the maximum angle ϕ_0 be

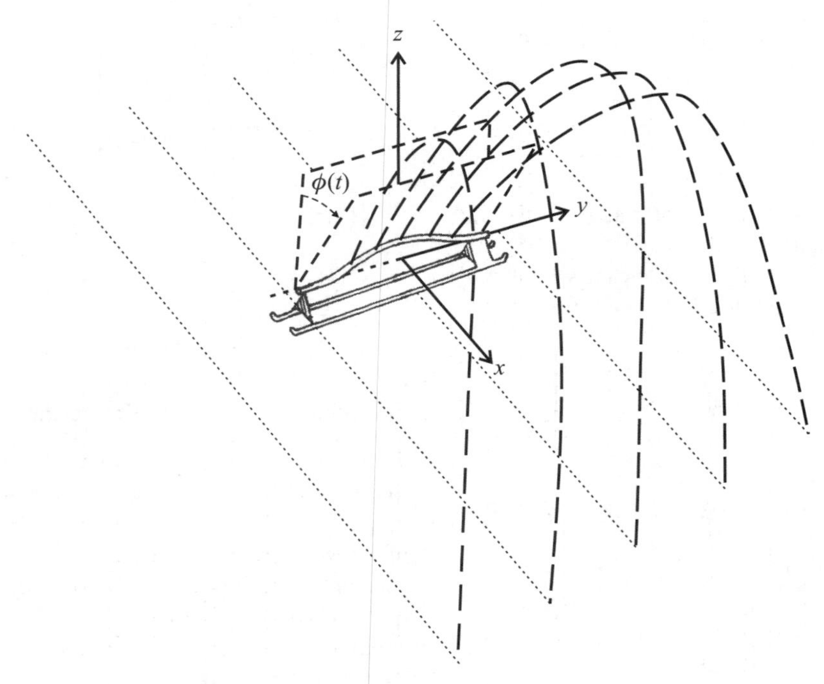

Figure 3.

denoted by T. (Thus $2T$ is the time required for one pass of the sprinkler over the lawn, and $4T$ is the period of the complete oscillation.) If we measure time so that $\phi(0) = -\phi_0$, then setting $t = 0$ in equation (5) gives $c = -\sin 2\phi_0$. Since $\phi(2T) = \phi_0$, we then get $\sin 2\phi_0 = 2KT - \sin 2\phi_0$, or $K = \frac{1}{T}\sin 2\phi_0$. Thus

$$\sin 2\phi(t) = \frac{t-T}{T}\sin 2\phi_0,$$

or, since $-\pi/2 \le 2\phi(t) \le \pi/2$,

$$\phi(t) = \frac{1}{2}\arcsin\left[\frac{t-T}{T}\sin 2\phi_0\right]. \tag{6}$$

The oscillatory motion of the sprinkler arm is therefore uniquely determined (once choices of ϕ_0 and T have been made) by the requirement that the **central** stream cover the ground uniformly.

Remark: Since the maximum range of the central stream occurs when $\phi = \pi/4$, one might think the ideal value for ϕ_0 would be $\pi/4$. However, we will show later that the shape of the region covered by the sprinkler will be more nearly rectangular if ϕ_0 is somewhat smaller than $\pi/4$.

It remains to describe a mechanism that will produce the desired oscillation, given by (6). (It was by observing my own sprinkler, the Nelson 'dial-a-rain', which appears to use the design described below, that I was led to the questions considered in this paper.)

Mechanical design of the sprinkler

The stream of water entering the sprinkler from a hose can be used to turn an impeller (waterwheel), which is then geared down to turn a cam with a constant angular velocity ω. A cam follower linkage converts the uniform rotational motion of the cam into an oscillatory motion of the sprinkler arm—as the cam makes a half-revolution, the sprinkler arm is pushed from $\phi = -\phi_0$ to $\phi = +\phi_0$; and on the next half-revolution of the (bilaterally symmetric) cam, the sprinkler arm makes the return sweep.

What shape of cam will cause the oscillation of the sprinkler arm to be that given by (6)? We may describe the shape of a cam in polar coordinates by $r(\theta) = r_0 + f(\theta)$, where the function $f(\theta)$ describes the 'eccentricity' of the cam, i.e., its deviation from the circle $r(\theta) = r_0$. The pole of our coordinate system is placed at the center around which the cam rotates, so it is the eccentricity $f(\theta)$ that produces the motion of the sprinkler arm. As the cam follower moves to the right or left of the point labelled P in Figure 4, by the amount $f(\theta)$, the other end of the connecting rod moves the sprinkler arm the same distance along a circular arc of radius ℓ. Denoting

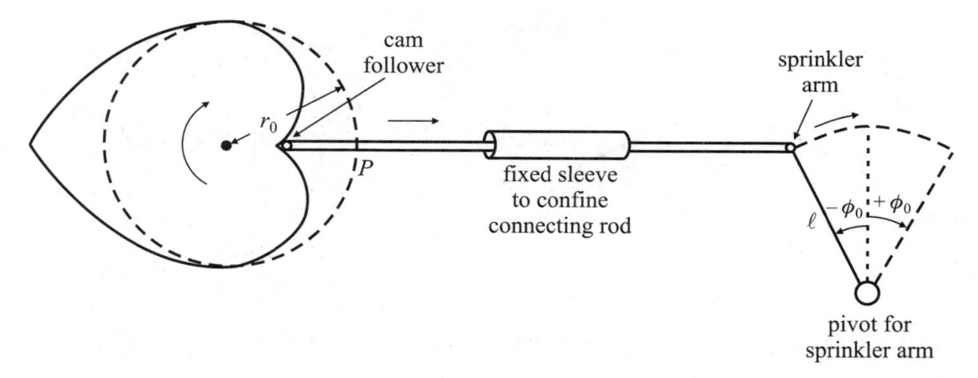

Figure 4. A typical cam mechanism.

the arclength by s and using the relation $s/\ell = \phi$, we have $f(\theta)/\ell = \phi$. Since $\theta(t) = \omega t$, we want

$$f(\omega t) = \frac{\ell}{2}\arcsin\left[\frac{t-T}{T}\sin 2\phi_0\right], \quad 0 \le t \le 2T.$$

Our goal is to find a formula for the eccentricity $f(\theta)$, so we must express t and T in terms of θ and ω, using the relation $\theta = \omega t$. This is easy: as the cam turns a half-revolution, from $\theta = 0$ to $\theta = \pi$, the sprinkler arm moves from $\phi = -\phi_0$ to $\phi = \phi_0$; so equating the times required gives $\pi/\omega = 2T$. Thus

$$\frac{t-T}{T} = \frac{2\omega t}{\pi} - 1,$$

so

$$f(\omega t) = \frac{\ell}{2}\arcsin\left[\frac{2}{\pi}\left(\omega t - \frac{\pi}{2}\right)\sin 2\phi_0\right],$$

$0 \le t \le 2T$. Replacing ωt by θ, we conclude that

$$f(\theta) = \frac{\ell}{2}\arcsin\left[\frac{2}{\pi}\left(\theta - \frac{\pi}{2}\right)\sin 2\phi_0\right], \quad (7)$$

$0 \le \theta \le \pi$. As θ goes from π to 2π we want the sprinkler arm to perform the same motion in reverse, i.e., the cam should be symmetric about the polar axis:

$$f(\theta) = f(2\pi - \theta) \quad \text{for } \pi \le \theta \le 2\pi. \quad (8)$$

The polar curve $r(\theta) = r_0 + f(\theta)$, where the eccentricity $f(\theta)$ is given by (7) and (8), is the cam shape that will produce the desired oscillatory motion of the sprinkler arm (see Figure 5). (Note that r_0 is arbitrary, provided $r_0 > \ell\phi_0$.) This curve has an interesting geometric property, described in the following definition.

Definition. A simple closed curve C is said to be of **constant diameter** d if there is a point O inside C such that every chord of C through O has the same length, d. Any chord through this 'center' point O is called a **diameter** of C.

N. B. This class of curves should not be confused with 'curves of constant width,' a family of convex curves which appears frequently in the literature, e.g. [1].

It is easy to verify using (7) and (8), that our cam curve $r(\theta) = r_0 + f(\theta)$ has constant diameter $2r_0$.

Proof. Let O be the pole of our coordinate system. Then the diameter of our curve which makes an angle θ with the polar axis is $r(\theta) + r(\theta + \pi)$, or $2r_0 + f(\theta) + f(\theta + \pi)$. Thus it must be shown that $f(\theta) + f(\theta + \pi) = 0$. Without loss of generality we may assume $0 \le \theta \le \pi$; then by (8),

$$\begin{aligned} f(\theta + \pi) &= f(2\pi - (\theta + \pi)) = f(\pi - \theta) \\ &= \frac{\ell}{2}\arcsin\left[\frac{2}{\pi}\left(\pi - \theta - \frac{\pi}{2}\right)\sin 2\phi_0\right] \\ &= \frac{\ell}{2}\arcsin\left[\frac{-2}{\pi}\left(\theta - \frac{\pi}{2}\right)\sin 2\phi_0\right] \\ &= -f(\theta). \end{aligned}$$

Examining this proof, we discover a simple construction for all curves of constant diameter. Given $d > 0$, take any continuous function $r(\theta)$ such that $r(O) + r(\pi) = d$ and $0 < r(\theta) < d$ for $0 \le \theta \le \pi$. If we extend the domain to $[\pi, 2\pi]$ by defining $r(\theta + \pi) = d - r(\theta)$, as in (8), the polar curve $r = r(\theta)$ will have constant diameter d.

For curves symmetric with respect to the polar axis, i.e., with $r(2\pi - \theta) = r(\theta)$ for $0 \le \theta \le \pi$, the constant diameter condition is simply that $r(\pi - \theta) = d - r(\theta)$ for $0 \le \theta \le \pi/2$. So $d = 2r(\pi/2)$. Thus an arbitrary continuous function $r(\theta)$ defined for $0 \le \theta \le \pi/2$, for which $0 < r(\theta) < 2r(\pi/2)$, can be extended uniquely to produce a simple closed curve of constant diameter $d = 2r(\pi/2)$ which is

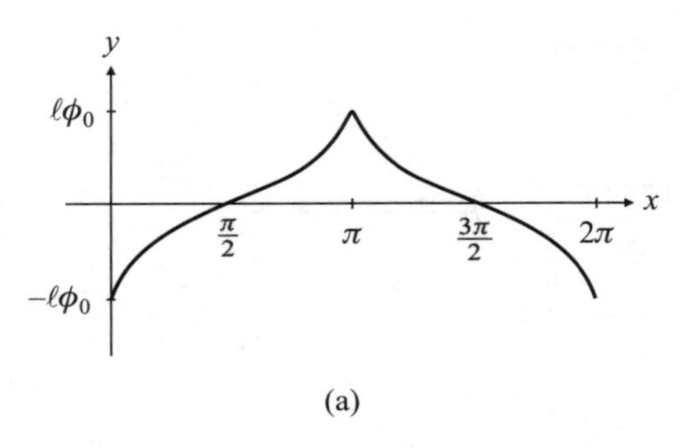

Figure 5. (a) The Cartesian graph of $f(x)$. (b) The polar graph of $r = r_0 + f(\theta)$.

symmetric with respect to the polar axis.

The observation that the cam curve for our sprinkler has constant diameter $2r_0$ suggests a particularly simple mechanical design for the cam follower linkage: a post fixed to the center of the cam, sliding in a slot in the connecting rod, with rollers fixed on the rod separated by the distance $2r_0$. As the cam turns, the rollers remain in contact with it at opposite ends of a diameter, and the connecting rod is alternately pushed and pulled along the line of its slot (see Figure 6). If the cam did not have constant diameter, a more complicated mechanical linkage would be required to keep the cam follower in contact with the cam, and to confine the motion of the connecting rod to one dimension.

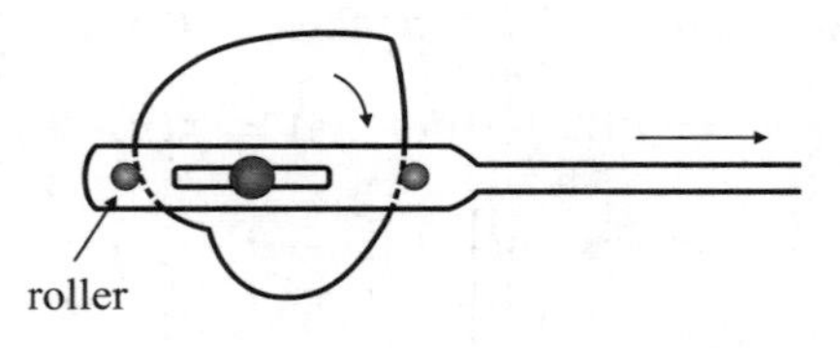

Figure 6.

The complete sprinkler pattern

In determining the curve of the sprinkler arm we considered only the situation in which the arm is vertical, and found that the requirement of uniform spacing of the streams on the ground then determines the curve uniquely. Similarly, in analyzing the oscillation of the sprinkler arm we considered only the central stream, and found that the requirement of uniform coverage by this single stream along its path uniquely determines the motion $\phi(t)$. It remains to be seen whether the streams from the other holes will move along the lawn at constant speeds, and whether these streams will remain equally spaced as the sprinkler arm rocks back and forth.

Suppose there are $2n + 1$ holes in the sprinkler arm: one in the center and n more spaced at equal intervals on each side. By symmetry we need only consider the streams from one half of the sprinkler arm. Using the coordinate system described earlier (see Figure 3), denote the angles between the vertical and the streams as they leave the sprinkler arm (when the plane of the arm is vertical) by $\alpha_0, \alpha_1, \ldots, \alpha_n$, where $0 = \alpha_0 < \alpha_1 < \cdots < \alpha_n \le \pi/4$. The streams will strike the ground at distances $d_i = (v^2/g)\sin 2\alpha_i$, and since the streams are equally spaced along the ground, $d_i = (i/n)d_n$. That is,

$$\frac{v^2}{g}\sin 2\alpha_i = \frac{i}{n}\left(\frac{v^2}{g}\sin 2\alpha_n\right),$$

or

$$\alpha_i = \frac{1}{2}\arcsin\left[\frac{i}{n}\sin 2\alpha_n\right].$$

The direction vectors of the streams as they leave the sprinkler arm are $N_i = \langle 0, \sin\alpha_i, \cos\alpha_i\rangle$, $0 \le i \le n$.

When the plane of the sprinkler arm is tilted at an angle ϕ, the direction vectors N_i are rotated through the angle ϕ around the y-axis, so the streams issue from the holes in the directions $N_i(\phi) = \langle \cos\alpha_i \sin\phi, \sin\alpha_i, \cos\alpha_i \cos\phi\rangle$. The angle θ_i between $N_i(\phi)$ and the vertical is given by

$$\cos\theta_i = N_i(\phi)\cdot\langle 0,0,1\rangle = \cos\alpha_i\cos\phi,$$

so the ith stream strikes the ground at a distance

$$\begin{aligned} d_i(\phi) &= \frac{v^2}{g}\sin 2\theta_i \\ &= \frac{2v^2}{g}\cos\theta_i\sin\theta_i \\ &= \frac{2v^2}{g}\cos\alpha_i\cos\phi\sqrt{1-\cos^2\alpha_i\cos^2\phi}. \end{aligned}$$

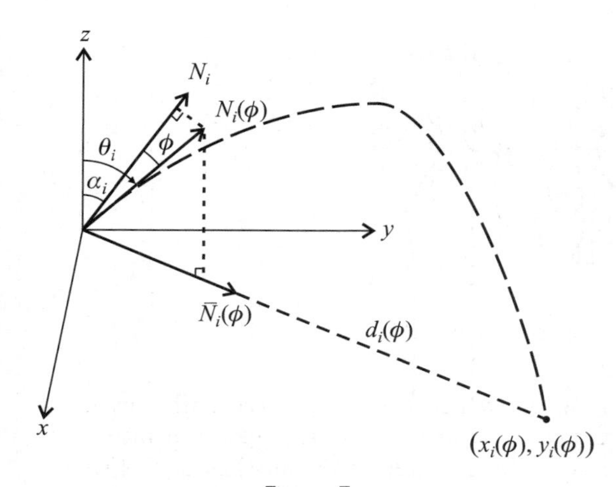

Figure 7.

The point of impact is $d_i(\phi)\overline{N}_i(\phi)$, where

$$\overline{N}_i(\phi) = \frac{\langle \cos\alpha_i \sin\phi, \sin\alpha_i, 0\rangle}{\sqrt{\cos^2\alpha_i \sin^2\phi + \sin^2\alpha_i}}$$

the unit vector in the direction of the projection of $N_i(\phi)$ on the xy plane. Thus parametric equations for the path of the ith stream as it moves over the lawn are

$$x_i(\phi) = \frac{2v^2\cos^2\alpha_i \cos\phi \sin\phi \sqrt{1-\cos^2\alpha_i \cos^2\phi}}{g\sqrt{\cos^2\alpha_i \sin^2\phi + \sin^2\alpha_i}},$$

$$y_i(\phi) = \frac{2v^2\cos^2\alpha_i \sin\alpha_i \cos\phi \sqrt{1-\cos^2\alpha_i \cos^2\phi}}{g\sqrt{\cos^2\alpha_i \sin^2\phi + \sin^2\alpha_i}}$$

Evidently $y_i(\phi)$ is not constant, so the path of the ith stream of water is not the straight line parallel to the central stream which one might have expected.

To parametrize the paths by time, we can simply replace the parameter ϕ by the expression for $\phi(t)$ in (6). Computer plots of the resulting family of curves are shown in Figures 8 and 9. In each of these figures, the curves running approximately parallel to the x-axis are the paths of the streams from one side of the sprinkler arm. Time is indicated by the curves nearly parallel to the y-axis, which are polygonal arcs connecting the points on the eight stream paths at six equally spaced instants $T + (k/6)T$, $1 \leq k \leq 6$. (Recall that as t runs through the interval $T \leq t \leq 2T$, the plane of the sprinkler arm turns through the interval $0 \leq \phi \leq \phi_0$.) Thus each of the resulting 'squares' receives the same amount of water on each pass of the sprinkler.

As seen in Figure 8, where $\phi_0 \approx 40°$ and $\alpha_7 \approx 30°$, the outer streams curve in significantly, making the 'squares' near the outside comer smaller in area. Since each 'square' receives the same amount of water per pass of the sprinkler, the sprinkler shown would overwater the four corners of the region it sprinkled.

In Figure 9, by reducing ϕ_0 to about 29° (and decreasing α_7 to 26° to keep similar proportions to the region watered), not only is the non-uniformity of coverage reduced, but at the same time the region covered is more nearly rectangular.

We conclude with some observations which could not be followed up here; their investigation is left to the proverbial interested reader.

1. No attempt has been made to define an optimal shape for the region watered. Evidently, decreasing the angle parameters ϕ_0 and $\alpha(L)$ will make the coverage more uniform, but at the cost of decreasing the area watered. An interesting question, suggested by a reviewer, might be to design a sprinkler to maximize the area covered without exceeding a stipulated amount of variation in the water applied per unit area. This would mean introducing some non-uniformity along the x and y axes (i.e., unequal spacing of the streams as they strike the ground when the plane of the sprinkler arm is vertical, and non-uniform speed of the central stream along its path),

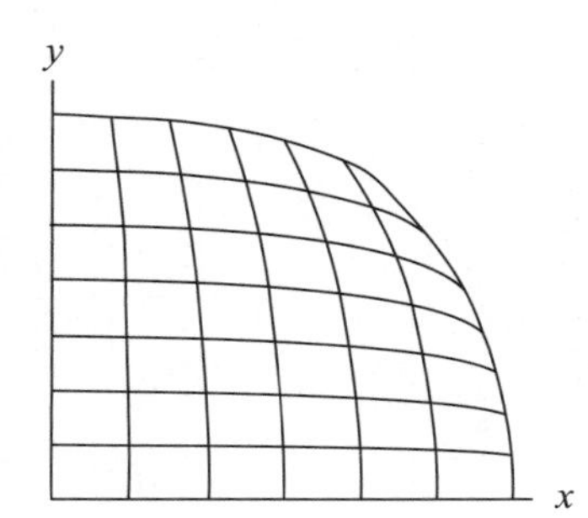

Figure 8.

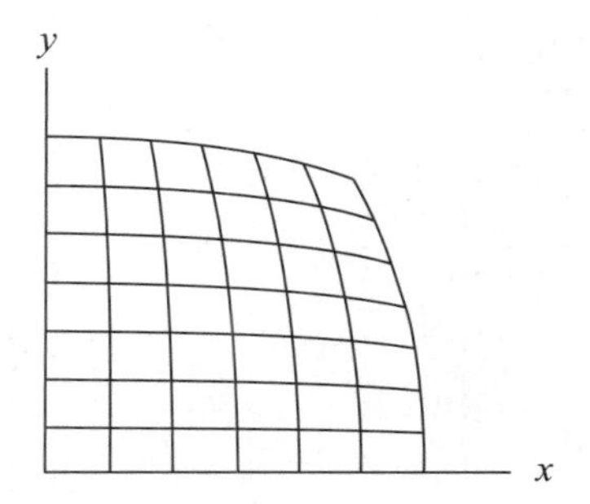

Figure 9.

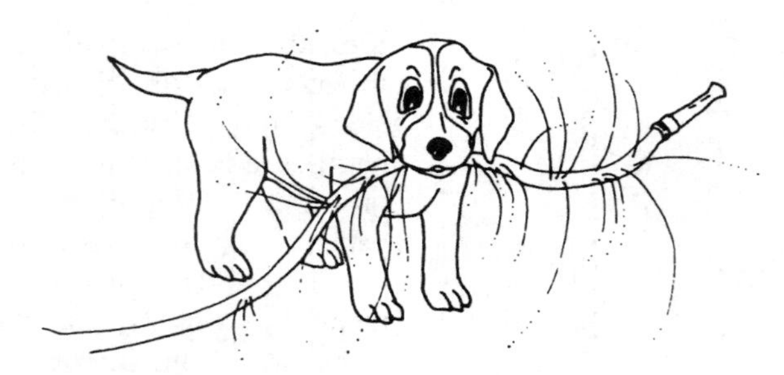

...of course, other sprinkler designs are possible.

to compensate for the overwatering of the corners observed in Figures 8 and 9 with our sprinkler.

2. My Nelson 'dial-a-rain' sprinkler has an additional feature of interest. On the sprinkler arm support is a dial which, when turned, changes the radius ℓ of the arc on which the sprinkler arm moves. The effect of doubling ℓ, for instance, can be shown to be to cut in half the region watered. The coverage of this smaller area is slightly less uniform, however.

I find it remarkable that not only are the curve of the sprinkler arm and the motion $\phi(t)$ of the arm unique, but even the mechanical design of the sprinkler is essentially determined by the requirement that water should be spread uniformly along the two coordinate axes. The wealth of mathematical questions raised in the analysis of this simple mechanism gives me a new respect for mechanical engineering, and greater confidence in the importance of classical mathematics to students in this field.

Reference

1. G. D. Chakerian, A characterization of curves of constant width, *Amer. Math. Monthly*, 81 (1974) 153–155.

The Centrality of Mathematics in the History of Western Thought

Judith V. Grabiner

Mathematics Magazine
61 (1988), 220–30

Editors' Note: This essay was reprinted in the *Magazine* from the Proceedings of the International Congress of Mathematicians (American Mathematical Society, 1987), the congress held in Berkeley, California, 1986.

An undergraduate in mathematics at the University of Chicago, Grabiner went on to take her PhD at Harvard in the history of mathematics. Her books, *The Calculus as Algebra/J. L. Lagrange 1736–1813* (Garland, 1990) and *The Origins of Cauchy's Rigorous Calculus* (MIT, 1981), are among the growing number of pieces she has written on the development of analytic geometry and calculus, with journal articles on Descartes, Fermat, Lagrange, Maclaurin, Weierstrass, and Cauchy. For her paper delivered at the Berkeley Congress she was awarded the Carl B. Allendoerfer Award by the MAA in 1989. This she added to her two other Allendoerfer Awards (1984, 1996) and her two Lester R. Ford Awards (1984, 1998). She has been given more awards for expository writing by the MAA than any other author.

Grabiner is the Flora Sanborn Pitzer Professor of Mathematics at Pitzer College, one of the Claremont colleges in California.

1. Introduction

Since this paper was first given to educators, let me start with a classroom experience. It happened in a course in which my students had read some of Euclid's *Elements of Geometry*. A student, a social science major, said to me, "I never realized mathematics was like this. Why, it's like philosophy!" That is no accident, for philosophy is like mathematics. When I speak of the centrality of mathematics in western thought, it is this student's experience I want to recapture—to reclaim the context of mathematics from the hardware store with the rest of the tools and bring it back to the university. To do this, I will discuss some major developments in the history of ideas in which mathematics has played a central role.

I do not mean that mathematics has by itself caused all these developments; what I do mean is that mathematics, whether causing, suggesting, or reinforcing, has played a key role; it has been there, at center stage. We all know that mathematics has been the language of science for centuries. But what I wish to emphasize is the crucial role of mathematics in shaping views of man and the world held not just by scientists, but by everyone educated in the western tradition.

Given the vastness of that tradition, I will give many examples only briefly, and be able to treat only a few key illustrative examples at any length. Sources for the others may be found in the bibliography. (See also [26].)

Since I am arguing for the centrality of mathematics, I will organize the paper around the key features of mathematics which have produced the effects I will discuss. These features are the certainty of mathematics and the applicability of mathematics to the world.

2. Certainty

For over two thousand years, the certainty of mathematics, particularly of Euclidean geometry, has had to be addressed in some way by any theory of knowledge. Why was geometry certain? Was it because of the subject matter of geometry, or because of its method? And what were the implications of that certainty?

Even before Euclid's monumental textbook, the philosopher Plato saw the certainty of Greek geometry—a subject which Plato called "knowledge of that which always is" [41, 527b]—as arising from the eternal, unchanging perfection of the objects of mathematics. By contrast, the objects of the physical world were always coming into being or passing away. The physical world changes, and is thus only an approximation to the higher ideal reality. The philosopher, then, to have his soul drawn from the changing to the real, had to study mathematics. Greek geometry fed Plato's idealistic philosophy; he emphasized the study of Forms or Ideas transcending experience: the idea of justice, the ideal state, the idea of the Good. Plato's views were used by philosophers within the Jewish, Christian, and Islamic traditions to deal with how a divine being, or souls, could interact with the material world [46, pp. 382–3] [51, pp. 17–40] [34, p. 305ff] [23, p. 46–67]. For example, Plato's account of the creation of the world in his *Timaeus*, where a god makes the physical universe by copying an ideal mathematical model, became assimilated in early Christian thought to the Biblical account of creation [29, pp. 21-22]. One finds highly mathematicized cosmologies, influenced by Plato, in the mystical traditions of Islam and Judaism as well. The tradition of Platonic Forms or Ideas crops up also in such unexpected places as the debates in eighteenth- and nineteenth-century biology over the fixity of species. Linnaeus in the eighteenth, and Louis Agassiz in the nineteenth century seem to have thought of species as ideas in the mind of God [16, p. 34] [13, pp. 36-7]. When we use the common terms "certain" and "true" outside of mathematics, we use them in their historical context, which includes the long-held belief in an unchanging reality—a belief stemming historically from Plato, who consistently argued for it using examples from mathematics.

An equally notable philosopher, who lived just before Euclid, namely Aristotle, saw the success of geometry as stemming, not from perfect eternal objects, but instead from its method [*Posterior Analytics*, I 10–11; I 1–2 (77a5, 71b ff)] [19, vol. I, Chapter IX]. The certainty of mathematics for Aristotle rested on the validity of its logical deductions from self-evident assumptions and clearly-stated definitions. Other subjects might come to share that certainty if they could be understood within the same logical form; Aristotle, in his *Posterior Analytics*, advocated reducing all scientific discourse to syllogisms, that is, to logically-deduced explanations from first principles. In this tradition, Archimedes proved the law of the lever, not by experiments with weights, but from deductions *à la* Euclid from postulates like "equal weights balance at equal distances" [18, pp. 189–194]. Medieval theologians tried to prove the existence of God in the same way. This tradition culminates in the 1675 work of Spinoza, *Ethics Demonstrated in Geometrical Order*, with such axioms as "That which cannot be conceived through another must be conceived through itself," definitions like "By substance I understand that which is in itself and conceived through itself" (compare Euclid's "A point is that which has no parts"), and such propositions as "God or substance consisting of infinite attributes... necessarily exists," whose proof ends with a QED [48, pp. 41–50]. Isaac Newton called his famous three laws "Axioms, or Laws of Motion." His *Principia* has a Euclidean structure, and the law of gravity appears as Book III, Theorems VII and VIII [37, pp. 13–14, pp. 414-17]. The Declaration of Independence of the United States is one more example of an argument whose authors tried to inspire faith in its certainty by using the Euclidean form. "We hold these truths to be self-evident..." not that all right angles are equal, but "that all men are created equal." These self-evident truths include that if any government does not obey these postulates, "it is the right of the people to alter or abolish it." The central section begins by saying that they will "prove" King George's government does not obey them. The conclusion is "We, *therefore*... declare, that these United Colonies are, and of right ought to be, free and independent states." (My italics) (Jefferson's mathematical education, by the way, was quite impressive by the standards of his time.)

Thus a good part of the historical context of the common term "proof" lies in Euclidean geometry-which was, I remind you, a central part of Western education.

However, the certainty of mathematics is not limited to Euclidean geometry. Between the rise of Islamic culture and the eighteenth century, the paradigm governing mathematical research changed from a geometric one to an algebraic, symbolic one. In algebra even more than in the Euclidean model of reasoning, the method can be considered independently of the subject-matter involved. This view looks at the method of mathematics as finding truths by manipulating symbols. The approach first enters the western world with the introduction of the Hindu-Arabic number system in the twelfth-century translations into Latin of Arabic mathematical works, notably al-Khowarizmi's algebra. The simplified calculations using the Hindu-Arabic numbers were called the "method of al-Khowarizmi" or as Latinized "the method of algorism" or algorithm.

In an even more powerful triumph of the heuristic power of notation, François Viète in 1591 introduced literal symbols into algebra: first, using letters in general to stand for any number in the theory of equations; second, using letters for any number of unknowns to solve word problems [4, pp. 59–63, 65]. In the seventeenth century, Leibniz, struck by the heuristic power of arithmetical and algebraic notation, invented such a notation for his new science of finding differentials—an algorithm for manipulating the d and integral symbols, that is, a calculus (a term which meant to him the same thing as "algorithm" to us). Leibniz generalized the idea of heuristic notation in his philosophy [30, pp. 12–25]. He envisioned a symbolic language which would embody logical thought just as these earlier symbolic languages enable us to perform algebraic operations correctly and mechanically. He called this language a "universal characteristic," and later commentators, such as Bertrand Russell, see Leibniz as the pioneer of symbolic logic [45, p. 170]. Any time a disagreement occurred, said Leibniz, the opponents could sit down and say "Let us calculate," and—mechanically—settle the question [30, p. 15]. Leibniz's appreciation of the mechanical element in mathematics when viewed as symbolic manipulation is further evidenced by his invention of a calculating machine. Other seventeenth-century thinkers also stressed the mechanical nature of thought in general: for instance, Thomas Hobbes wrote "Words are wise men's counters, they do but reckon by them" [21, Chapter 4, p. 143]. Others tried to introduce heuristically powerful notation in different fields: consider Lavoisier's new chemical notation which he called a "chemical algebra" [14, p. 245].

These successes led the great prophet of progress, the Marquis de Condorcet, to write in 1793 that algebra gives "the only really exact and analytical language yet in existence.... Though this method is by itself only an instrument pertaining to the science of quantities, it contains within it the principles of a universal instrument, applicable to all combinations of ideas" [9, p. 238]. This could make the progress of "every subject embraced by human intelligence... as sure as that of mathematics" [9, pp. 278–9]. The certainty of symbolic reasoning has led us to the idea of the certainty of progress. Though one might argue that some fields had not progressed one iota beyond antiquity, it was unquestionably true by 1793 that mathematics and the sciences had progressed. To quote Condorcet once more: "the progress of the mathematical and physical sciences reveals an immense horizon... a revolution in the destinies of the human race" [9, p. 237]. Progress was possible; why not apply the same method to the social and moral spheres as well?

No account of attempts to extend the method of mathematics to other fields would be complete without discussing René Descartes, who in the 1630's combined the two methods we have just discussed—that of geometry and that of algebra—into analytic geometry. Let us look at his own description of how to make such discoveries. Descartes depicted the building-up of the deductive structure of a science—proof—as a later task than analysis or discovery. One first needed to analyze the whole into the correct "elements" from which truths could later be deduced. "The first rule," he wrote in his *Discourse on Method*, "was never to accept anything as true unless I recognized it to be evidently such... The second was to divide each of the difficulties which I encountered into as many parts as possible, and as might be required for an easier solution... " Then, "the third [rule] was to [start]... with the things which were simplest and easiest to understand, gradually and by degrees reaching toward more complex knowledge" [10, Part II, p. 12]. Descartes presented his method as the key to his own mathematical and scientific discoveries. Consider, for instance, the opening line of his *Geometry*: "All problems in geometry can easily be reduced to such terms that a knowledge of the lengths of certain straight lines suffices for their construction. Just as arithmetic is composed of only four or five operations..., so in geometry" [11, Book I, p. 3]. Descartes's influence on subsequent philosophy, from Locke's empiricism to Sartre's existentialism, is well known and will not be reviewed here. But for our purposes it is important to note that the thrust of Descartes's argument is that emulating the method successful in mathematical discovery will lead to successful discoveries in other fields [10, Part Five].

Descartes' method of analysis fits nicely with the Greek atomic theory, which had been newly revived in the seventeenth century: all matter is the sum of atoms; analyze the properties of the whole as the sum of these parts [17, Chapter VIII, esp. p. 217]. Thus the idea of studying something by "analysis" was doubly popular in seventeenth- and eighteenth-century thought. I would like to trace just one line of influence of this analytic method. Adam Smith in his 1776 *Wealth of Nations* analyzed [47, p. 12] the competitive success of economic systems by means of the concept of division of labor. The separate elements, each acting as efficiently as possible, provided for the overall success of the manufacturing process; similarly, each individual in the whole economy, while striving to increase his individual advantages, is "led as if by an Invisible Hand to promote ends which were not part of his original intention" [47, p. 27]—that is, the welfare of the whole of society. This Cartesian method of studying a whole system by analyzing it into its elements, then synthesizing the elements to produce the whole, was especially popular in France. For instance, Gaspard François de Prony had the job of calculating, for the French Revolutionary government, a set of logarithmic and trigonometric tables. He, himself, said he did it by applying Adam Smith's ideas about the division of labor. Prony organized a group of people into a hierarchical system to compute these tables. A few mathematicians decided which functions to use; competent technicians then reduced the job of calculating the functions to a set of simple additions and subtractions of pre-assigned numbers; and, finally, a large number of low-level human "calculators" carried out the additions and subtractions. Charles Babbage, the early nineteenth-century pioneer of the digital computer, applied the Smith-Prony analysis and embodied it in a machine [1, Chapter XIX]. The way Babbage's ideas developed can be found in a chapter in his *Economy of Machinery* entitled "On the Division of Mental Labour" [1, Chapter XIX]. Babbage was ready to convert Prony's organization into a computing machine because Babbage had long been impressed by the arguments of Leibniz and his followers on the power of notation to make such mathematical calculation mechanical, and Babbage, like Leibniz, accounted for the success of mathematics by "the accurate simplicity of its language" [22, p. 26]. Since Babbage's computer was designed to be "programmed" by punched cards, Hollerith's later invention of punched-card census data processing, twentieth-century computing, and other applications of the Cartesian "divide-and-solve" approach, including top-down programming, are also among the offspring of Descartes's mathematically-inspired method.

Whatever view of the *cause* of the certainty of mathematics one adopts, the *fact* of certainty in itself has had consequences. The "fact of mathematical certainty" has been taken to show that there exists *some* sort of knowledge, and thus to refute skepticism. Immanuel Kant in 1783 used such an argument to show that metaphysics is possible [25, Preamble, Section IV]. If metaphysics exists, it is independent of experience. Nevertheless, it is not a complex of tautologies. Metaphysics, for Kant, had to be what he called "synthetic," giving knowledge based on premises which is not obtainable simply by analyzing the premises logically. Is there such knowledge? Yes, said Kant, look at geometry. Consider the truth that the sum of the angles of a triangle is two right angles. We do not get this truth by analyzing the concept of triangle—all that gives us, Kant says, is that there are three angles. To gain the knowledge, one must make a construction: draw a line through one vertex parallel to the opposite side. (I now leave the proof as an exercise.) The construction is essential; it takes place in space, which Kant sees as a unique intuition of the intellect. (This example [24, II "Method of Transcendentalism," Chapter I, Section I, p. 423] seems to require the space to be Euclidean; I will return to this point later on.) Thus synthetic knowledge independent of experience *is* possible, so metaphysics—skeptics like David Hume to the contrary—is also possible.

This same point—that mathematics is knowledge, so there *is* objective truth—has been made throughout history, from Plato's going beyond Socrates' agnostic critical method, through George Orwell's hero, Winston Smith, attempting to assert, in the face of the totalitarian state's overwhelming power over the human intellect, that two and two are four.

Moreover, since mathematics is certain, perhaps we can, by examining mathematics, find which properties *all* certain knowledge must have. One such application of the "fact of mathematical certainty" was its use to solve what in the sixteenth century was called the problem of the criterion [43, Chapter I]. If there is only one system of thought around, people might well accept that one as true—as many Catholics did about the teachings of the Church in the Middle Ages. But then the Reformation developed alternative religious systems, and the Renaissance rediscovered the thought of pagan antiquity. Now the problem of finding the criterion that identified the true system became acute. In the seventeenth and eighteenth centuries, many thinkers looked to mathematics to help find an answer. What was the sign of the certainty of the conclusions of mathematics? The fact that nobody disputed them [43, Chapter VII]. Distinguishing mathematics from religion and philosophy, Voltaire wrote, "There are no sects in geometry. One does not speak of a Euclidean, an Archimedean" [49, Article "Sect."]. What every reasonable person agrees upon—that is the truth. How can this be applied to religion? Some religions forbid eating beef, some forbid eating pork; therefore, since they disagree, they both are wrong. But, continues Voltaire, all religions agree that one should worship God and be just; that must therefore be true. "There is but one morality," says Voltaire, "as there is but one geometry" [49, Miscellany, p. 225].

3. Applicability

Let us turn now from the certainty of mathematics to its applicability. Since applying mathematics to describe the world works so well, thinkers who reflect on the applicability of mathematics find that it affects their views not only about thought, but also about the world. For Plato, the applicability of mathematics occurs because this world is merely an approximation to the higher mathematical reality; even the motions of the planets were inferior to pure mathematical motions [41, 529d]. For Aristotle, on the other hand, mathematical objects are just abstracted from the physical world by the intellect. A typical mathematically-based science is optics, in which we study physical objects—rays of light—as though they were mathematical straight lines [*Physics* II, Chapter 2; 194a]. We can thus use all the tools of geometry in that science of optics, but it is the light that is real.

One might think that Plato is a dreamer and Aristotle a hard-headed practical man. But today's engineer steeped in differential equations is the descendant of the dreamer. From Plato—and his predecessors the Pythagoreans who taught that "all is number"—into the Renaissance, many thinkers looked for the mathematical reality beyond the appearances. So did Copernicus, Kepler, and Galileo [7, Chapters 3, 5, 6]. The Newtonian world-system that completed the Copernican revolution was embodied in a mathematical model, based on the laws of motion and inverse-square gravitation, and set in Platonically absolute space and time ([6]; cf. [7, Chapter 7]). The success of Newtonian physics not only strongly reinforced the view that mathematics was the appropriate language of science, but also strongly reinforced the emerging ideas of progress and of truth based on universal agreement.

Another consequence of the Newtonian revolution was Newton's explicit help to theology, strongly buttressing what was called the argument for God's existence from design. The mathematical perfection of the solar system—elliptical orbits nearly circular, planets moving all in the same plane and direction—could not have come about by chance, said Newton, but "from the counsel and dominion of an intelligent and powerful Being" [37, General Scholium, p. 544]. "Natural theology," as this doctrine was called, focussed on examples of design and adaptation in nature, inspiring considerable research in natural history, especially on adaptation, research which was to play a role in Darwin's discovery of evolution by natural selection [14, pp. 263–266].

Just as the "fact of mathematical certainty" made certainty elsewhere seem achievable, so the "fact of mathematical applicability" in physical science inspired the pioneers of the idea of social science, Auguste Comte and Adolphe Quételet. Both Comte and Quételet were students of mathematical physics and astronomy in the early nineteenth century; Comte, while a student at the École polytechnique in Paris, was particularly inspired by Lagrange, Quételet, and Laplace. Lagrange's great *Analytical Mechanics* was an attempt to reduce all of mechanics to mathematics. Comte went further: if physics was built on mathematics, so was chemistry built on physics, biology on chemistry, psychology on biology, and finally his own new creation, sociology (the term is his) would be built on psychology [8, Chapter II]. The natural sciences were no longer (as they had once been) theological or metaphysical; they were what Comte called "positive"—based only on observed connections between things. Social science could now also become positive. Comte was a reformer, hoping for a better society through understanding what he called "social physics." His philosophy of positivism influenced twentieth-century logical positivism, and his ideas on history—"social dynamics"—influenced Feuerbach and Marx [32, Chapter 4]. Still, Comte only prophesied but did not create quantitative social science; this was done by Quételet.

For Quételet's conception of quantitative social science, the fact of applicability of mathematics was crucial. "We can judge of the perfection to which a science has come," he wrote in 1828, "by the ease with which it can be approached by calculation" (quoted in [27, p. 250]). Quételet noted that Laplace had used probability and statistics in determining planetary orbits; Quételet was especially impressed by what we call the normal curve of errors. Quételet found empirically that many human traits—height, for instance—gave rise to a normal curve. From this, he defined the statistical concept, and the term, "average man" (*homme moyen*). Quételet's work demonstrates that, just as the Platonic view that geometry underlies reality made mathematical physics possible, so having a statistical view of data is what makes social science possible.

Quételet found also that many social statistics—the number of suicides in Belgium, for instance, or the number of murders—produced roughly the same figures every year. The constancy of these rates over time, he argued, dictated that murder or suicide had constant social causes. Quételet's discovery of the constancy of crime rates raised an urgent question: whether the individuals are people or particles, do statistical laws say anything about individuals? or are the individuals free?

Laplace, recognizing that one needed probability to do physics, said that this fact did not mean that the laws governing the universe were ultimately statistical. In ignorance of the true causes, Laplace said, people thought that events in the universe depended on chance, but in fact all is determined. To an infinite intelligence which could comprehend all the forces in nature and the "respective situation of the beings who composed it," said Laplace, "nothing would be uncertain" [28, Chapter II]. Similarly, Quételet held that "the social state prepares these crimes, and the criminal is merely the instrument to execute them" [27].

Another view was held by James Clerk Maxwell. In his work on the statistical mechanics of gases, Maxwell argued that statistical regularities in the large told you nothing about the behavior of individuals in the small [33, Chapter 22, pp. 315–16]. Maxwell seems to have been interested in this point because it allowed for free will. And this argument did not arise from Maxwell's physics; he had read and pondered the work of Quételet on the application of statistical thinking to society [44]. The same sort of dispute about the meaning of probabilistically-stated laws has of course recurred in the twentieth-century philosophical debates over the foundations of quantum mechanics.

Thus discussions of basic philosophical questions—is the universe an accident or a divine design? is there free will or are we all programmed?—owe surprisingly much to the applicability of mathematics.

4. More than one geometry?

Given the centrality of mathematics to western thought, what happens when prevailing views of the nature of mathematics change? Other things must change too. Since geometry had been for so long the canonical example both of the certainty and of the applicability of mathematics, the rise of non-Euclidean geometry was to have profound effects.

As is well known, in attempts to prove Euclid's parallel postulate and thus, as Saccheri put it in 1733, remove the single blemish from Euclid, mathematicians deduced a variety of surprising consequences from denying that postulate. Gauss, Bolyai, and Lobachevsky in the early nineteenth century each separately recognized that these consequences were not absurd, but rather were valid results in a consistent, non-Euclidean (Gauss's term) geometry.

Recall that Kant had said that space (by which he meant Euclidean 3-space) was the form of all our perceptions of objects. Hermann von Helmholtz, led in mid-century to geometry by his interest in the psychology of perception, asked whether Kant might be wrong: could we imagine ordering our perceptions in a non-Euclidean space? Yes, Helmholtz said. Consider the world as reflected in a convex mirror. Thus, the question of which geometry describes the world is no longer a matter for intuition—or for self-evident assumptions—but for *experience* [20].

What did this view—expressed as well by Bernhard Riemann and W. K. Clifford, among others—do to the received accounts of the relation between mathematics and the world? It detached mathematics from the world. Euclidean and non-Euclidean geometry give the first clear-cut historical example of two mutually contradictory mathematical structures, of which at most one can actually represent the world. This seems to indicate that the choice of mathematical axioms is one of intellectual freedom, not empirical constraint; this view, reinforced by Hamilton's discovery of a non-commutative algebra, suggested that mathematics is a purely formal structure, or as Benjamin Peirce put it, "Mathematics is the science which draws necessary conclusions," [40]—*not* the science of number (even symbolic algebra had been just a generalized science of number) or the science of space. Now that the axioms were no longer seen as necessarily deriving from the world, the applicability of mathematics to the world became turned upside down. The world is no longer, as it was for Plato, an imperfect model of the true mathematical reality; instead, mathematics provides a set of different models for one empirical reality. In 1902 the physicist Ludwig Boltzmann expressed a view which had become widely held: that models, whether physical or mathematical, whether geometric or statistical, had become the means by which the sciences "comprehend objects in thought and represent them in language" [3]. This view, which implies that the sciences are no longer claiming to speak directly about reality, is now widespread in the social sciences as well as the natural sciences, and has transformed the philosophy of science. As applied to mathematics itself—the formal model of mathematical reasoning—it has resulted in Gödel's demonstration that one can never prove the consistency of mathematics, and the resulting conclusion among some philosophers that there is no certainty anywhere, not even in mathematics [2, p. 206].

5. Opposition

The best proof of the centrality of mathematics is that every example of its influence given so far has provoked strong and significant opposition. Attacks on the influence of mathematics have been of three main types. Some people have simply favored one view of mathematics over other views; other people have granted the importance of mathematics but have opposed what they consider its overuse or extension into inappropriate domains; still others have attacked mathematics, and often all of science and reason, as cold, inhuman, or oppressive.

Aristotle's reaction against Platonism is perhaps the first example of opposition to one view of mathematics (eternal objects) while championing another (deductive method). Another example is Newton's attack on Descartes's attempt to use nothing but "self-evident" assumptions to figure out how the universe worked. There are many mathematical systems God could have used to set up the world, said Newton. One could not decide *a priori* which occurs; one must, he says, observe in nature which law actually holds. Though mathematics is the tool one uses to discover the laws, Newton concludes that God set up the world by free choice, not mathematical necessity [35, pp. 7–8] [36, p. 47]. This point is crucial to Newton's natural theology: that the presence of order in nature proves that God exists.

Another example of one view based on mathematics attacking another can be found in Malthus's *Essay on Population* of 1798. He accepts the Euclidean deductive model—in fact he begins with two "postulata": man requires food, and the level of human sexuality remains constant [31, Chapter I]. His consequent analysis of the growth of population and of food supply rests on mathematical models. Nonetheless, one of Malthus's chief targets is the predictions by Condorcet and others of continued human progress modelled on that of mathematics and science. As in Newton's attack on Descartes, Malthus applied one view of mathematics to attack the conclusions others claimed to have drawn from mathematics.

Our second category of attacks—drawing a line that mathematics should not cross—is exemplified by the seventeenth-century philosopher and mathematician Blaise Pascal. Reacting against Cartesian rationalism, Pascal contrasted the "esprit géometrique" (abstract and precise thought) with what he called the "esprit de finesse" (intuition) [39, *Pensée* 1] holding that each had its proper sphere, but that mathematics had no business outside its own realm. "The heart has its reasons," wrote Pascal, "which reason does not know" [39, *Pensée* 277]. Nor is this contradicted by the fact that Pascal was willing to employ mathematical thinking for theological purposes—recall his "wager" argument to convince a gambling friend to try acting like a good Catholic [39, *Pensée* 233]; the point here was to use his friend's own probabilistic reasoning style in order to convince him to go on to a higher level.

Similarly, the mathematical reductionism of men like Lagrange and Comte was opposed by men like Cauchy. Cauchy, whom we know as the man who brought Euclidean rigor to the calculus, opposed both Lagrange's attempt to reduce mechanics to calculus and calculus to formalistic algebra [15, pp. 51–54], and opposed the positivists' attempt to reduce the human sciences to an ultimately mathematical form. "Let us assiduously cultivate the mathematical sciences," Cauchy wrote in 1821, but "let us not imagine that one can attack history with formulas, nor give for sanction to morality theorems of algebra or integral calculus" [5, p. vii]. Analogously, in our own day, computer scientist Joseph Weizenbaum attacks the modern, computer-influenced view that human beings are nothing but processors of symbolic information, arguing that the computer scientist should "teach the limitations of his tools as well as their power" [50, p. 277].

Finally, we have those who are completely opposed to the method of analysis, the mathematization of nature, and the application of mathematical thought to human affairs. Witness the Romantic reaction against the Enlightenment: Goethe's opposition to the Newtonian analysis of white light, or, even more extreme, William Wordsworth in *The Tables Turned*:

> Sweet is the lore which Nature brings;
> Our meddling intellect
> Mis-shapes the beauteous forms of things:—
> We murder to dissect.

Again, Walt Whitman, in his poem "When I heard the learn'd astronomer," describes walking out on the lecture on celestial distances, having become "tired and sick," going outside instead to look "up in perfect silence at the stars."

Reacting against statistical thinking on behalf of the dignity of the individual, Charles Dickens in his 1854 novel *Hard Times* satirizes a "modern school" in which a pupil is addressed as "Girl number twenty" [12, Book I, Chapter II]; the schoolmaster's son betrays his father, justifying himself by pointing out that in any given population a certain percentage will become traitors, so there is no occasion for surprise or blame [12, Book III, Chapter VII]. In a more political point, Dickens through his hero denounces the analytically-based efficiency of industrial division of labor, saying it regards workers as though they were nothing but "figures in a sum" [12, Book II, Chapter V].

The Russian novelist Evgeny Zamyatin, in his early-twentieth-century antiutopian novel *We* (a source for Orwell's *1984*), envisions individuals reduced to being numbers, and mathematical tables of organization used as instruments of social control. Though the certainty of mathematics, and thus its authority, has sometimes been an ally of liberalism, as we have seen in the cases of Voltaire and Condorcet, Zamyatin saw how it could also be used as a way of establishing an unchallengeable authority, as philosophers like Plato and Hobbes had tried to use it, and he wanted no part of it.

6. Conclusion

As the battles have raged in the history of Western thought, mathematics has been on the front lines. What does it all (to choose a phrase) add up to?

My point is not that what these thinkers have said about mathematics is right, or is wrong. But this history shows that the nature of mathematics has been—and must be—taken into account by anyone who wants to say anything important about philosophy or about the world. I want, then, to conclude by advocating that we teach mathematics *not* just to teach quantitative reasoning, *not* just as the language of science—though these are very important—but that we teach mathematics to let people know that one cannot fully understand the humanities, the sciences, the world of work, and the world of man without understanding mathematics in its central role in the history of Western thought.

Acknowledgements. I thank the students in my Mathematics 1 classes at Pomona and Pitzer Colleges for stimulating discussions and suggestions on some of the topics covered in this paper, especially David Bricker, Maria Camarena, Marcelo D'Asero, Rachel Lawson, and Jason Gottlieb. I also thank Sandy Grabiner for both his helpful comments and his constant support and encouragement.

References

1. Charles Babbage, *On the Economy of Machinery*, Charles Knight, London, 1832.
2. William Barrett, *Irrational Man: A Study in Existential Philosophy*, Doubleday, New York, 1958. Excerpted in William L. Schaaf, ed., *Our Mathematical Heritage*, Collier, New York, 1963.
3. Ludwig Boltzmann, Model, *Encyclopedia Britannica*, 1902.
4. Carl Boyer, *History of Analytic Geometry*, Scripta Mathematica, New York, 1956.
5. A.-L. Cauchy, *Cours d'analyse*, 1821. Reprinted in A.-L. Cauchy, *Oeuvres*, series 2, vol. 3, Gauthier-Villars, Paris, 182.
6. I. Bernard Cohen, *The Newtonian Revolution*, Cambridge University Press, Cambridge, 1980.
7. ——, *The Birth of a New Physics*, 2d edition. W. W. Norton, New York and London, 1985.
8. Auguste Comte, *Cours de philosophic positive*, vol. I, Bachelier, Paris, 1830.
9. Marquis de Condorcet, *Sketch for a Historical Picture of the Progress of the Human Mind*, 1793, tr, June Barraclough. In Keith Michael Baker, ed., *Condorcet: Selected Writings*. Bobbs-Merrill, Indianapolis, 1976.
10. René Descartes, *Discourse on Method*, 1637, tr. L. J. Lafleur, Liberal Arts Press, New York, 1956.
11. ——, *La géométrie*, 1637, tr., D. E. Smith and Marcia L. Latham, *The Geometry of René Descartes*, Dover, NY, 1954.
12. Charles Dickens, *Hard Times*, 1854. Norton Critical Edition edited by George Ford and Sylvere Monod, W. W. Norton, New York and London, 1966.
13. Neal C. Gillespie, *Charles Darwin and the Problem of Creation*, University of Chicago Press, Chicago and London, 1979.
14. Charles C. Gillespie, *The Edge of Objectivity*, Princeton University Press, Princeton, NJ, 1960.
15. Judith V. Grabiner, *The Origins of Cauchy's Rigorous Calculus*, MIT Press, Cambridge, Mass., 1981.

16. John C. Greene, *Science, Ideology, and World View: Essays in the History of Evolutionary Ideas*, University of California Press, Berkeley, 1981.

17. A. Runert Hall, *From Galileo to Newton, 1630–1720*. Harper and Row. New York and Evanston. 1963.

18. T. L. Heath (Editor), The Works of Archimedes with the Method of Archimedes, New York, n.d.

19. ——, *The Thirteen Books of Euclid's Elements*, Volume I, Cambridge University Press, 1925, reprinted by Dover, New York, 1956.

20. Hermann von Helmholtz, On the origin and significance of geometrical axioms, 1870, reprinted in Hermann von Helmholtz, *Popular Scientific Lectures*, edited by Morris Kline, Dover, NY, 1962.

21. Thomas Hobbes, *Leviathan, or the Matter, Form, and Power of a Commonwealth, Ecclesiastical and Civil*, 1651, in E. Burtt, ed., *The English Philosophers from Bacon to Mill*, Modem Library, New York, 1939.

22. Anthony Hyman, *Charles Babbage: Pioneer of the Computer*. Princeton University Press, Princeton, NJ, 1982.

23. Werner Jaeger, *Early Christianity and Greek Paideia*, Oxford University Press, London, Oxford, and New York, 1961.

24. Immanuel Kant, *Critique of Pure Reason*, 1781, tr. F. Max Müller, Macmillan, New York, 1961.

25. ——, *Prolegomena to Any Future Metaphysics*, 1783, ed. Lewis W. Beck, Liberal Arts Press, New York, 1951.

26. Morris Kline, *Mathematics in Western Culture*, Oxford University Press, New York, 1953.

27. D. Landau and P. Lazarsfeld, Quételet, *International Encyclopedia of the Social Sciences*, Vol. 13, Macmillan, New York, 1968, pp. 247-257.

28. Pierre Simon Laplace, *A Philosophical Essay on Probabilities*, 1819 tr. F. W. Truscott and F. L. Emory, Dover, NY, 1951.

29. Desmond Lee, 1965 Introduction, in Desmond Lee, ed., *Plato: Timaeus and Critias*, Penguin Books of Great Britain, London, 1971.

30. Leibniz, *Preface to the General Science and Towards a Universal Characteristic*; 1677; reprint, Selections, Philip P. Weiner, editor, 1951, pp. 12–17, 17–25.

31. Thomas R. Malthus, An Essay on the Principle of Population, as it Affects the Future Improvement of Society: with Remarks on the Speculations of Mr. Godwin, M. Condorcet, and other writers, 1798, in Garrett Hardin, ed., *Population, Evolution and Birth Control*, Freeman, San Francisco, 1964, pp. 4–16.

32. Maurice Mandelbaum, The Search for a Science of Society: From Saint-Simon to Marx and Engels, Chapter 4 in Maurice Mandelbaum, *History, Man, and Reason*, Johns Hopkins University Press, Baltimore and London, 1971, pp. 63-76.

33. James Clerk Maxwell, *Theory of Heat*, Longmans, Green, & Co., London, 1871.

34. Seyyed Hossein Nasr, *Science and Civilization in Islam*, New American Library, New York, Toronto and London, 1968.

35. Isaac Newton, Letter to Henry Oldenburg, July, 1672, in [38].

36. ——, Letter to Richard Bentley, December 10, 1692, in [38].

37. ——, *Sir Isaac Newton's Mathematical Principles of Natural Philosophy and His System of the World*, tr. Andrew Motte, revised and edited by Florian Cajori, University of California Press, Berkeley, 1934.

38. ——, *Newton's Philodophy of Nature: Selections from His Writings*, ed. H. S. Thayer, Hafner, New York, 1951.

39. Blaise Pascal, *Pensées*, E. P. Dutton, New York, 1958.

40. Benjamin Peirce, Linear Associative Algebra, *Amer. J. Math.*, 4 (1881).

41. Plato, *Republic*, tr. A. D. Lindsay, E. P. Dutton, New York, 1950.

42. ——, *Timaeus*, in [29].

43. Richard H. Popkin, *The History of Scepticism from Erasmus to Spinoza*, University of California Press, Berkeley and Los Angeles, 1979.

44. Theodore M. Porter, A Statistical Survey of Gases: Maxwell's Social Physics, *Historical Studies in the Physical Sciences* 12 (1981), 77–116.

45. A. Quételet, *Instructions populaires sur le calcul des probabilités*, Tarlier, Brussels, 1828.

46. Bertrand Russell, *A Critical Exposition of the Philosophy of Leibniz*, 2d edition, Allen and Unwin, London, 1937.

47. William G. Sinnigen and Arthur E. R. Boak, *A History of Rome to A. D. 565*, sixth edition, Macmillan, New York, 1977.

48. Andrew Skinner, Introduction, Adam Smith, *The Wealth of Nations*, (1776), Penguin Books, London, 1974, pp. 11–97.

49. Benedict de Spinoza, *Ethics*, preceded by On the Improvement of the Understanding, ed. James Gutmann, Hafner, New York, 1953.

50. François-Marie Arouet de Voltaire, *The Portable Voltaire*, The Viking Press, New York, 1949, selections from *Dictionnaire Philosophique* (1764), pp. 53–228.

51. Joseph Weizenbaum, *Computer Power and Human Reason: From Judgment to Calculation*, Freeman, San Francisco, 1976.

52. Harry Austryn Wolfson, What is New in Philo?, in Harry A. Wolfson, *From Philo to Spinoza: Two Studies in Religious Philosophy*, Behrman House, New York, 1977.

Geometry Strikes Again

Branko Grünbaum

Mathematics Magazine
58 (1985), 12–18

Editors' Note: This article by Professor Grünbaum of the University of Washington caused the MAA some embarrassment: the logo of the MAA—a regular icosahedron—that appeared on publications and letterhead was mathematically wrong. At least it's wrong under the obvious assumption that the polyhedron was intended to be regular. You can read the details here.

Branko Grünbaum was born in 1929 in Osijek, Croatia (then part of Yugoslavia). He earned his PhD at the Hebrew University in Jerusalem in 1958. After a two-year stay at the Institute for Advanced Study, Princeton, he moved to the University of Washington, where he has remained throughout his career, with the exception of occasional short visits elsewhere. His books include *Convex Polytopes* (Interscience, 1967), *Arrangements and Spreads* (American Mathematical Society, 1972), and (with G. D. Shephard) *Tilings and Patterns* (Freeman, 1987). Grünbaum and Shephard won an Allendoerfer Award in 1977 for an article, "Tiling by regular polygons."

We append here a short note by Doris Schattschneider outlining the history of the MAA logo. With Grünbaum's discovery, the logo was corrected and one might have assumed that only a correct logo would be used in the future. Alas, the incorrect logo surfaced again on the cover of the *Monthly* in 1996. It was eventually corrected yet again in 2000. It's hard to keep a bad illustration down. For more details on this figure and others see William Casselman's article "Pictures and proofs," *Notices of the American Mathematical Society* 47 (November 2000), 1257–66.

Last Sunday I was leisurely reading the May 1984 issue of *Mathematics Magazine*. Coming to the "Philatelists take note" item (on page 187) about the stamp issued by East Germany to mark the bicentennial of Euler's death, I remembered that the same stamp was used on a recent request for reprints I received from East Germany. So I retrieved that postcard and washed off the stamp. It turned out that I was lucky, and the stamp was much less marred by cancellation marks than the one reproduced in *Mathematics Magazine*. Contemplating the stamp I noticed that the drawing of the regular icosahedron shown in it (and I have no doubt that it was meant to show a regular icosahedron) is wrong. This is not a question of which kind of perspective or projection was used, it is just a logical error (repeated twice): if three of the five vertices of a plane pentagon project into one line, then the other two vertices must project into the same line. In Figure 1 the offending drawing is enlarged and the misaligned vertices are easily picked out. Recalling how highly precision draftsmanship used to be valued in European education, I could not help snickering at the low level to which the East Germans have fallen. Full of condescension I closed my issue of the *Magazine*.

But the heavier outer cover of the issue opened by itself, and revealed to my horrified and unbelieving eyes another fallacious rendition of the regular icosahedron, in the upper left comer of the title page, IN THE LOGO OF THE MATHEMATICAL ASSOCIATION OF AMERICA!!! Under any reasonable projection, parallel segments either remain parallel or lie on concurrent lines-but in the logo three easily discernible pairs (see Figure 2) of such segments

Figure 1. An enlarged view of the East German stamp commemorating the bicentennial of the death of Leonhard Euler. Note the two regular pentagons which are seen "edge on," each of which shows two vertices not lying in the plane determined by its other three vertices.

(a)

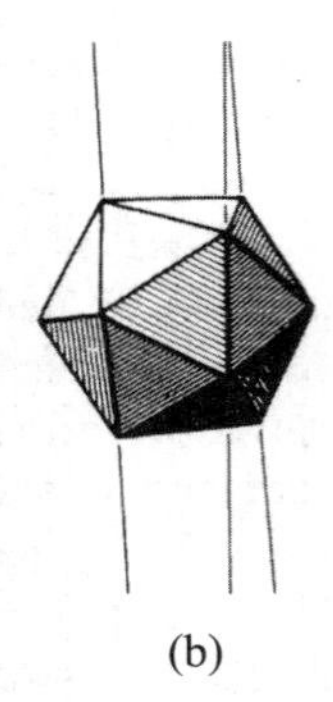

(b)

Figure 2. (a) The emblem of the Mathematical Association of America, from the title page of the May 1984 issue of *Mathematics Magazine*. (b) The icosahedron from (a) with three lines determined by it; the lines should be parallel to each other (if parallel projection is used), or concurrent at one point (if the projection is central).

do neither. Quickly grabbing the recent issue of *FOCUS*, and then the last issue of the *Monthly*, I saw that the drawing in the *Magazine* is no isolated distortion. The official symbol of the Association, appearing on all its publications, proclaims louder than words that we have eyes and look but do not see, that those who shouted "Down with Euclid" achieved their goal much faster and much more thoroughly than they had any right to expect—that good old Geometry is dead! Thumbing through back issues of the *Monthly*, the time of death can be determined with precision. The previous logo was discarded after the December 1971 issue, and the defective new one has been gracing the Monthly since January 1972. (Possibly this can help in determining who was responsible for the foul deed—it certainly does not seem to me that the departure of the old dear was through natural causes, or accidental.)

Since revival seems impossible, at this stage I decided to try to extract at least a measure of revenge by writing this short note, and illustrating it with some of the gruesome pictures which I have collected over the last few years. They document, better than thousands of words could, the low level of common knowledge of Geometry and the lack of commonsense feeling for graphical rendition of spatial figures. In contrast to the examples given above—in which the lack of accuracy could be blamed on the graphic artist, who might have "corrected for better artistic effect" the original sketches—in the illustrations that follow the responsibility for the grotesque must squarely rest on the authors.

Since (for the uninitiated) looking at icosahedra can become tedious after the first few dozen, let

(a)

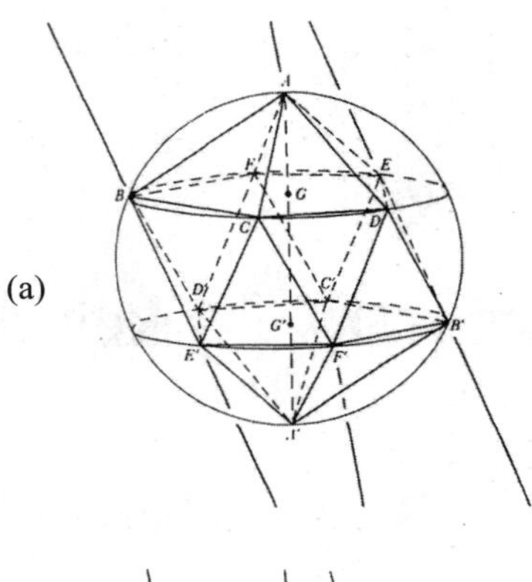

(b)

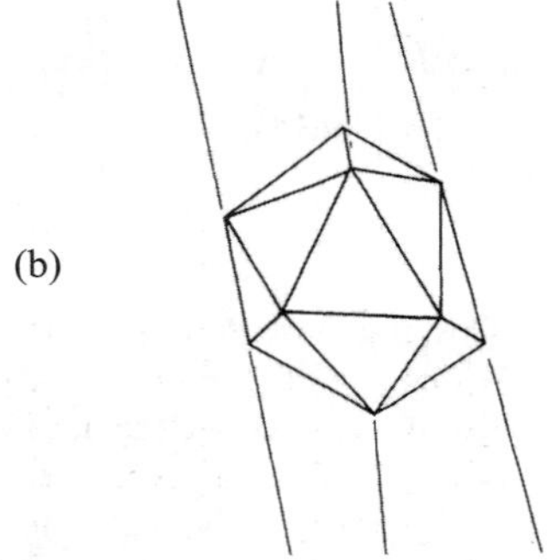

Figure 3. The icosahedra from (a) Mueller [101, page 257, and (b) Firby & Gardiner [51, page 84. As in Figure 2, the lines outside the icosahedra should be either parallel, or else concurrent.

me show here just two from my rich harvest of juicy "plums" (see Figure 3). From looking at the representations of icosahedra in various books one could easily conjecture a metatheorem to the effect that "regular icosahedra cannot be drawn correctly"; however, that conjecture is readily disproved (see p. 252). The truly frightening aspect of the situation is not that icosahedra are often grossly misrepresented and that hardly anybody notices, but that such fallacious images *pervade* our books—even those texts that are in other respects quite good. This goes for materials meant for elementary or high schools and their teachers, as much as for those intended for college students, research, or the general public (Figure 4)—and the number of examples could be increased almost indefinitely. I would like to stress that I have *not* been spending my time looking for errors, but just tried (with only partial success) to remember those that I encounter and notice.

It may be worth mentioning that it is not just polyhedra which get such cavalier treatment (although it is easiest to detect there). From beginning calculus through foundations of geometry to linear algebra and beyond—our texts are full of pictures (see Fig-

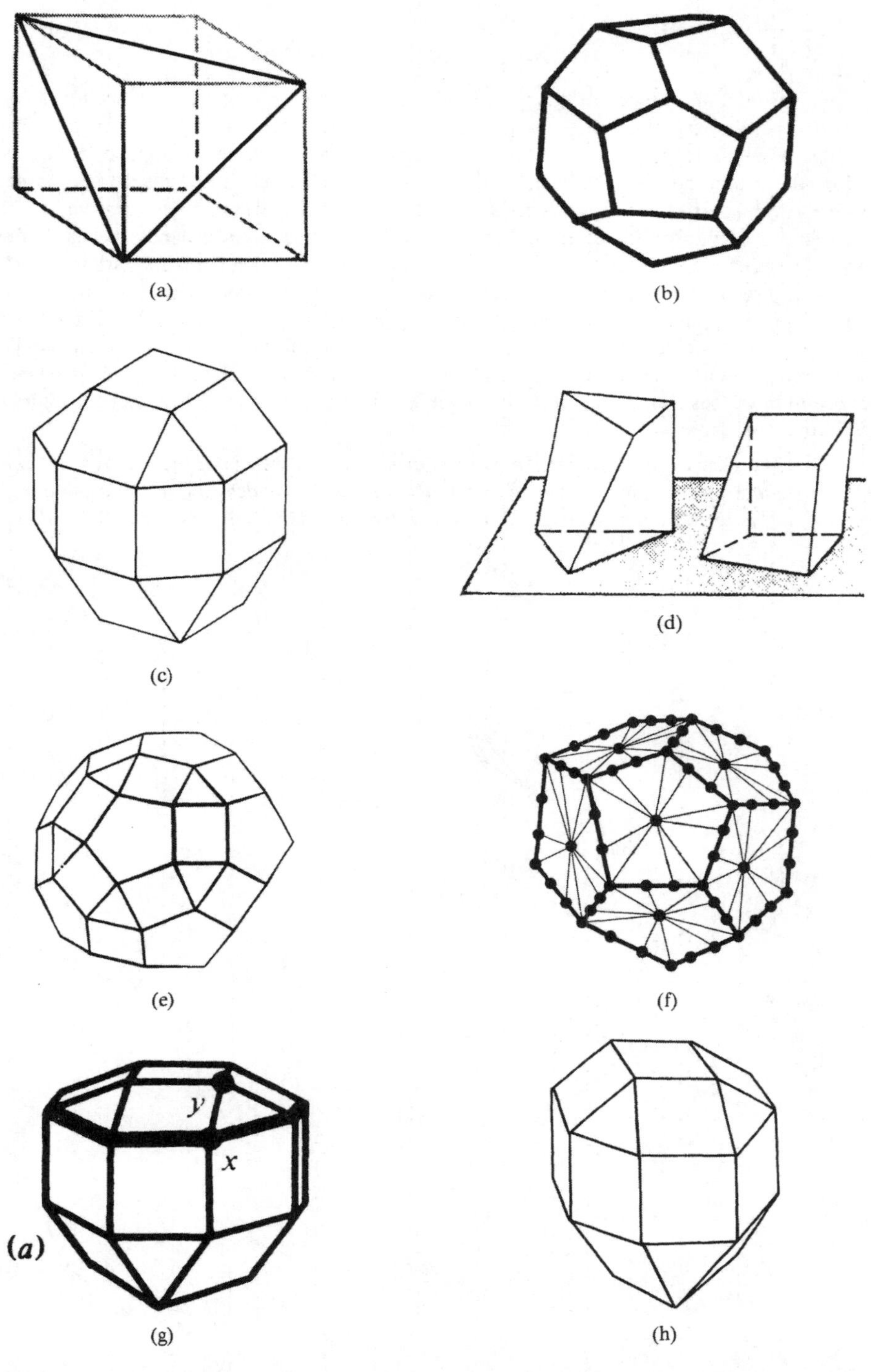

Figure 4. (a) A cube, and the equilateral triangle determined by three of its vertices. In the drawing, the edges of the triangle which are supposed to be foreshortened are actually longer than the edge seen full size. From Jacobs 171, page 85. (b) A view of the regular dodecahedron, from Young & Bush [14], page 101. (c) The rhombicuboctahedron, from O'Daffer & Clemens [111, page 141. (d) Two prisms, from Peterson 1121, page 77. (e) The rhombicosidodecahedron, from Fejes Tóth [4], page 111. (f) A graph derived from that of the regular dodecahedron, from Lovasz [8], page 434. (g) and (h) Two pseudo-rhombicuboctahedra, from Baglivo & Graver [1], page 74, and Martin 19], page 208.

Drawing a Regular Icosahedron

In her comments on a first version of this note, Prof. Schattschneider suggested that I include a *correct* drawing of the regular icosahedron. While it is not hard to find quite accurate drawings in various books, very few of them divulge how this can be done. My favorite method (which can easily be applied in various projections) is based on the observation that the vertices of a regular icosahedron can be placed one on each edge of a regular octahedron so as to divide the edge in the golden ratio $\tau : 1$, where $\tau = (\sqrt{5} + 1)/2 = 1.618034\ldots$. Since projections of the regular octahedron are easily drawn, this gives a handy means of finding various representations of the icosahedron. (It also can be used as an illustration of the various "dry" theorems of linear algebra, providing very useful practice material—especially if turned into a program on a programmable calculator or a computer, and connected to a graphics display or plotter.) The following are several views of the regular icosahedron obtainable as orthogonal projections in different directions. Skew parallel projections (which are frequently used in mathematical and other texts) can also be obtained in this manner. But although they are useful in various situations and easily drawn, I believe that they are only a poor substitute for a properly executed orthogonal projection.

Editor's note: The view of the icosahedron on our cover was obtained from the orthogonal projection shown in the lower right corner, below. The simple directions to an artist for drawing it are: draw a regular hexagon and then locate the three points (such as B) for which the ratio OB/OA equals $\tau = 1.618\ldots$.

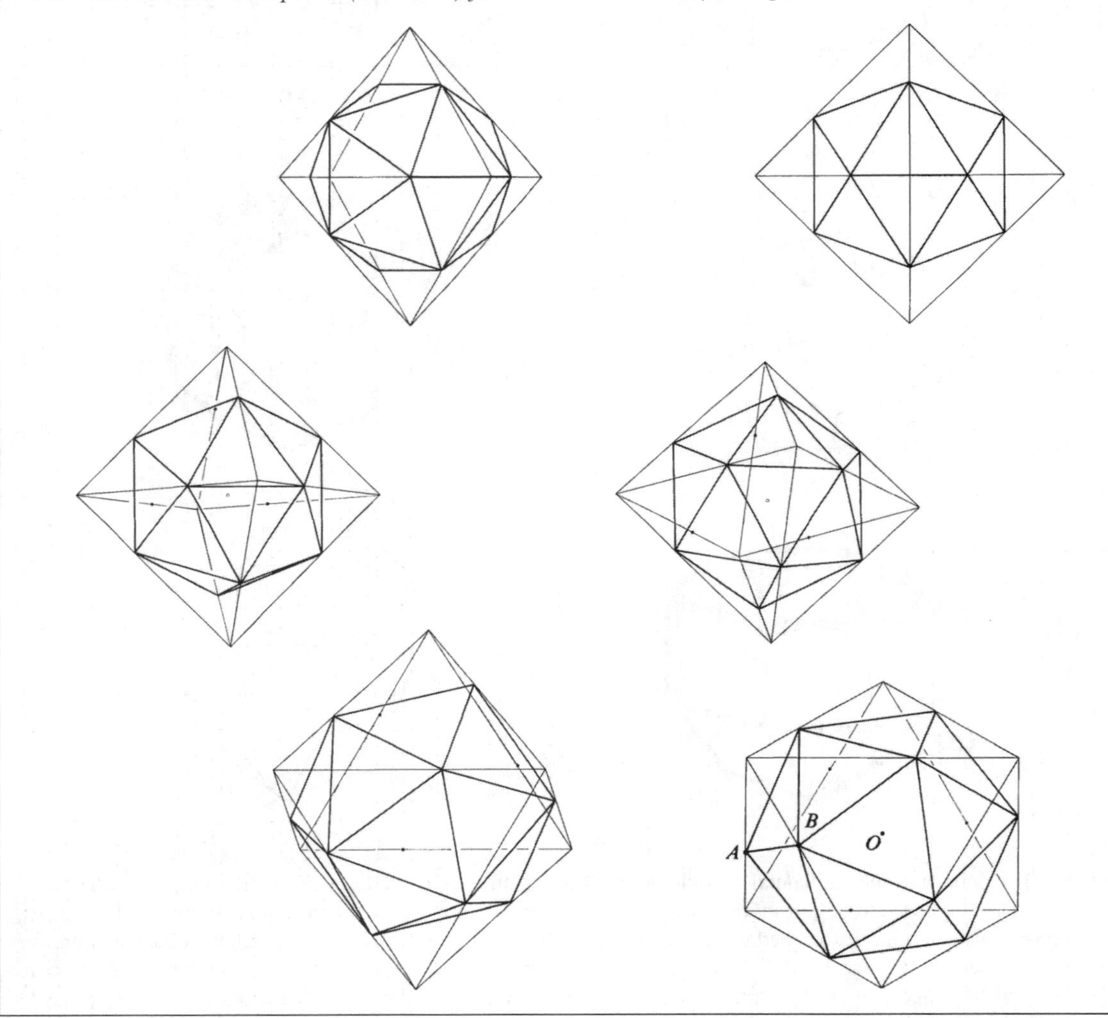

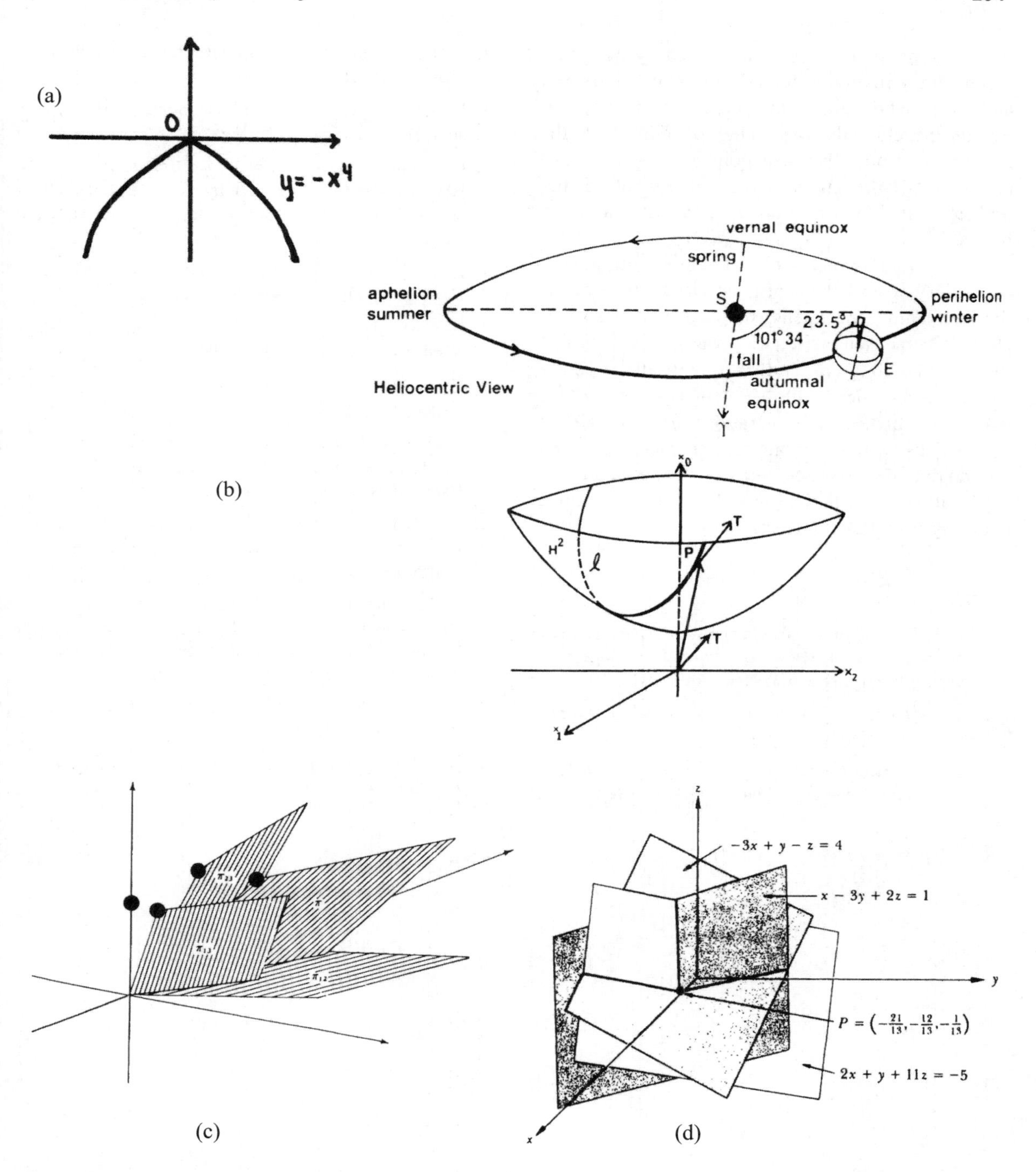

Figure 5. (a) The graph of a very smooth curve, from Gill [61, page 75. (b) Two ellipses, from Faber [3], pp. 84, 263. (c) A parallelogram π spanned by two vectors in 3-space, and its parallel projections on the coordinate planes, from Banchoff & Wermer 121, page 123. I used the solid dots to mark points which should be the vertices of a parallelogram. (d) The remarkable intersection of three unremarkable planes, from Shields [13], page 6.

ure 5) that should not have been permitted to see the light of the day and should be banned from U.S. mails. (Some of the effects are almost amusing; had it been created on purpose, the illustration in Figure 5(d) would deserve a place with the Penrose bolts and Escher's staircase.)

Concerning a search for causes, I think that the pervasive phenomenon of such misleading illustrations can probably be best explained by assuming a conspiracy involving the adherents of axiomatics, and those who believe that geometry is only linear algebra clumsily done. They certainly have the motive—and now they can point to this little note and say: "See, even the dyed-in-the-wool geometers say you cannot rely on the pictures you see in books."

Time is running out. Unless there is a strong general outcry against the continued visual abuse of the few remaining specimens, the endangered species Homo Geometricus will surely vanish. That will be the end of a long era, and YOU will all be poorer for it.... But meanwhile—if you plan to write a book with abominable illustrations, or if you aim to outsmart us by having your next geometry text contain no drawings—watch out: we may be on our last legs, but we are still kicking and will try to continue exposing visual skulduggery.

References

1. J. A. Baglivo and J. E. Graver, *Incidence and Symmetry in Design and Architecture*, Cambridge University Press, London & New York, 1983.
2. T. Banchoff and J. Wermer, *Linear Algebra Through Geometry*, Springer-Verlag, New York, 1983.
3. R. L. Faber, *Foundations of Euclidean and Non-Euclidean Geometry*, Marcel Dekker, New York, 1983.
4. L. Fejes Tóth, *Regular Figures*, Pergamon Press, New York, 1964.
5. P. A. Firby and C. F. Gardiner, *Surface Topology*, John Wiley & Sons, New York, 1982.
6. G. S. Gill, *Solutions Manual for "Calculus and Analytic Geometry"* (5th ed.) by G. B. Thomas, Jr. and R. L. Finney, Addison-Wesley, Reading, Mass., 1979.
7. H. R. Jacobs, *A Teacher's Guide to Geometry*, W. H. Freeman and Co., San Francisco, 1974.
8. L. Lovász, *Combinatorial Problems and Exercises*, North-Holland, Amsterdam, 1979.
9. G. E. Martin, *Transformation Geometry*, Springer-Verlag, New York, 1982.
10. I. Mueller, *Philosophy of Mathematics and Deductive Structure in Euclid's Elements*, The MIT Press, Cambridge, Mass. and London, 1981.
11. P. G. O'Daffer and S. R. Clemens, *Geometry: An Investigative Approach*, Addison-Wesley, Menlo Park, California, 1976.
12. J. C. Peterson, Informal geometry in grades 7–14, *Geometry in the Mathematics Curriculum*, 36th Yearbook, National Council of Teachers of Mathematics, 1973, pp. 52–91.
13. P. C. Shields, *Elementary Linear Algebra*, 3rd ed. Worth, New York, 1980.
14. J. E. Young and G. A. Bush, *Geometry for Elementary Teachers*, Holden-Day, San Francisco, 1971.

The Mystery of the MAA Logo

Several questions naturally arose in the correspondence concerning the rendering of the icosahedron in the MAA logo; *When* was the logo (the seal on the contents page of this *Magazine*) adopted? *Why* was the icosahedron chosen (and not, for instance, the pentagram of the Pythagoreans, or the cone and cylinder of Archimedes)? Was the icosahedron *ever* accurately rendered? Simple research has turned up virtually no answers|–it would appear that mathematicians are uninterested in leaving records to satisfy our curiosity.

The MAA was founded on December 31, 1915, and was incorporated in September 1920. The reasons for its establishment, as well as its constitution, by-laws, and list of charter members can all be found in issues of the *American Mathematical Monthly* for the years which surround 1915. Kenneth May's history, *The Mathematical Association of America: It's First Fifty Years*, MAA, 1972, and the special 50th anniversary issue of the *Monthly* (v. 74 II (1967)) are also good references, but none of these accounts mentions the adoption of the now-familiar logo with an icosahedron surrounded by a circular band containing the official name THE MATHEMATICAL ASSOCIATION OF AMERICA. (The November 1920 issue of the *Monthly* records the incorporation of the MAA, in Illinois, and states that the corporate seal of the Association "shall have inscribed thereon the name of the Association and the words 'Corporate Seal-Illinois'"|this is not the same as the MAA logo.)

The first occurrence of the logo with the icosahedron that we could find was on the cover and title page of the first Cams Monograph, *Calculus of Variations*, by C. A. Bliss, published in 1925 by the MAA. It continued to appear in that manner on all subsequent Carus Monographs. (A. Willcox and M. Callanan, at the Washington office of the MAA, searched early minutes of the meetings in which the Carus Monograph series was planned, but found no mention of the creation of the logo to appear on the title page.) Most of us think of the logo today as "always" being part of the cover of the *Monthly*, but it made its first appearance there in January, 1942, when Lester B. Ford assumed the editorship. These are the best answers that we could find to the "when?" question; there was not even a htnt of an answer to the "why?" question.

Accuracy? Although the logo was redrawn around 1971 (because the old plates were in bad shape), and the rendering (by an artist) may have become 'worse' from the standpoint of descriptive geometry, careful scrutiny of an enlargement of the logo on the 1925 Carus Monograph reveals the same flaw pointed out hy Grünbaum in Figure 2. The carefully executed representation of an icosahedron which is featured on our cover will now become the master for all new renderings of the MAA logo.

The best irony in our attempt to find answers to the various questions about the logo was to discover in the *Monthly* (v. 31, 1924. pp. 157–158), in the year in which the first Carus Monograph was prepared, the following account of a talk entitled "The nature and function of descriptive geometry," given by Professor W. H. Roever at the annual MAA meeting held in Cincinnati in 1923;

> In the paper by Professor Roever the principal purposes of the subject [descriptive geometry] were defined to be (1) representation of the objects of space by means of figures which lie in a plane (or upon a surface), and (2) solution of the problems of space by means of constructions which can be executed in the plane. It was shown that the requirement for a picture to be adequate, i.e., capable of producing a retinal image differing but little from that produced by the object itself, naturally leads to the use of central or parallel (orthographic or oblique) projection as a means of representation; and also that the requirement for an unambiguous correspondence between space and the plane results in the need for two projections or for one projection with information concerning the object (such, for instance, as the perpendicularity of edges). ...
>
> Finally, need for the production of good pictures in books on mathematics was stressed, and the value of the study of descriptive geometry as a means of developing the power of space visualization was emphasized. The reader interested in a brief account of the subject is referred to the author's paper entitled "Descriptive geometry and its merits as a collegiate as well as an engineering subject" published in the *Monthly* (1918, 145–159).
>
> Professor Bradshaw spoke of a recent book on calculus in which is found a so-called "Standard figure of the ellipsoid," a critical examination of which from the standpoint of descriptive geometry reveals three different directions of projection. ...

Plus ça change, plus c'est la même chose!

—Doris Schattschneider

Why Your Classes Are Larger Than "Average"

David Hemenway

Mathematics Magazine
55 (1982), 162–64

Editors' Note: Admissions officers often proudly publicize an institution's "average class size," leaving the teaching staff wondering "Why don't I ever get any classes like that?" Here we learn why.

Professor Hemenway is a Professor in the Department of Health Policy and Management at Harvard University. An economist by training, he is currently director of the Harvard Injury Control Research Center and the Harvard Youth Violence Prevention Center. He has written extensively on firearms injuries and the economics of health care.

Most schools advertise their "average class size," yet most students find themselves in larger classes most of the time. Here is a typical example.

In the first quarter of the 1980–81 academic year, 111 courses including tutorials, were given at Harvard School of Public Health. These ranged in size from one student to 229. The **average class** size, from the administration's and professors' perspective, was 14.5. The **expected class size** for a typical student was over 78! This huge discrepancy was due to the existence of a few very large classes. Indeed, only three courses had more than 78 students. One enrolled 105, another 171, and there were 229 in Epidemiology.

Given one class of the size of Epidemiology, an expected class size of approximately 78 for a typical student can be achieved in various ways. Four possible configurations for the rest of the classes are: (i) 450 individual tutorials, (ii) 50 courses of size 10, (iii) 25 courses of size 30, (iv) 25 courses of size 50. The administration's "average class size" for these four cases would be 1.5, 14.3 (close to the advertised figure), 38, and 57 respectively.

The discrepancy between average class size and expected class size for a typical student is explained by a simple computation. Suppose we have a population of M individuals divided into N groups, and we let X_i denote the size of the ith group, $1 \leq i \leq N$. Then the expected number of people in a randomly selected group ("average class size") is given by

$$\bar{X} = (\Sigma X_i)/N = M/N,$$

and the expected size of a group containing a randomly selected individual is given by

$$X^* = \Sigma(X_i/M)X_i = (\Sigma X_i^2)/M.$$

Hence

$$\begin{aligned} X^* - \bar{X} &= \left[N\Sigma X_i^2 - (\Sigma X_i^2)\right]/MN \\ &= \left[\frac{N\Sigma X_i^2 - (\Sigma X_i)^2}{N^2}\right]\frac{N}{M} \\ &= \sigma^2/\bar{X}, \end{aligned}$$

where σ^2 is the variance in group sizes.

The difference between the two means $\bar{X}$ and X^* is directly proportional to the variance in sizes of groups and inversely proportional to average group size. It follows that $X^* \geq \bar{X}$, with equality only when all the groups are the same size.

Here are additional examples from everyday life of the differences between $\bar{X}$ and X^*.

The Nationwide Personal Transportation Survey indicated that average car occupancy ($\bar{X}$) for "home-to-work" trips in metropolitan areas in 1969 was 1.4 people. The table below gives the data.

Number of Occupants	"Home-to-Work" Trips
1	73.5
2	18.2
3	4.7
4	1.9
5	1.1
6	.5
7	.1

Calculating X^* from these statistics we find that the average number of occupants in the car of a typical commuter was 1.9.

To eliminate most congestion problems in U.S. cities would only require raising the average number of people per car ($\bar{X}$) to 2. This doesn't sound impossible. But suppose this were accomplished by inducing some drivers of single-occupant vehicles to join together in five-person car pools. The percentage of single-occupant cars would need to fall to 58.7%; five-occupant cars would rise to 15.9%. The percentage of people in single-occupant cars would fall below 30%. If X^* is calculated for this situation, one finds that the typical commuter would be in a car carrying more than three people.

I often buy dinner at a fast-food restaurant near my home. Although most customers order "to go," the place is almost always crowded, and I consider it quite a success. One evening about 6:30 I went in and there was no one in line. The manager was serving me, so I asked, "Where is everyone?" "It often gets quiet like this," he said, "even at dinnertime. The customers always seem to come in spurts. Wait fifteen minutes and it will be crowded again." I was surprised that I had never before seen the restaurant so empty. But I probably shouldn't have been. If I am a typical customer, I am much more likely to be there during one of the spurts, so my estimate of the popularity of the restaurant (X^*) is likely to be much greater than its true popularity ($\bar{X}$).

The average number of people at the beach on a typical *day* will always be less than the average number of people the typical *beach-goer* finds there. This is because there are lots of people at the beach on a crowded day, but few people are ever there when the beach is practically deserted.

If the waiting time at a health clinic increases with the number of patients, the average waiting time for a typical *day* will always be less than the average waiting time for a typical *patient*. This is because there are more patients waiting on those days when the waiting time is especially long.

The expected size of a typical generation will be smaller than the expected number of contemporaries for a randomly chosen individual from one of those generations.

Figures for the population density of any region will understate the actual degree of crowding for the average inhabitant.

This Note distinguishes mathematically between two types of means. It does not report any original findings about human behavior. Yet it does indicate something about perceptions—especially my own. I was surprised at the restaurant. I was also surprised when the courses I took in college were larger than advertised. And I was surprised to realize how many commuters had to carpool to reach an average of even two people per car. If you are similarly surprised by any of these observations, your perceptions and perhaps even your behavior may be affected.

Helpful comments were received from Frederick Mosteller and an anonymous referee.

The New Polynomial Invariants of Knots and Links

W. B. R. Lickorish and
Kenneth C. Millett

Mathematics Magazine
61 (1988), 3–23

Editors' Note: William Bernard Raymond Lickorish is a Fellow of Pembroke College and Professor of Mathematics at the University of Cambridge, where he was a student. During a short stay between his student years and his return to Cambridge, he taught at the University of Sussex. At Cambridge he was head of the Department of Pure Mathematics and Mathematical Statistics, 1997–2002. A topologist, he has published a book on the topic of this paper, *An Introduction to Knot Theory* (Springer, 1997). He was awarded the Whitehead Prize by the London Mathematical Society.

Kenneth Cary Millett did his undergraduate work at MIT and took his PhD at the University of Wisconsin-Madison in 1964. He is Professor of Mathematics at the University of California, Santa Barbara, an institution he joined in 1969. The American Mathematical Society gave him their Award for Distinguished Public Service in 1998.

This piece from the Magazine is a result of a conversation over tea one summer in the Mathematics Commons Room at Cambridge when Professor Millett was returning home after a yearlong sabbatical at the Institut des Hautes Études Scientifiques outside Paris. Lickorish and Millett discovered a similarity between the new (at that time) Jones polynomials and the earlier Alexander polynomials used in the study of knots.

For this article the authors were given an Allendoerfer Award in 1989 and this was followed by the Chauvenet Prize in 1991. The Chauvenet Prize is the most venerable and prestigious of the MAA's awards for distinguished expository writing and is the only MAA writing award that is not restricted to publications in MAA journals. This was only the second time that an article in *Mathematics Magazine* was honored with the Chauvenet Prize. The list of recipients of the Chauvenet, given since 1925, includes such stellar names as G. H. Hardy, Paul R. Halmos, Mark Kac, S. S. Chern, Carl Pomerance, Stephen Smale, Barry Mazur, and Donald G. Saari.

* Supported in part by National Science Grant DMS8503733.

The theory of knots and links is the analysis of disjoint simple closed curves in ordinary 3-dimensional space. It is the consideration of a collection of pieces of string in 3-space, the two ends of each string having vanished by being fastened together as in a necklace. Many examples can be seen in the diagrams that follow. If the strings can be moved around from one position to another those two positions are the same link or 'equivalent' links. Of course, during the movement no part of a string is permitted to pass right through another part in some supernatural fashion; the string is regarded as being extremely thin and pliable; it can stretch and there is no friction nor rigidity to be considered. As an example, Figure 1 shows two pictures of the same link, a famous link called the Whitehead link. Thus the problem of understanding knots and links is one of geometry and topology, and within those disciplines the subject has received considerable study during the last hundred or more years. Knot theory has been a real inspiration to both algebraic and geometric topology, and, conversely, the theoretical machinery of topology has been used to make vigorous attacks on knot theory. The principal problem has always been to find ways of deciding whether or not two links are equivalent. Confronted with two heaps of intertwined strings, how is one to know if one can move the first to the configuration of the second (without cheating and breaking the strings)? Algebraic topology provides some 'invariants,' but recently some entirely new methods have been discovered which are extraordinarily effective (though not infallible),

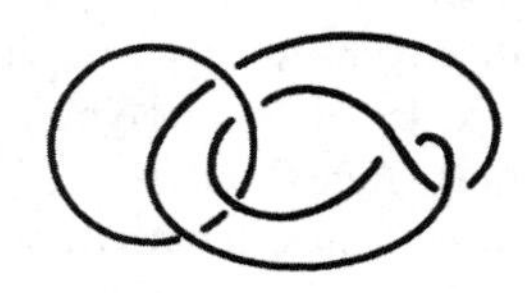

Figure 1.

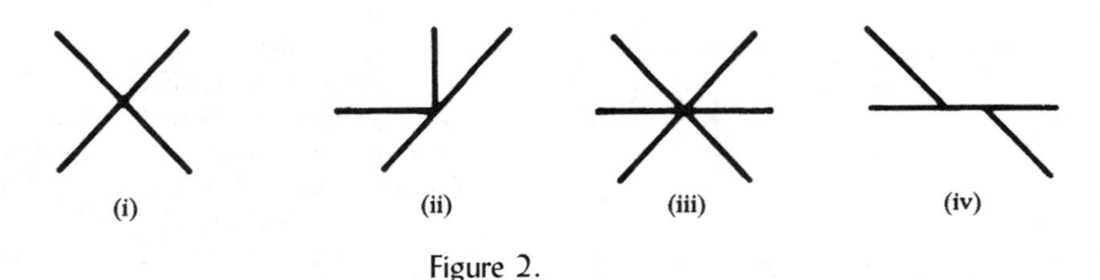

Figure 2.

and which, judged by the standards of most modern mathematics, are breathtakingly simple.

The story begins in the spring of 1984. Professor V. F. R. Jones, now of the University of California at Berkeley, had for some years been studying operator algebras and trace functions on these algebras. It was pointed out to him that some of the formalism of his work closely resembled that of the well-known *braid group* of E. Artin [3]. This braid group can be used to study knots, and eventually Jones realized that, using his trace functions, he could define polynomials for knots and links which are *invariants* [9]. This means that to each configuration of pieces of string is associated a polynomial, and that if the string is moved (as described above) to a new position it still has the same polynomial. Thus, if calculation shows two heaps of string have two distinct polynomials, then it is not possible to move the strings from one position to the other. To get the idea of an invariant, consider what is probably the easiest of them all, namely the number of strings that make up a link; a link of two strings can never be deformed to one of three strings. Another polynomial invariant for links (discovered in about 1926 by J. Alexander [1]) was well known so, for a while, it was suspected that Jones' polynomial might be but some elementary manipulation of that polynomial. Soon however it was established that the Jones polynomial was entirely new, independent of all other known invariants. Strenuous efforts to understand the Jones polynomial have since been made by many mathematicians scattered around the world. The most amazing things about it are its simplicity and the fact that it exists at all. In retrospect it seems that several mathematicians during the last thirty years came exceedingly close to discovering Jones' polynomial and would surely have done so had they dreamt there was anything there to discover. A very simple complete proof of the existence of this polynomial appears in §3 below. By now, Jones' polynomial has been generalized two or three times, lengthy computer generated tabulations of examples have been produced, proofs have been explored and simplified, some correlations with algebraic topology have been found and a few geometric applications have been produced. Nevertheless, at the time of writing, there is still a feeling that these new ideas are not really understood, that they do not really fit in with more established theories, and that more generalisations and applications may be possible. Intense investigation continues.

1. Basic Background

A little basic information about knots and links may allay some misunderstandings, but the confident will proceed to the next section. There are several excellent surveys of the subject prior to Jones' discovery; [2], [15], [17], [16] and [5] are accounts in (approximate) order of increasing mathematical sophistication. As already stated, a link is a finite collection of disjoint simple closed curves in 3-dimensional space $\mathbb{R}^3$, the individual simple closed curves being called the *components* of the link. A link of just one component is a *knot*. It is tacitly assumed that the closed curves are *piecewise linear*, that is that they consist of a finite number (probably very large) of straight line segments placed end to end. This is a technical restriction best ignored in practice; it does however ensure that an infinite number of kinks of any sort, possibly converging to zero size, never occurs. Restricting the components to being differentiable would do equally well. The orthogonal projection of $\mathbb{R}^3$ onto a plane $\mathbb{R}^2$ in $\mathbb{R}^3$ maps a link to a diagram of the type seen frequently in the pages that follow. The direction of that projection is always chosen so that, when in $\mathbb{R}^2$ projections of two distinct parts of the link meet, they do so transversally at a *crossing* as in Figure 2(i), never as in Figure 2(ii), (iii), or (iv). At a crossing it is indicated which of the two arcs corresponds to the upper string, and which to the lower, by breaking the line of the lower one at the crossing. Such a planar diagram will be called a *projection* of the link.

A movement of a link from one position in $\mathbb{R}^3$ to another is called an *ambient isotopy*; that idea defines when two links are the same or '*equivalent*.' Such a movement changes the planar projections of

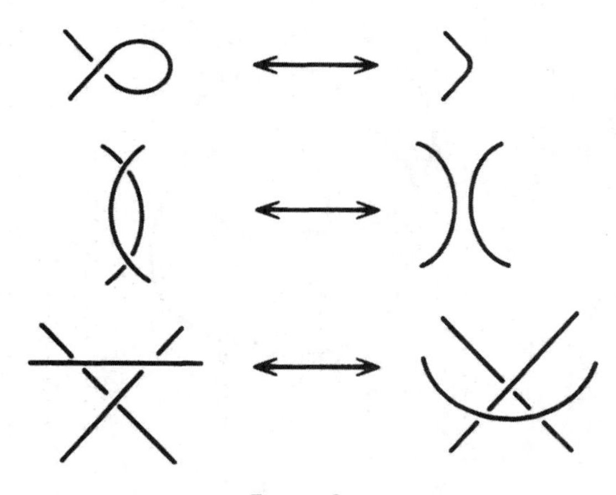

Figure 3.

a link. An established theorem states that two links are equivalent if and only if (any of) their projections differ by a sequence of *Reidemeister moves* [16]. These moves, of types I, II and III, are those shown in Figure 3 (and their reflections), where, for each type, a small part of the projection is shown before and after the move; the remainder of the projection remains unchanged. It is clear that if two link projections so differ then the links are equivalent; the converse is established by a routine and inelegant proof. Thus to show that the two diagrams of Figure 1 represent the same link one can construct a sequence of these moves that allow the first diagram to evolve to the second. However, for two general link projections one has no idea whether many million moves may be required.

An oriented link is a link with a direction (usually indicated by an arrow) assigned to each component, so that each acquires a preferred way of travelling around it. Thus a link with n components has 2^n possible orientations. The two oriented links of Figure 4 are distinct, for one cannot be moved to the other sending the directions on the one link to those on the other (this is proved in what follows).

A knot is *unknotted* if it is equivalent to a knot that has a projection with zero crossings. Two oriented knots can be *summed* together as indicated in Figure 5; they are placed some way apart and a 'straight' band joins one to the other so that in the resultant sum the orientations match up. A knot, other than the unknot, is *prime* if it cannot be expressed as a sum of two knots neither of which is unknotted. Note that the sum of two oriented links of more than one component is not well defined unless it is specified which two components are to be banded together. As it is known that any knot is uniquely expressible as the sum of prime knots, listings of knots usually include only the prime knots.

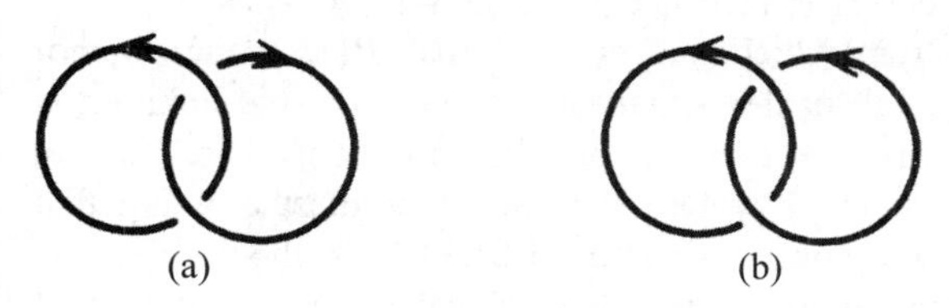

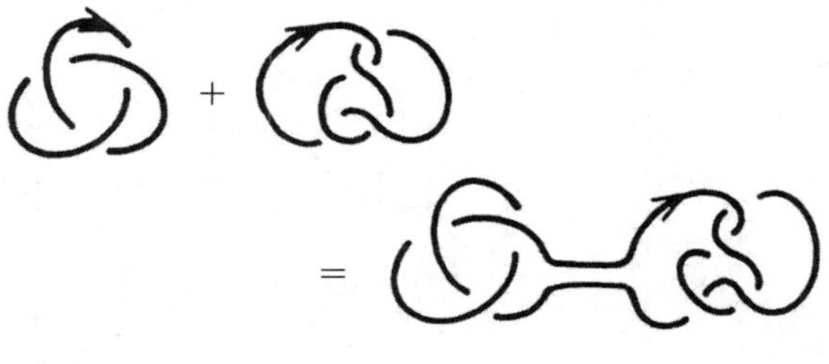

Figure 4.

Figure 5.

If L is an oriented link, let ρL denote L with all its directions reversed, and let $\overline{L}$ be the reflection of L. When considering projections, this reflection is usually thought of as reflection in the plane of the paper, so that the projection of $\overline{L}$ comes from that of L by changing all underpasses to overpasses and *vice versa*. Thus from L can be created ρL, $\overline{L}$, and $\rho\overline{L} = \overline{\rho L}$, and these may be four distinct links, they may be the same in pairs, or all four may be the same. The trefoil knot 3_1 (see Figure 6) creates two pairs in this way, the figure of eight knot 4_1 has all four the same whilst 9_{32} has all four distinct.

Inherent in the idea that reflection can change a knot is the convention that the enveloping three-dimensional space $\mathbb{R}^3$ is oriented; it is equipped with a distinction between left-hand and right-hand screwing motions. Knot tables have traditionally listed prime knots according to the minimum number of crossings in a projection of the knot. Thus 7_4 denotes the fourth knot, in some traditional order, that needs seven and no more than seven crossings. The tables have deliberately ignored reflections and reversals, so that an entry may stand for as many as four knots if these orientations are taken into account. With these conventions knots have been classified up to thirteen crossings [19] with the help of computers and the following table has been produced.

#crossings	3	4	5	6	7	8	9	10	11	12	13
#knots	1	1	2	3	7	21	49	165	552	2176	9988

This totals 12965 knots.

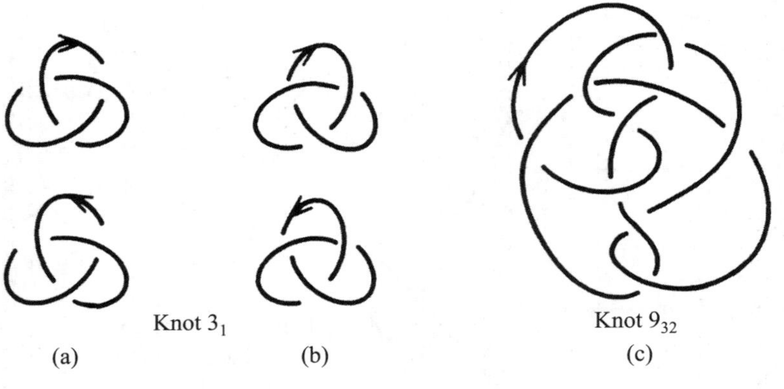

Figure 6.

There now follows a discussion of the new polynomial invariants of knots and links. It is not possible to restrict the discussion to knots alone, for many-component links are an integral part of the theory. The ideas will not here be developed in the order of their discovery but in an order that now seems simpler to understand.

2. The Oriented Polynomial

The polynomials to be considered here are Laurent polynomials with *two* variables ℓ and m and with integer coefficients. A Laurent polynomial differs from the usual polynomials of high school only inasmuch as negative as well as positive powers of the variables may occur. One such polynomial $P(L)$ will be associated to each oriented link L. For example, to the link of Figure 1 will be assigned the polynomial

$$(-\ell^{-1} - \ell)m^{-1} + (\ell^{-1} + 2\ell + \ell^3)m - \ell m^3.$$

The result that encapsulates this was discovered almost simultaneously by four sets of authors [7] in the wake of Jones' first announcement ($P(L)$ *generalises* Jones' polynomial, see §3). It can be stated as follows:

Theorem 1 *There is a unique way of associating to each oriented link L a Laurent polynomial $P(L)$, in the variables ℓ and m, such that equivalent oriented links have the same polynomial and*

(i) $P(unknot) = 1$,

(ii) *if L_+, L_-, and L_0 are any three oriented links that are identical except near a point where they are as in Figure 7, then*

$$\ell P(L_+) + \ell^{-1} P(L_-) + mP(L_0) = 0.$$

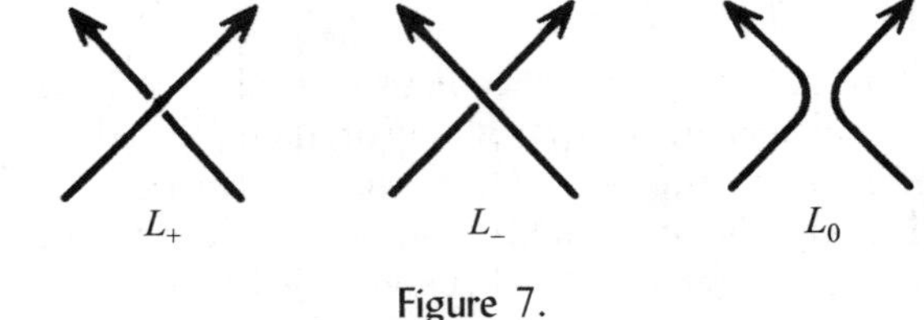

Figure 7.

In a projection of an oriented link the crossings are of two types; that of L_+ in Figure 7 is called positive, that of L_- is negative. This idea will be exceedingly important. In L_+, the direction of one segment can be thought of as pointing in the direction dictated by a right-hand screw motion along the direction of the other segment. All the orientations are needed to make this important distinction. Of course, the choice of which type of crossing is given which sign is but a convention.

The meaning of Theorem 1 becomes apparent as one uses it to make a few calculations. Shown in Figure 8 is a very elementary example of a triple of links L_+, L_-, and L_0. The first two links are just

Figure 8.

pictures of the unknot, the first with a single positive crossing, the second with just a negative one. Formulae (i) and (ii) imply that $\ell 1 + \ell^{-1} 1 + mP(L_0) = 0$, from which one deduces that $P(L_0)$, the polynomial for the trivial link of two separated unknots, is $-(\ell^{-1} + \ell)m^{-1}$. Consider now the triple of links in Figure 9 (where it is the uppermost crossing that is to be considered). Here L_+ is the link whose polynomial has just been calculated, or at least it is equivalent to it and so has that same polynomial;

Figure 9.

L_0 is the unknot. Thus

$$-\ell(\ell^{-1}+\ell)m^{-1}+\ell^{-1}P(L_-)+m1=0,$$

so that the polynomial for the simple link L_-, the link of Figure 4(a), is $(\ell+\ell^3)m^{-1}-\ell m$. Figure 10 shows a third triple (look at the right-hand crossing). This yields the equation

$$\ell 1+\ell^{-1}P(L_-)+m\{(\ell+\ell^3)m^{-1}-\ell m\}=0.$$

Hence the polynomial of the left-hand version of the trefoil knot 3_1 (seen also in Figure 6(a)) is $-2\ell^2-\ell^4+\ell^2m^2$. This proves that the trefoil is indeed

Figure 10.

knotted, for were it equivalent to the unknot it would, by Theorem 1, have 1 for its polynomial. Similarly the link of Figure 4(a) is not equivalent to the link consisting of two separated unknots.

Consider the method of the preceding calculation of the trefoil's polynomial. Attention was given to one crossing. Switching that crossing produced the unknot, nullifying it (to get 'L_0') produced a link with fewer crossings that had already been considered. This procedure works in general. Suppose that one is confronted with an oriented link L of n crossings and c components. Assume that one has already calculated the polynomials of all (relevant) oriented links of $n-1$ crossings; there are only finitely many of them. Then formula (ii) calculates $P(L)$ in terms of the polynomial of a modified L, namely L with some chosen crossing switched. However it is always possible to change L to U^c, the unlink of c unknots, by switching a subset of some s of the crossings. Thus, performing the switches one by one, using formula (ii) each time, $P(L)$ is calculated in terms of the polynomials of s links of fewer crossings (the 'L_0's') and of $P(U^c)$. However, it is an easy exercise to show, by induction on c, that $P(U^c)=\left(-(\ell^{-1}+\ell)m^{-1}\right)^{c-1}$. This method of calculation will always work, and it is essentially the only known method of calculating these polynomials. It is nevertheless not a very welcome method for the length of calculation increases exponentially with the number of crossings of the link presentation.

Note that if *all* the diagrams in the above calculations were reflected in the plane of the paper this would change each positive crossing to a negative crossing and vice versa. Each L_+ would become an L_-. This would simply exchange the roles played in the calculation by ℓ and ℓ^{-1}. Thus reflecting a link has the effect on its polynomial of interchanging ℓ and ℓ^{-1}. The left-hand trefoil of Figure 6(a) has polynomial $-2\ell^2-\ell^4+\ell^2m^2$, so the right-hand trefoil's polynomial (Figure 6(b)) is $-2\ell^{-2}-\ell^{-4}+\ell^{-2}m^2$. These polynomials are obviously different so, by Theorem 1, the trefoils are inequivalent (this fact was tricky to prove until 1984 when Jones produced a version of this proof). Similarly the two oriented links of Figure 4 are distinct. For any polynomial P, let $\overline{P}$ denote P with ℓ and ℓ^{-1} interchanged (c.p. complex conjugation). The above discussion has demonstrated the following result.

Proposition 1 *For any oriented link L,* $\overline{P(L)}=P(\overline{L})$.

Consider now the triple of Figure 11. This yields

$$\ell P(4_1)+\ell^{-1}1+m\{(\ell+\ell^3)m^{-1}-\ell m\}=0$$

so that

$$P(4_1)=-\ell^{-2}-1-\ell^2+m^2$$

Here then the polynomial *is* symmetric with respect

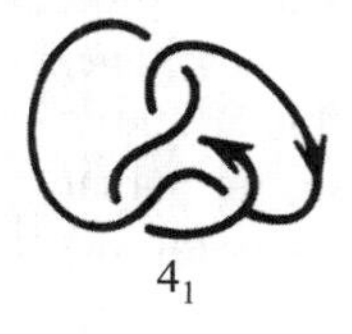

Figure 11.

to ℓ and ℓ^{-1} and so it does not show 4_1 and $\overline{4}_1$ to be different, and, in fact, a little experimentation shows them to be the same. In practice the P-polynomial provides a very good test as to whether or not $L=\overline{L}$, but any hope that it might be an infallible test is dashed by the knot 9_{42} shown in Figure 12. It is known that $9_{42}\neq\overline{9}_{42}$ because a

Figure 12.

certain 'signature' invariant, from algebraic topology, is nonzero. However,

$$P(9_{42}) = (-2\ell^{-2}-3-2\ell^2)+(\ell^{-2}+4+\ell^2)m^2-m^4$$

and this is a self-conjugate polynomial.

It should be remarked that other *notations* can be used in the whole of this theory of polynomials for knots and links. For example, $P(L)$ can be taken to be a polynomial in three variables x, y, z with the vital defining formula being $xP(L_+) + yP(L_-) + zP(L_0) = 0$. However, the three variables are homogeneous variables as in projective planar geometry (there are still really only two variables), and the balance between ℓ and ℓ^{-1} is lost. Some authors also have a strong preference for some negative signs in the defining formula.

Recall that ρL is obtained from L by reversing all its arrows. Unfortunately, $P(\rho L) = P(L)$, for changing L to ρL leaves the signs of all its crossings unchanged. Thus any calculation for $P(L)$ induces exactly the same calculation for $P(\rho L)$. (This means that for a knot, a link of one component, it is not really necessary to specify an orientation at all when thinking about the polynomial.) If, however, the directions of some, but not all, of the components of L are changed, then $P(L)$ can change in a rather drastic way that is not well understood. Examples occur in some of the polynomials of two-component links listed in the table at the end of this article.

A result concerning the behaviour of the polynomial under sums and 'distant' unions is as follows:

Proposition 2
(i) $P(L_1 + L_2) = P(L_1)P(L_2)$;
(ii) $P(L_1 \cup L_2) = -(\ell + \ell^{-1})m^{-1}P(L_1)P(L_2)$.

In (i) $L_1 + L_2$ denotes the sum (see Figure 5) of oriented links using *any* component of L_1 to add to any component of L_2. As different choices may be made for these components, this leads easily to examples of distinct links having the same polynomial. For example, the two links in Figure 13 are

Figure 13.

distinct as their individual components are different knots. However, by Proposition 2(i) they both have polynomial

$$\{-2\ell^2 - \ell^4 + \ell^2m^2\}\{-\ell^{-2} - 1 - \ell^2 + m^2\}$$
$$\{(\ell^{-1} + \ell^{-3})m^{-1} - ell^{-1}m\}.$$

In (ii) $L_1 \cup L_2$ denotes the union of L_1 and L_2 placed some distance apart from each other so that no part of L_1 crosses over or under part of L_2. Proposition 2 is significant because it relates simple geometry to the *product* of polynomials. This uses the multiplicative structure of polynomials; $P(L)$ is not just an array of coefficients but is a polynomial that may be used to multiply another polynomial! The proposition is easy to prove from Theorem 1.

There is another way in which it is known that two oriented links will have the same polynomial. Deep in the geometric structure of link theory is the simple idea of decomposing a link using spheres that cut the link at four points. The dotted sphere of Figure 14 is an example. If the inside of that sphere is rotated through angle π (about the polar axis) the second diagram results. Such an operation is called *mutation*. Mutation never changes the P-polynomial of a link though it can well change the link, as indeed it does in Figure 14.

For a further example consider a pretzel knot as shown in Figure 15. The ith circle contains a twist of c_i crossings as indicated, each c_i being odd (to side-step any orientation difficulties). Mutation, with respect to the indicated ellipse, interchanges c_2 and

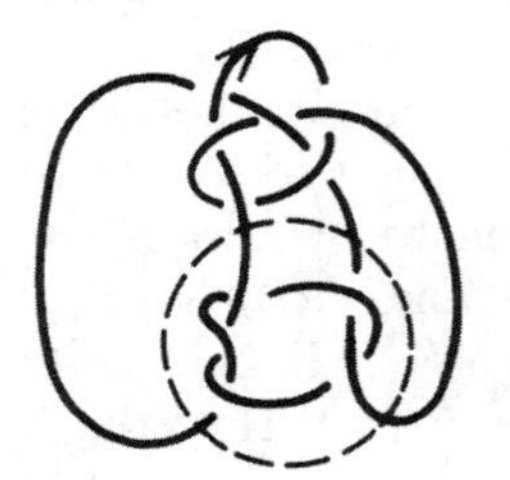

Figure 14.

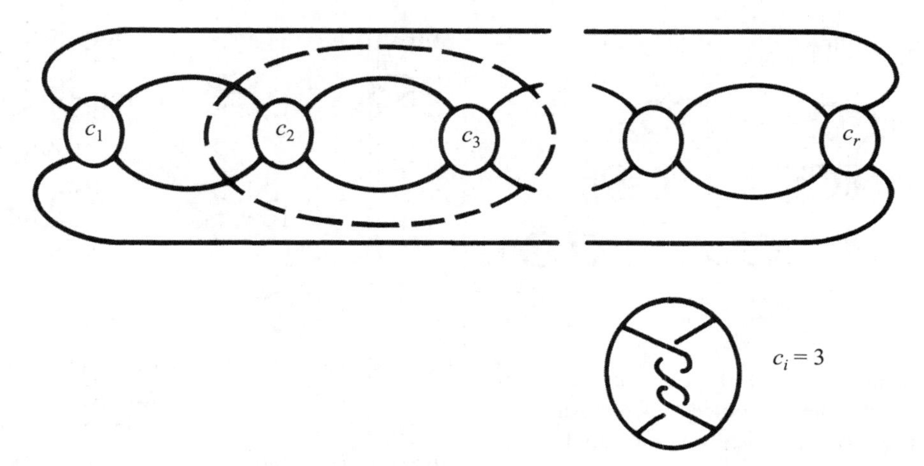

Figure 15.

c_3. As any permutation is the result of a sequence of such adjacent interchanges, the P-polynomial of the pretzel knot is independent of the ordering of the c_i's; in general the knot does change if that ordering changes.

So far nothing has been said about a proof for Theorem 1. The proof in [13] consists of defining the polynomial with a lengthy argument of induction on the number of crossings of a presentation, showing that however a calculation (like those already discussed) is made the same polynomial results, and checking invariance under the Reidemeister moves. It is thus entirely combinatoric, but the induction argument needs delicate handling. Although other proofs differ in style and emphasis they all seem to use essentially the same combinatorics.

3. The Jones Polynomial

Whereas the proof that $P(L)$ exists is a little arduous, an almost trivial proof of the existence of the polynomial of Jones has been found by L. H. Kauffman [11]. This proof, which must, in recent years, be one of the most remarkable discoveries of readily accessible mathematics, is outlined below.

The original polynomial of V. F. R. Jones associated with an oriented link L is denoted $V(L)$. It is a Laurent polynomial in the variable $t^{1/2}$, that being simply a symbol whose square is the symbol t. It satisfies

$$V(\text{unknot}) = 1$$

$$t^{-1}V(L_+) - tV(L_-) + (t^{-1/2} - t^{1/2})V(L_0) = 0,$$

where L_+, L_-, and L_0 are oriented links related as before. Thus $V(L)$ is obtained from $P(L)$ by the substitution

$$(\ell, m) = \left(it^{-1}, i(t^{-1/2} - t^{1/2})\right),$$

where $i^2 = -1$. As mentioned before, $P(L)$ was conceived as a generalisation of $V(L)$.

Begin all over again by considering *projections* (pictures) of *unoriented* links. For each such projection L define a Laurent polynomial $\langle L \rangle$ in one variable A by the following three rules that will shortly be explained:

(a) $\langle \bigcirc \rangle = 1$,

(b) $\langle L \cup \bigcirc \rangle = -(A^{-2} + A^2)\langle L \rangle$,

(c) $\langle \times \rangle = A\langle \asymp \rangle + A^{-1}\langle)\,(\rangle$.

This $\langle L \rangle$ is called the *bracket polynomial* of L. Rule (a) states that 1 is the polynomial of the particular projection of the unknot that has no crossing at all. In rule (b) $L \cup \bigcirc$ denotes the projection that consists of L plus an extra component that contains no crossing. Rule (c) refers to three projections exactly the same, except near one point where they are as shown. The first projection of this triple shows a crossing, and in the other two that crossing has been destroyed. It should be noted that, given the picture of the crossing, one can distinguish between the two other pictures using the orientation of space: If, when moving along the underpass towards the crossing one swings to the right, up on to the overpass, one creates the picture of the link whose polynomial is multiplied by A in rule (c). No arrows are required for that. A simple example involving the use of all three rules is as follows:

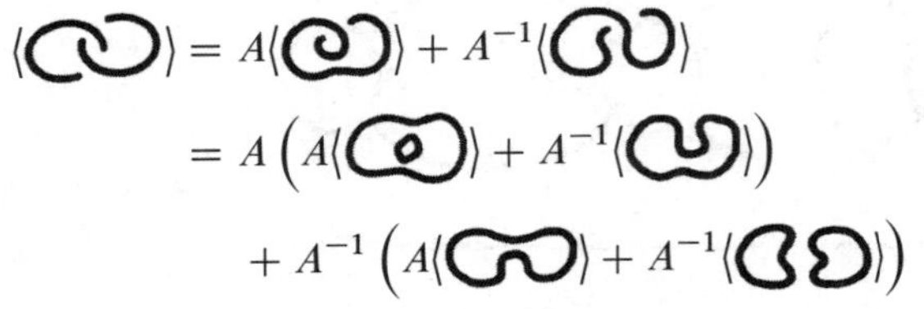

$$\langle\ \rangle = A\langle\ \rangle + A^{-1}\langle\ \rangle$$
$$= A\left(A\langle\ \rangle + A^{-1}\langle\ \rangle\right)$$
$$+ A^{-1}\left(A\langle\ \rangle + A^{-1}\langle\ \rangle\right)$$
$$= -(A^2 + A^{-2})^2 + 2.$$

Here when calculating $\langle L\rangle$ there are no problems about making judicious choices of crossings to switch in order to maneuver towards an unknotted situation (as there were with the P-polynomial). Each use of rule (c) *reduces* the number of crossings in the projections until there are no crossings at all; then rules (b) and (a) finish the job of calculation. It is evident that the choice of the order in which the crossings are attacked is irrelevant, so that these rules do indeed define unambiguously a polynomial for each unoriented link projection. What remains to be done is to see if $\langle L\rangle$ is unchanged by the Reidemeister moves I, II and III of §1; if it is, then it is an invariant of real links in $\mathbb{R}^3$:

Move I.

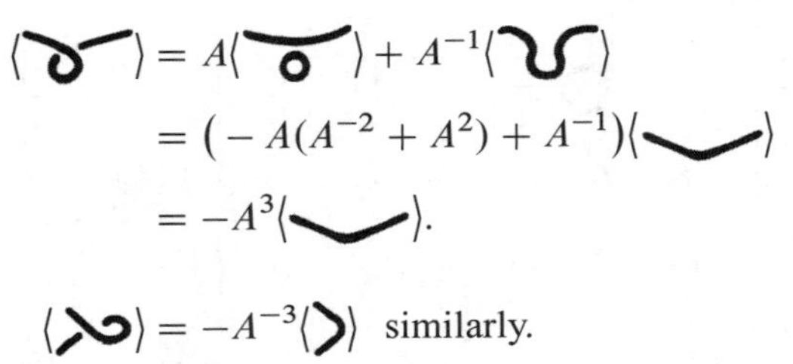

$$\langle\ \rangle = A\langle\ \rangle + A^{-1}\langle\ \rangle$$
$$= (-A(A^{-2} + A^2) + A^{-1})\langle\ \rangle$$
$$= -A^3\langle\ \rangle.$$

$$\langle\ \rangle = -A^{-3}\langle\ \rangle \text{ similarly.}$$

Thus the bracket polynomial *fails* to be invariant under Move I, and that is an exceedingly important observation.

Move II.

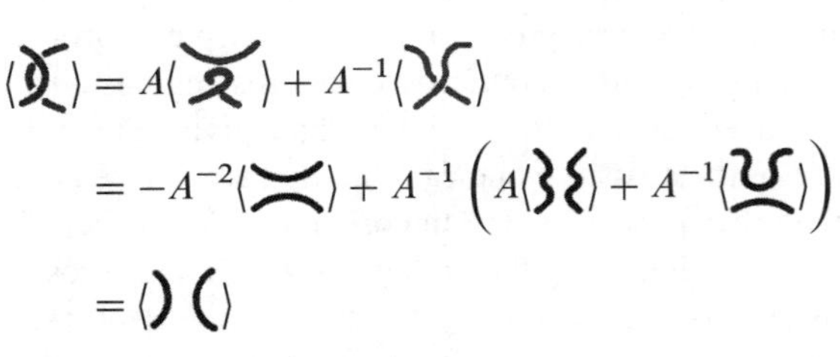

$$\langle\ \rangle = A\langle\ \rangle + A^{-1}\langle\ \rangle$$
$$= -A^{-2}\langle\ \rangle + A^{-1}\left(A\langle\ \rangle + A^{-1}\langle\ \rangle\right)$$
$$= \langle\ \rangle$$

Hence the bracket polynomial *is* invariant under Move II.

Move III.

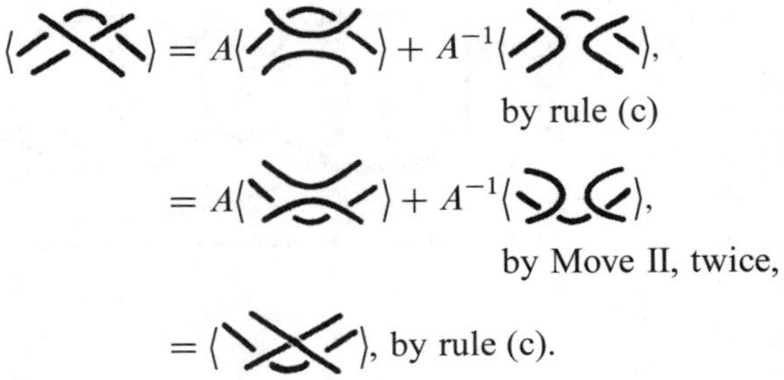

$$\langle\ \rangle = A\langle\ \rangle + A^{-1}\langle\ \rangle,$$
by rule (c)
$$= A\langle\ \rangle + A^{-1}\langle\ \rangle,$$
by Move II, twice,
$$= \langle\ \rangle, \text{ by rule (c).}$$

Hence there is also invariance under Move III.

Now give L an *orientation*. Let $w(L)$, the *writhe* of L, be the algebraic sum of the crossings of L, counting $+1$ for a positive crossing, and -1 for a negative crossing (for example, w(left-hand trefoil) $= -3$). Move I adds or subtracts one to $w(L)$, so $w(L)$ is certainly not invariant under that move, but it is (clearly) invariant under Moves II and III. Thus any combination of $w(L)$ and $\langle L\rangle$ will be invariant under Moves II and III, and their non-invariant behaviours under Move I *cancel* in the expression

$$X(L) = (-A)^{-3w(L)}\langle L\rangle.$$

The above is a *complete proof* that $X(L)$ is a well defined invariant of oriented links.

For projections related in the usual way, rule (c) gives

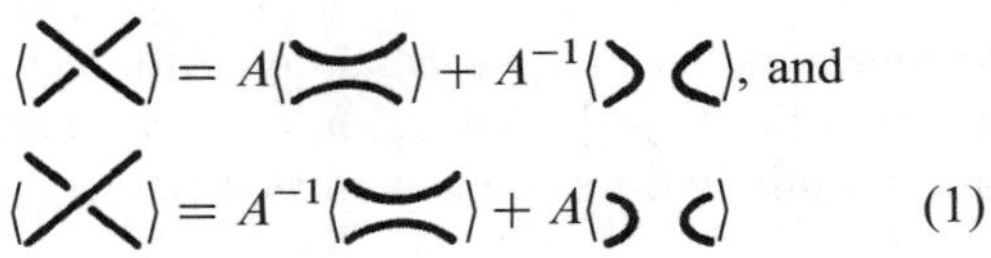

$$\langle\ \rangle = A\langle\ \rangle + A^{-1}\langle\ \rangle, \text{ and}$$
$$\langle\ \rangle = A^{-1}\langle\ \rangle + A\langle\ \rangle \tag{1}$$

Thus

$$A\langle\ \rangle - A^{-1}\langle\ \rangle = (A^2 - A^{-2})\langle\ \rangle.$$

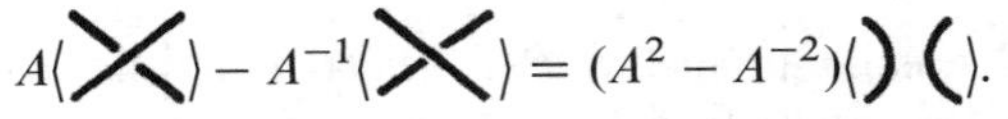

Suppose that orientations can be chosen for these last three projections so that the arrows point approximately upwards (c.f. Figure 7); call them L_+, L_- and L_0. Then $w(L_\pm) = w(L_0) \pm 1$. Hence, substitution gives

$$A(-A)^3X(L_+) - A^{-1}(-A)^{-3}X(L_-)$$
$$= (A^2 - A^{-2})X(L_0).$$

Writing $t^{-1/4} = A$ this becomes

$$t^{-1}X(L_+) - tX(L_-) + (t^{-1/2} - t^{1/2})X(L_0) = 0,$$

so that, under the substitution $A = t^{-1/4}$, $X(L)$ is the original Jones polynomial $V(L)$ for they satisfy the same defining formula.

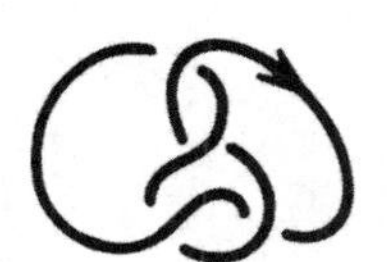
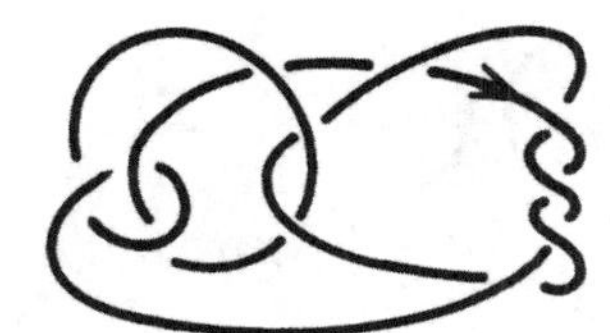

Figure 16.

No analogously simple proof is known for the existence of the P-polynomial; in a proof, the difficulty is to show that different chains of calculations *never* give different polynomials. In terms of distinguishing links the P-polynomial is more powerful than is the V-polynomial; two variables are better than one. For example, the knots in Figure 16 have the same V-polynomial but different P-polynomials.

There is a property of the Jones polynomial, a *reversing result*, that *seems* to have no analogue in terms of the P-polynomial. Suppose that k is one component of an oriented link L and that a new oriented link L^* is formed from L by reversing just the orientation of k. Of course, $\langle L \rangle = \langle L^* \rangle$, for the bracket polynomial disregards all. orientations. Thus $V(L)$ and $V(L^*)$ are the same up to multiplication by some power of t (for each is $\langle L \rangle$ multiplied by a power of $A = t^{-1/4}$). The precise result is:

Proposition 3 *$V(L^*) = t^{-3\lambda} V(L)$, where 2λ is the sum of the signs of the crossings of k with the other components of $L - k$.*

This λ is called the *linking number* of k and $L - k$. It is noteworthy that, though this result follows trivially from the approach of the bracket polynomial, it is by no means obvious when working from the (L_+, L_-, L_0)-definition of the V-polynomial.

The simplicity of Kauffman's approach to the V-polynomial has led to a much better understanding of that polynomial and to a most pleasing application ([11], [14], [18]) concerning alternating knots. A projection of a link is alternating if, when travelling along any part of the link, the crossings are encountered alternately over, under, over, under,.... In Figure 16 the four-crossing projection is alternating, the other is not alternating. The first thirty-one knots in the classical knot tables have alternating projections. A crossing in a link projection will be called *removable* if it is like the crossing in Figure 17; it could be removed by rotating the part of the link in the box labelled Y.

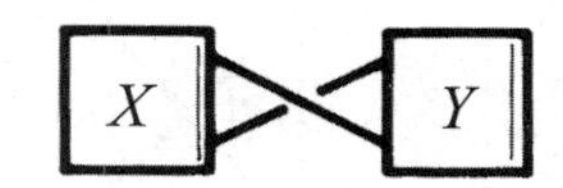

Figure 17.

Proposition 4 (See [11], [14] and [18].) *Let L be a connected oriented link projection of n crossings, then*

(i) *$n \geq$ Spread $V(L)$, where spread $V(L)$ is the difference between the maximum and the minimum degrees of t that appear in $V(L)$;*

(ii) *$n =$ Spread $V(L)$ if L is also alternating and has no removable crossing.*

It is easy to see, by inspection, if a knot projection is alternating and has no removable crossing. If it has these properties, and n crossings, the proposition implies that the knot can have no projection with fewer crossings. If there were a projection with $n-1$ crossings, then, by (i), $n - 1 >$ Spread $V(L)$, which contradicts (ii). This solves a very old problem in knot theory; it has always been suspected that an alternating projection was the simplest available.

For the record, this section should include mention of the *Alexander polynomial* $\Delta(L)$ of an oriented link L. Like the Jones polynomial it is a Laurent polynomial in $t^{1/2}$, it can be defined in a similar way by $\Delta(\text{unknot}) = 1$, and

$$\Delta(L_+) - \Delta(L_-) + (t^{1/2} - t^{-1/2})\Delta(L_0) = 0.$$

Thus $\Delta(L)$ is obtained from $P(L)$ by a substitution for the variables (though it was the similarity between this formula and that defining $V(L)$ that lead to the discovery of the P-polynomial). The Alexander polynomial has been known and developed for about sixty years [1]. It is discussed in the text books of knot theory, usually being defined in terms of the determinant of a certain matrix (though see [6]). The value of $\Delta(L)$ when $t = -1$ is called the determinant of the link, and this integer was one of the first link invariants to be studied. Although the Alexander polynomial is quite good at distinguishing knots, there do exist knots that it cannot distinguish from the unknot; an example is the pretzel knot (see Figure 15) for which $(c_1, c_2, c_3) = (-3, 5, 7)$; for this the Jones polynomial is certainly non-trivial. The Alexander polynomial is fairly well understood in terms of the machinery of algebraic topology (homology groups, fundamental groups, covering spaces, etc.). The same cannot be said for the Jones polynomial and its generalisations. Is there a knot K, other than the unknot, for which $V(K) = 1$? The answer to this is not known. If there is no such K then the Jones polynomial is

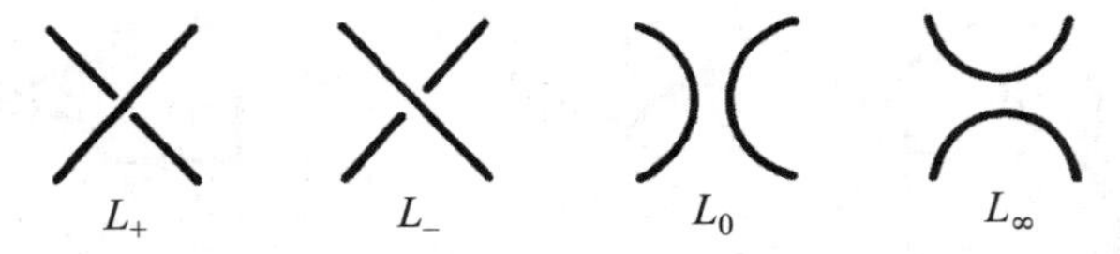

Figure 18.

the long sought elementary method of determining knottedness. There is no reason to suppose that it is so powerful an invariant, but computer searches have revealed no example of such a K, neither has understanding given any clue to finding a method by which such a K might be constructed.

4. The Semioriented Polynomial

Although it may seem that the preceding sections contain many polynomials, only the P-polynomial and some specialisations of it occur. There has been discovered, however, another polynomial, the F-polynomial, that is similar in concept to the P-polynomial though the two are quite distinct. The way to define this F-polynomial is rather like the way in which $V(L)$ was derived from $\langle L\rangle$.

First, for a *projection* L of an *unoriented* link, define a Laurent polynomial $\Lambda(L)$ in two variables a and x by the rules

(a) $\Lambda(\bigcirc) = 1$;

(b) $\Lambda(\boxed{L}\ldots) = a\Lambda(L)$, $\Lambda(\boxed{L}\ldots) = a^{-1}\Lambda(L)$, and $\Lambda(L)$

does not change when L is changed by a Reidemeister move of type II or type III;

(c) $\Lambda(L_+) + \Lambda(L_-) = x\big(\Lambda(L_0) + \Lambda(L_\infty)\big)$

where L_+, L_-, L_0 and L_∞ are projections of unoriented links that are exactly the same except near a point where they are as shown in Figure 18.

Notes.

(a) This means that Λ is 1 for the projection of just one component which has no crossing.

(b) If a positive kink is removed, the Λ-polynomial is multiplied by a (or by a^{-1} for a negative kink). Thus the Λ-polynomial is not invariant under Reidemeister Move I.

(c) In the absence of orientations it is not clear which picture in Figure 18 should be called L_+ and which L_-, nor which is L_0 and which L_∞. Those ambiguities are irrelevant in the light of the symmetry of the formula of Rule (c).

It should be clear that the methods of calculation developed for the P-polynomial will work equally well in this new situation. The new situation is easier in that orientations do not (yet) appear, but more troublesome in that Rule (c) uses four pictures instead of three. As an exercise check that the Λ-polynomial of the two-component link projection with no crossing is $\big((a^{-1}+a)x^{-1}-1\big)$ while that of the usual projection of the left-hand trefoil knot is

$$-2a^{-1} - a + (1+a^2)x + (a^{-1}+a)x^2.$$

A proof that, for a given projection, different schemes of calculation always give the same polynomial requires a more complicated version of the inductive method mentioned at the end of §2 for the P-polynomial.

The Λ-polynomial is, as stated, invariant under the second and third of Reidemeister's moves. Its failure to be invariant under move I can easily be corrected (as was done for $\langle L\rangle$) if L is *now oriented.* As before let $w(L)$ be the sum of the signs of the crossings of the oriented link L.

Theorem 2 *For any oriented link L, let*

$$F(L) = a^{-w(L)}\Lambda(L).$$

This $F(L)$ is a well-defined invariant of oriented links in 3-space.

Tables of the P-polynomial and of this second two-variable polynomial have been produced by M. B. Thistlethwaite for the 12,965 knots in his tabulation of knot projections up to thirteen crossings. He works, of course, with a computer, that being all the more desirable for the F-polynomial; the occurrence of four rather than three diagrams in the defining formula does make calculations for F much more arduous than for P. The F-polynomials contain very many more terms than do the P-polynomials; for example for either knot in Figure 14 the P-polynomial has 14 terms, the F-polynomial has 45 terms. A few more examples appear in the tables at the end of this paper. The greater number of terms seems to mean, in practice, that two knots are more likely to be distinguished by F than by P.

8_8

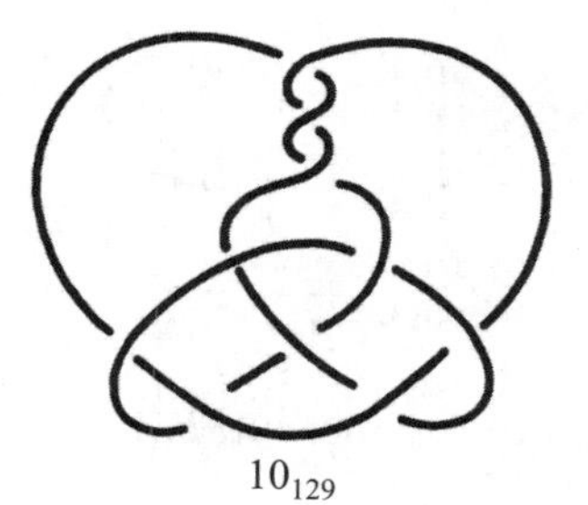

10_{129}

These have the same P but different Q (and F) polynomials.

11_{255}

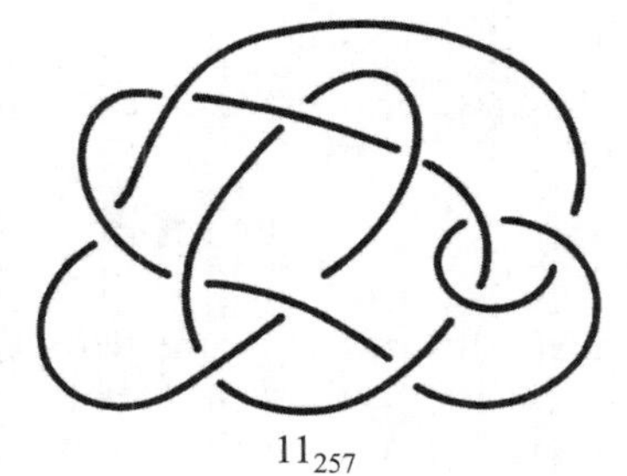

11_{257}

These have the same F but different Δ (and P) polynomials.

Figure 19.

The F-polynomial has a right to be called 'semioriented' because, although L must be oriented to define $F(L)$, changing the orientation only changes $F(L)$ by multiplication by a power of a. The relevant result is:

Proposition 5 *Suppose L^* is obtained from L by reversing the orientation of a component k, then $F(L^*) = a^{4\lambda}F(L)$, where λ is the linking number of k with the other components of $L - k$.*

Compare this with Proposition 3; the proof here is much the same.

An interesting specialisation of $F(L)$ is obtained by the substitution $a = 1$. The resultant Laurent polynomial $Q(L)$, in the one variable x, is called the *absolute* polynomial. This substitution makes all the subtleties of the above definition disappear, no orientations of any sort are required (hence the name 'absolute') and one can work entirely with links in $\mathbb{R}^3$ rather than projections in the plane. The Q-polynomial is simply defined by $Q(\text{unknot}) = 1$, and

$$Q(L_+) + Q(L_-) = x\big(Q(L_0) + Q(L_\infty)\big).$$

Chronologically this polynomial was discovered by Brandt, Lickorish, and Millett [4] and Ho [8] as an extension of the ideas of the P-polynomial, and Kauffman [10] explained how to insert the second variable 'a' to create the F-polynomial. The fact that the Q-polynomial uses no arrows, and $Q(L) = Q(\overline{L})$, makes this something of a recommended polynomial for beginners. Unfortunately the proofs that the Q- and F-polynomials are unambiguously defined are of almost the same form and complexity. In Figure 19 are examples that show that the P and F polynomials are independent in the sense that neither is hidden within the other, to be revealed by some subtle change of the variables.

As might be expected, some basic properties of the F-polynomial are similar to those of the P-polynomial. This is summarized in the next result; it uses some of the notations from §2.

Proposition 6 (i) *$\overline{F(L)} = F(\overline{L})$, where $\overline{a} = a^{-1}$ and $\overline{x} = x$;*

(ii) *$F(L_1 + L_2) = F(L_1)F(L_2)$;*

(iii) *$F(L_1 U L_2) = \big((a^{-1} + a)x^{-1} - 1\big)F(L_1)F(L_2)$;*

(iv) *F is unchanged by mutation.*

A result that came as something of a surprise [12] was that both the P-polynomial and the F-polynomial contain the original polynomial of Jones, the V-polynomial. The result, which now has an easy proof, is:

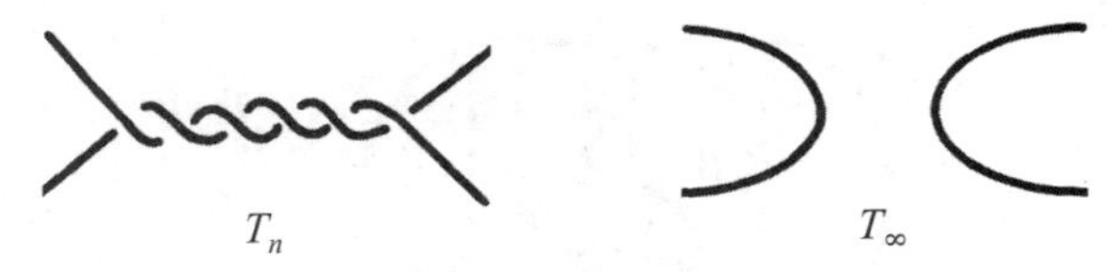

Figure 20.

Proposition 7 *For any oriented link L, the substitution*

$$(a, x) = (-t^{-3/4}, (t^{-1/4} + t^{1/4}))$$

reduces $F(L)$ to $V(L)$.

Proof. Adding together the two equations (1) from §3 gives

$$\langle \times \rangle + \langle \times \rangle = (A + A^{-1})\left(\langle)(\rangle + \langle \asymp \rangle\right)$$

Thus $\langle L \rangle$ satisfies exactly the same defining equations as $\Lambda(L)$ when $x = (A + A^{-1})$ and $a = -A^3$ (the latter arising from comparison of the effects of the first Reidemeister move on the two polynomials). Then just recall that it is the substitution $A = t^{-1/4}$ that produces the Jones polynomial.

5. Calculations, Problems and Tables

Calculations of any of the polynomials mentioned in previous sections can be performed 'by hand' for links with few crossings or for those with some simple pattern. Sometimes the linear nature of the formulae that define these polynomials can be exploited in a most pleasing way. That idea can be illustrated using the Q-polynomial to avoid orientation complications. Suppose that a link T_n contains as part of it the n-crossing twist as shown in Figure 20 (where by convention $n = \infty$ is also permitted, as illustrated); if n is negative the twist goes the other way.

Focussing on one of these crossings produces a quadruple of links L_+, L_-, L_0 and L_∞. (as in the defining formula for the Q-polynomial), namely, T_n, T_{n-2}, T_{n-1} and T_∞. The defining formula gives

$$Q(T_n) + Q(T_{n-2}) = x\big(Q(T_{n-1}) + Q(T_\infty)\big).$$

Thus

$$\begin{bmatrix} Q(T_n) \\ Q(T_{n-1}) \\ Q(T_\infty) \end{bmatrix} = M \begin{bmatrix} Q(T_{n-1}) \\ Q(T_{n-2}) \\ Q(T_\infty) \end{bmatrix} = M^n \begin{bmatrix} Q(T_0) \\ Q(T_{-1}) \\ Q(T_\infty) \end{bmatrix}$$

where M is the matrix

$$\begin{bmatrix} x & -1 & x \\ 1 & 0 & 0 \\ 0 & 0 & 1 \end{bmatrix}$$

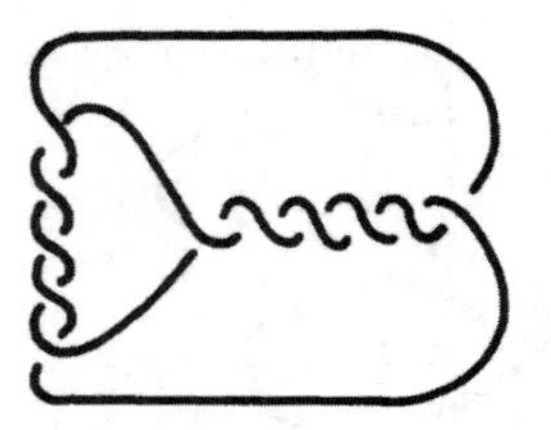

Figure 21.

As an exercise use this idea to calculate, in terms of the matrix M, the Q-polynomial of the link of Figure 21, where the twists have m and n crossings respectively.

In the following exercises L is an oriented link with $c(L)$ components. The proofs are all performed using, in the relevant defining formulae, induction on the number of crossings in a projection.

Exercises

(i) $P(L) = 1$ when $m = -(\ell + \ell^{-1})$.

(ii) In $P(L)$ the least power of m is $m^{1-c(L)}$.

(iii) $F(L) = (-2)^{c(L)-1}$ when $(a, x) = (1, -2)$.

(iv) $F(L) = (-1)^{c(L)-1}$ when $(a, x) = (i, x)$, $x \neq 0$.

(v) $V(L) = \Lambda(L)$ when $t = -1$.

(vi) If $c(L) = 1$, $F(L) = P(L)$ when $x = 0 = m$, and $a = \ell$.

Unanswered questions abound. Here are some of them.

Questions

(1) Can the new polynomials be defined without reference to diagrams in the plane?

(2) Can the P- and F-polynomials be defined in as simple a way as the bracket polynomial?

(3) Can any of the new polynomials for a link be calculated in one step (e.g., by means of a determinant) without working out many polynomials of simpler links?

(4) Is there some simple characterisation of what polynomials can arise as the P- or F- or V- or Q-polynomial of some link?

(5) Is there a nontrivial link of c components with the same polynomial (in the P, V, F, or Q sense) as the trivial unlink of c components?

(6) Does any of the new polynomials give information about the number of crossing switches needed to undo a knot (this is called its *unknotting number*)?

(7) Does there exist some grand master polynomial in which particular substitutions produce both P and F?

(8) Is there a "coloured" theory for P or F? This would be a theory that had more variables and which could distinguish, for example, a red trefoil linked with a blue unknot from a blue trefoil linked with a red unknot. There is such a variant of the Alexander polynomial.

(9) Are there polynomials other than P and F that can be defined along the same lines as they are defined? Several attempts have been made but all have turned out to be subtle variants of the original two polynomials.

Tables. On pages 272–274 are given tables of the P and F polynomials for a few knots and links of low numbers of crossings to give a feeling for what is involved.

References

1. J. W. Alexander, Topological invariants of knots and links, *Trans. Amer. Math. Soc.* 30 (1928), 275–306.
2. C. W. Ashley, *The Ashley Book of Knots*, Faber and Faber, 1947.
3. J. S. Birman, Braids, links and mapping class groups, *Ann. Math. Studies* 82 (1974).
4. R. D. Brandt, W. B. R. Lickorish, and K. C. Millett, A polynomial invariant for unoriented knots and links, *Invent. Math.* 84 (1986), 563–573.
5. G. Burde and H. Zieschang, *Knots*, de Gruyter, 1985.
6. J. H. Conway, An enumeration of knots and links, *Computational Problems in Abstract Algebra* (ed. J. Leech), Pergamon Press, 1969, 329–358.
7. P. Freyd, D. Yetter, J. Hoste, W. B. R. Lickorish, K. Millett, and A. Ocneanu, A new polynomial invariant of knots and links, *Bull. Amer. Math. Soc.* 12 (1985), 239–246.
8. C. F. Ho, A new polynomial for knots and links-preliminary report, *Abstracts Amer. Math. Soc.* 6, 4 (1985), 300.
9. V. F. R. Jones, A polynomial invariant for knots via von Neumann algebras, *Bull. Amer. Math. Soc.* 12 (1985), 103–111.
10. L. H. Kauffman, An invariant of regular isotopy, *Trans. Amer. Math. Soc.* 318 (1990), 417–471.
11. ——, State models and the Jones polynomial, *Topology* 26 (1987), 395–407.
12. W. B. R. Lickorish, A relationship between link polynomials, *Math. Proc. Cambridge Philos. Soc.*, (1986), 100, 109–112.
13. W. B. R. Lickorish and K. C. Millett, A polynomial invariant of oriented links, *Topology* 26 (1987), 107–141.
14. K. Murasugi, Jones polynomials and classical conjectures in knot theory, *Topology* 26 (1987), 187–194.
15. L. Neuwirth, The theory of knots, *Scientific American* (June 1979), 84–96.
16. K. Reidemeister, *Knotentheorie* (reprint), Chelsea, New York, 1948.
17. D. Rolfsen, *Knots and Links*, Publish or Perish, Wilmington, Delaware, 1976.
18. M. B. Thistlethwaite, A spanning tree expansion of the Jones polynomial, *Topology* 26 (1987), 296–309.
19. ——, Knot tabulations and related topics, In *Aspects of Topology* (Ed. I. M. James and E. H. Kronheimer) L.M.S. Lecture Notes 93 (1985), 1–76.

F & P for knots $\leqslant 7$ crossings, for links $\leqslant 6$ crossings.

		P	F
3_1		$(-2\ell^2-\ell^4)+\ell^2m^2$	$(-2a^2-a^4)+(a^3+a^5)x+(a^2+a^4)x^2$
4_1		$(-\ell^{-2}-1-\ell^2)+m^2$	$(-a^{-2}-1-a^2)+(-a^{-1}-a)x+(a^{-2}+2+a^2)x^2+(a^{-1}+a)x^3$
5_1		$(3\ell^4+2\ell^6)+(-4\ell^4-\ell^6)m^2+\ell^4m^4$	$(3a^4+2a^6)+(-2a^5-a^7+a^9)x+(-4a^4-3a^6+a^8)x^2+(a^5+a^7)x^3+(a^4+a^6)x^4$
5_2		$(-\ell^2+\ell^4+\ell^6)+(\ell^2-\ell^4)m^2$	$(-a^2+a^4+a^6)+(-2a^5-2a^7)x+(a^2-a^4-2a^6)x^2+(a^3+2a^5+a^7)x^3+(a^4+a^6)x^4$
6_1		$(-\ell^{-2}+\ell^2+\ell^4)+(1-\ell^2)m^2$	$(-a^{-2}+a^2+a^4)+(2a+2a^3)x+(a^{-2}-4a^2-3a^4)x^2+(a^{-1}-2a-3a^3)x^3$ $+(1+2a^2+a^4)x^4+(a+a^3)x^5$
6_2		$(2+2\ell^2+\ell^4)+(-1-3\ell^2-\ell^4)m^2+\ell^2m^4$	$(2+2a^2+a^4)+(-a^3-a^5)x+(-3-6a^2-2a^4+a^6)x^2+(-2a+2a^3)x^3$ $+(1+3a^2+2a^4)x^4+(a+a^3)x^5$
6_3		$(\ell^{-2}+3+\ell^2)+(-\ell^{-2}-3-\ell^2)m^2+m^4$	$(a^{-2}+3+a^2)+(-a^{-3}-2a^{-1}-2a-a^3)x+(-3a^{-2}-6-3a^2)x^2+(a^{-3}+a^{-1}+a+a^2)x^3$ $+(2a^{-2}+4+2a^2)x^4+(a^{-1}+a)x^5$

7_1

$(-4\ell^6-3\ell^8)+(10\ell^6+4\ell^8)m^2$
$\quad+(-6\ell^6-\ell^8)m^4+\ell^6m^6$

$(-4a^6-3a^8)+(3a^7+a^9-a^{11}+a^{13})x+(10a^6+7a^8-2a^{10}+a^{12})x^2+(-4a^7-3a^9+a^{11})x^3$
$\quad+(-6a^6-5a^8+a^{10})x^4+(a^7+a^9)x^5+(a^6+a^8)x^6.$

7_2 

$(-\ell^2-\ell^6-\ell^8)+(\ell^2-\ell^4+\ell^6)m^2$

$(-a^2-a^6-a^8)+(3a^7+3a^9)x+(a^2+3a^6+4a^8)x^2+(a^3-a^5-6a^7-4a^9)x^3+(a^4-3a^6-4a^8)x^4$
$\quad+(a^5+2a^7+a^9)x^5+(a^6+a^8)x^6.$

7_3 

$(-2\ell^{-8}-2\ell^{-6}+\ell^{-4})+(\ell^{-8}+3\ell^{-6}-3\ell^{-4})m^2$
$\quad+(-\ell^{-6}+\ell^{-4})m^4$

$(-2a^{-8}-2a^{-6}+a^{-4})+(-2a^{-11}+a^{-9}+3a^{-7})x+(-a^{-10}+6a^{-8}+4a^{-6}-3a^{-4})x^2$
$\quad+(a^{-11}-a^{-9}-4a^{-7}-2a^{-5})x^3+(a^{-10}-3a^{-8}-3a^{-6}+a^{-4})x^4$
$\quad+(a^{-9}+2a^{-7}+a^{-5})x^5+(a^{-8}+a^{-6})x^6.$

7_4° 

$(-\ell^{-8}+2\ell^{-4})+(\ell^{-6}-2\ell^{-4}+\ell^{-2})m^2$

$(-a^{-8}+2a^{-4})+(4a^{-9}+4a^{-7})x+(2a^{-8}-3a^{-6}-4a^{-4}+a^{-2})x^2+(-4a^{-9}-8a^{-7}-2a^{-5}+2a^{-3})x^3$
$\quad+(-3a^{-8}+3a^{-4})x^4+(a^{-9}+3a^{-7}+2a^{-5})x^5+(a^{-8}+a^{-6})x^6.$

7_5 

$(2\ell^4-\ell^8)+(-3\ell^4+2\ell^6+\ell^8)m^2$
$\quad+(\ell^4-\ell^6)m^4$

$(2a^4-a^8)+(-a^5+a^7+a^9-a^{11})x+(-3a^4+a^8-2a^{10})x^2+(-a^5-4a^7-2a^9+a^{11})x^3$
$\quad+(a^4-a^6+2a^{10})x^4+(a^5+3a^7+2a^9)x^5+(a^6+a^8)x^6.$

7_6 

$(1+\ell^2+2\ell^4+\ell^6)+(-1-2\ell^2-2\ell^4)m^2$
$\quad+\ell^2m^4$

$(1+a^2+2a^4+a^6)+(a+2a^3-a^7)x+(-2-4a^2-4a^4-2a^6)x^2+(-4a-6a^3-a^5+a^7)x^3$
$\quad+(1+a^2+2a^4+2a^6)x^4+(2a+4a^3+2a^5)x^5+(a^2+a^4)x^6.$

7_7

$(\ell^{-4}+2\ell^{-2}+2)+(-2\ell^{-2}-2-\ell^2)m^2+m^4$

$(a^{-4}+2a^{-2}+2)+(2a^{-3}+3a^{-1}+a)x+(-2a^{-4}-6a^{-2}-7-3a^2)x^2+(-4a^{-3}-8a^{-1}-3a+a^3)x^3$
$\quad+(a^{-4}+2a^{-2}+4+3a^2)x^4+(2a^{-3}+5a^{-1}+3a)x^5+(a^{-2}+1)x^6.$

		P	F
0_1^2		$(-\ell^{-1}-\ell)m^{-1}$	$(a^{-1}+a)x^{-1}-1$
2_1^2		$(\ell+\ell^3)m^{-1}-\ell m$	$(-a-a^3)x^{-1}+a^2+(a+a^3)x$
4_1^2		$(-\ell^3-\ell^5)m^{-1}+(3\ell^3+\ell^5)m-\ell^3m^3$	$(a^3+a^5)x^{-1}-a^4+(-3a^3-2a^5+a^7)x+(a^4+a^6)x^2+(a^3+a^5)x^3$
4_1^2		$(-\ell^{-5}-\ell^{-3})m^{-1}+(\ell^{-3}-\ell^{-1})m$	$(a^{-5}+a^{-3})x^{-1}-a^{-4}+(-3a^{-5}-2a^{-3}+a^{-1})x+(a^{-4}+a^{-2})x^2+(a^{-5}+a^{-3})x^3$
5_1^2	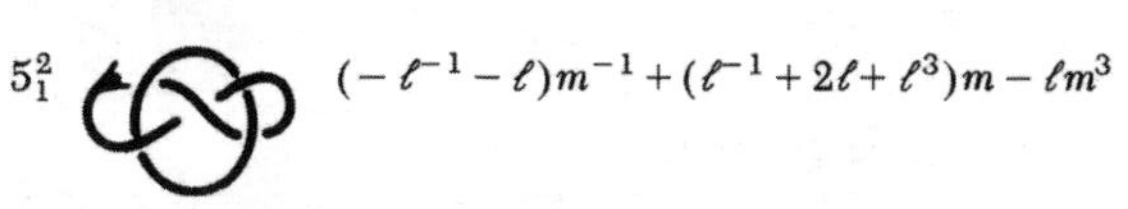	$(-\ell^{-1}-\ell)m^{-1}+(\ell^{-1}+2\ell+\ell^3)m-\ell m^3$	$(a^{-1}+a)x^{-1}-1+(-2a^{-1}-4a-2a^3)x+(-1+a^4)x^2+(a^{-1}+3a+2a^3)x^3+(1+a^2)x^4$
6_1^2		$(\ell^5+\ell^7)m^{-1}+(-6\ell^5-3\ell^7)m+(5\ell^5+\ell^7)m^3$ $-\ell^5m^5$	$(-a^5-a^7)x^{-1}+a^6+(6a^5+4a^7-a^9+a^{11})x+(-3a^6-2a^8+a^{10})x^2+(-5a^5-4a^7+a^9)x^3$ $+(a^6+a^8)x^4+(a^5+a^7)x^5$
6_1^2	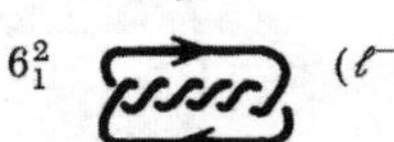	$(\ell^{-7}+\ell^{-5})m^{-1}+(-\ell^{-5}+\ell^{-3}-\ell^{-1})m$	$(-a^{-7}-a^{-5})x^{-1}+a^{-6}+(6a^{-7}+4a^{-5}-a^{-3}+a^{-1})x+(-3a^{-6}-2a^{-4}+a^{-2})x^2$ $+(-5a^{-7}-4a^{-5}+a^{-3})x^3+(a^{-6}+a^{-4})x^4+(a^{-7}+a^{-5})x^5$
6_2^2	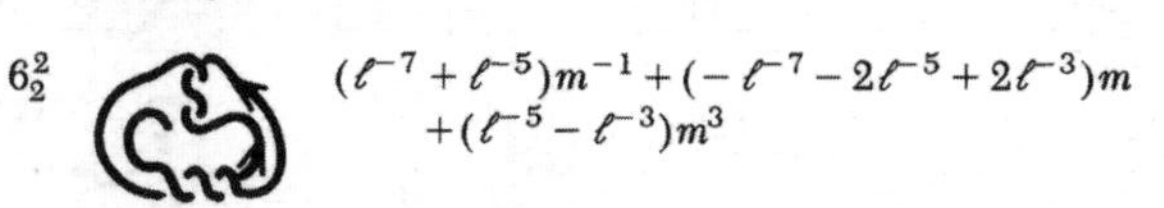	$(\ell^{-7}+\ell^{-5})m^{-1}+(-\ell^{-7}-2\ell^{-5}+2\ell^{-3})m$ $+(\ell^{-5}-\ell^{-3})m^3$	$(-a^{-7}-a^{-5})x^{-1}+a^{-6}+(-2a^{-9}+3a^{-7}+3a^{-5}-2a^{-3})x+(-a^{-8}-2a^{-6}-a^{-4})x^2$ $+(a^{-9}-2a^{-7}-2a^{-5}+a^{-3})x^3+(a^{-8}+2a^{-6}+a^{-4})x^4+(a^{-7}+a^{-5})x^5.$
6_3^2		$(-\ell^3-\ell^5)m^{-1}+(2\ell^3-\ell^5-\ell^7)m$ $+(-\ell^3+\ell^5)m^3$	$(a^3+a^5)x^{-1}-a^4+(-2a^3-a^5-a^9)x+(-3a^6-3a^8)x^2+(a^3+a^9)x^3$ $+(a^4+3a^6+2a^8)x^4+(a^5+a^7)x^5$
6_3^2		$(-\ell^{-5}-\ell^{-3})m^{-1}+(2\ell^{-3}+\ell^{-1}+\ell)m-\ell^{-1}m^3$	$(a^{-5}+a^{-3})x^{-1}-a^{-4}+(-2a^{-5}-a^{-3}-a)x+(-3a^{-2}-3)x^2+(a^{-5}+a)x^3$ $+(a^{-4}+3a^{-2}+2)x^4+(a^{-3}+a^{-1})x^5$

Briefly Noted

Some articles and notes we found to be of considerable interest, but with limited space we decided not to include the whole text. Here we give short descriptions of some and suggest that the reader check them out in the archives if they appear to be of interest. For further descriptions of outstanding pieces in the *Magazine*, we suggest reading our article "Twentieth-Century Gems from *Mathematics Magazine*" that appears in the *Magazine* 78 (2005), 110–123.

In "The theory of numbers for undergraduates" in 10 (1935/36), 53–57, Emory P. Starke of Rutgers University made a case for inclusion of number theory in the undergraduate curriculum, a rather natural suggestion. His opening paragraph includes the following quote: "A month's intelligent instruction in the theory of numbers ought to be twice as instructive, twice as useful, and at least ten times as entertaining as the same amount of 'calculus for engineers'." It is no surprise that these are the words of G. H. Hardy. What is curious is that these remarks are taken from Hardy's 1928 address to the American Mathematical Society in New York, his Josiah Willard Gibbs lecture. These are lectures given every year (with five exceptions) since 1923 in the field of applied mathematics! Was Hardy so prescient that he could predict the developments in cryptography that would make some parts of number theory into applied mathematics?

• Benjamin F. Finkel, founder of the *American Mathematical Monthly*, wrote a series of eighteen articles for the *Magazine* between 1940 and 1942 on the early history of American mathematical journals. How many readers are even aware of the existence of *The Mathematical Companion*, *The Mathematical Miscellany*, the *Cambridge Miscellany* (referring to Cambridge, Massachusetts, not England), and the *Mathematical Monthly* (not the *American Mathematical Monthly* founded by Finkel over fifty years later)? All of these predated the *American Journal of Mathematics*, founded by Sylvester in 1876. These very early publications were not research journals, as Sylvester's was, but they appear not to have been entirely pedagogical or expository either, and some significant research mathematicians were involved. For example, the *Cambridge Miscellany* was edited by Benjamin Peirce, known for his work on linear associative algebras. He was a major contributor to the development of algebra, and one of the first American mathematicians to be recognized abroad. These eighteen articles provide a fascinating glimpse into

19th century American mathematics.

Finkel's founding of the *Monthly* led indirectly to the founding of the MAA. He established the *Monthly* in 1893 and when it encountered hard times in the first part of the 20th century, the American Mathematical Society was approached as a possible sponsor. The AMS declined and a group in the Midwest got together at Ohio State in 1915 to organize the MAA in order to sponsor the *Monthly*.

• Robert H. Cameron of MIT won the Chauvenet Prize for a long article entitled, "Some introductory exercises in the manipulation of Fourier transforms," 15 (1940/41), 331–56. The paper shows how to use Fourier transforms to solve difficult differential or integral equations such as the nonlinear integral equation

$$2f(x)+\int_{-\infty}^{\infty} f(x-t)f(t)\,dt = \frac{4x^2+10}{\pi(x^4+5x^2+4)},$$

which is to be solved for the unknown function $f(x)$. The author focuses on *how* to operate with Fourier transforms, stating that for simplicity, "we shall in this introductory paper lay aside all ideas of mathematical rigor." This includes not specifying the type of integral, for example, Lebesgue, or the space, such as L_1 or L_2, in which the functions lie. (Would such an article even be nominated for a Chauvenet Prize today?) The paper does give arguments for most of the nine formal properties of Fourier transforms that are presented and used. And in its conclusion Cameron warns that while in elementary calculus lack of rigor and formal use of limit theorems "may lead to false conclusions, but usually not," purely formal work with Fourier transforms "is sure to lead one into difficulties, and rather soon at that." (About halfway through the article, after using improper integrals and defining the Fourier transform, the author defines the absolute value function $|x|$ in a footnote!)

• An attractive pedagogical article on "Normal curve areas and geometric transformations" by David Gans, 31 (1957/58), 205–6, compared area under an arbitrary normal curve (mean μ, standard deviation σ)

$$y = \frac{1}{\sigma\sqrt{2\pi}}e^{-\frac{1}{2}\left(\frac{x-\mu}{\sigma}\right)^2} \qquad (1)$$

with area under the standard normal curve (mean 0, standard deviation 1)

$$z = \frac{1}{\sqrt{2\pi}}e^{-\frac{1}{2}t^2}. \qquad (2)$$

The article explained geometrically why the area under (1) from x_1 to x_2 is equal to the area under (2) from t_1 to t_2, where $t_i = \frac{x_i-\mu}{\sigma}$. The explanation is based on decomposing the affine transformation of the plane given by $t = \frac{x-\mu}{\sigma}$, $z = \sigma y$ into three simpler transformations:

(a) a horizontal translation by μ,
(b) a horizontal scaling by $1/\sigma$,
(c) a vertical scaling by σ.

Since (a) preserves area, (b) multiplies area by $1/\sigma$, and (c) multiplies area by σ, then their composition preserves area.

• A problem in the *Monthly* in 1953 (E1054) asked for the probability that the roots of the quadratic equation $x^2+bx+c=0$, $-k<b$, $c<k$, are imaginary. A solution by MAA secretary, Harry M. Gehman, was published: the probability is $(12-k)/24$ if $k\le 4$ and $2/(3\sqrt{k})$ if $k\ge 4$. One of the solvers observed that something curious happens when $k\to\infty$; this probability is zero, or the probability that the roots are real is 1. The *Magazine* probably asked the right question in an article by Joseph W. Andrushkiw: "The probability that the roots of a real quadratic equation $ax^2+bx+c=0$ lie inside or on the circumference of the unit circle in the complex plane," 32 (1958/59), 123–28. Note that here the equation is no longer monic. The author finds that the probability of the equation having real roots is $41/72+(\log 2)/12 = .6272\ldots$, and if transformed to the question in the title, the probability the roots lie inside or on the circumference of the unit circle is $7/24 = .291\overline{6}$.

• For fixed a the function defined by

$$f(n) = \int_a^t x^n\,dx = \frac{t^{n+1}-a^{n+1}}{n+1}$$

for $n\ne -1$, and

$$f(-1) = \int_a^t x^{-1}\,dx = \ln t - \ln a$$

appears to have a discontinuity at $n=-1$. In M. J. Pascual's "Note on $\int_a^x t^y\,dt$," 35 (1962), 175, a single application of l'Hôpital's rule as n approaches -1 shows that f is indeed continuous at $n=-1$.

• In the classic *Mathematical Snapshots* (Stechert-Hafner, New York, 1938, p. 132), H. Steinhaus gave a proof of the irrationality of $\sqrt{2}$ that used inequalities instead of properties of divisibility. In "A method of establishing certain irrationalities" 37 (1964), 208–10, Ivan Niven and Eugene A. Maier used a similar approach to generalize first to the nth root of any positive integer, and then to real algebraic integers. The latter result asserts that any algebraic integer that is a real number yet not an ordinary integer is irrational.

• In "The maximum diameter of a convex polyhedron," by E. Jocovič and J. W. Moon, 38 (1965), 31–32, 3-dimensional convex polyhedra are considered as graphs and the "distance" between two vertices is defined to be the length of the shortest path between them on the graph itself, where adjacent vertices are a distance 1 apart. The diameter of a polyhedron is then the greatest distance between any two vertices; a cube, for example, has diameter 3 and a regular dodecahedron has diameter 5. Using Menger's theorem from graph theory, the authors prove that the maximum diameter of all 3-dimensional, convex polyhedra with n vertices is $\lfloor \frac{n+1}{3} \rfloor$. For the case $n = 2 \pmod 3$ this bound is achieved by a triangular prism with tetrahedral caps at each end, as shown below.

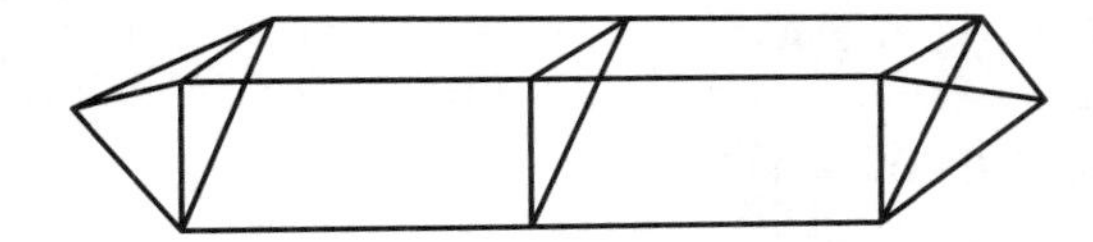

• In "Boolean matrices and switching nets" by Wai-kai Chen, 39 (1966), 1–8, a Boolean matrix is not just a matrix of 0s and 1s, as in some contemporary texts, but a matrix with entries from a general Boolean algebra B. The Boolean matrices of a fixed size themselves form a Boolean algebra, where the Boolean operations are defined entry-wise. Further, ordinary matrix multiplication, defined with the Boolean operations of B instead of $+$ and $\cdot$, yields some strange properties. For example, a Boolean matrix A is invertible if and only if it is orthogonal (meaning $A^{-1} = A^T$). The article reviews Boolean matrix algebra, including properties of the so-called or-determinant of a square Boolean matrix. The author then applies these properties in switching theory for the analysis and synthesis of combinatorial circuits.

• Mathematical folklore, Paul B. Yale observes in 39 (1966), 135–41, suggests that there are many automorphisms of the field of complex numbers $\mathbb{C}$. That is, many bijective maps of $\mathbb{C}$ preserve both addition and multiplication. The two obvious automorphisms are the identity map and the conjugation map; Yale calls every other automorphism of $\mathbb{C}$ a wild automorphism. Is this name appropriate? Yale proves that any wild automorphism maps the real line onto a dense subset of the complex plane, even though it leaves fixed all rational numbers! And, there really are many such automorphisms, since a Zorn's lemma argument shows that any automorphism of a subfield of $\mathbb{C}$ can be extended to an automorphism of $\mathbb{C}$ itself.

• The Malfatti problem, proposed in 1803, reduces to the problem of inscribing three nonoverlapping circles in a given triangle so that the sum of their areas is maximized. Malfatti and many others after him assumed that the solutions would consist of mutually tangent circles such as in Figure 1a, and such circles became known as Malfatti circles. It wasn't until 1930 that Lob and Richmond (On the solutions of Malfatti's problems for a triangle, *Proc. London Math. Soc.* 322 (1930), 287–304) noted that the Malfatti circles did not always solve the Malfatti problem. For an equilateral triangle an arrangement such as in Figure 1b beats the Malfatti circles, while for a wide, short triangle an arrangement such as in Figure 1c beats them.

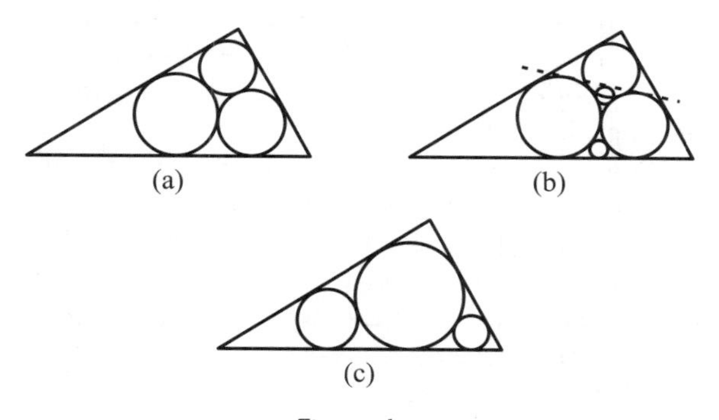

Figure 1.

In "On the original Malfatti problem," 40 (1967), 241–47, Michael Goldberg proves that the Malfatti circles never solve the Malfatti problem! Goldberg gives a construction, which he calls the "Lob-Richmond-Goldberg construction," that always yields a larger area than the Malfatti circles.

• A powerful technique that sometimes helps solve difficult combinatorial problems, Putnam-level problems, and so on, is the use of graph theory. An example of such a problem was the aptly-named puzzle

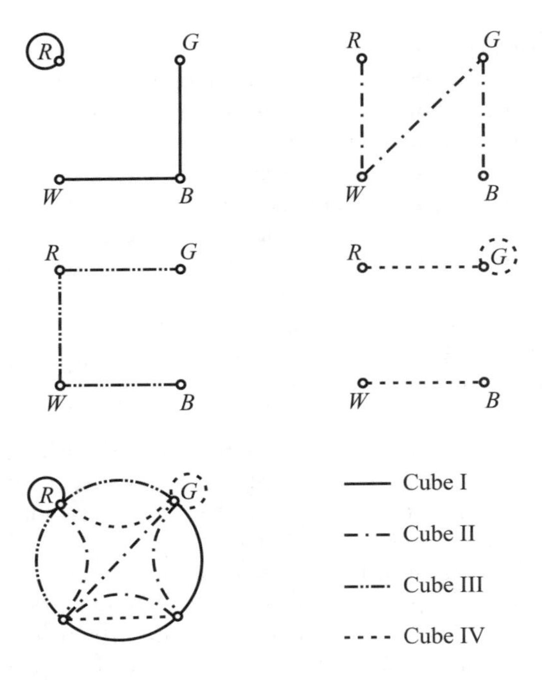

"Instant Insanity," that was common in school mathematics labs in the 1960s. With its 82,944 possible configurations of blocks, "Instant Insanity," as with the Rubik Cube a decade later, was not accessible to most trial-and-error problem-solving approaches. In "A diagrammatic solution to 'Instant Insanity' problem," 44 (1971), 119–124, A. P. Grecos and R. W. Gibberd represent the puzzle by a four-vertex connected multigraph where the vertices are the four colors of block faces and an edge connects colors of the opposite faces of a block.

Holistic inspection of the graph almost immediately yields both solutions to the "Instant Insanity" puzzle. The authors use the same procedure to solve a more difficult six-block puzzle that was given as a challenge exercise in an earlier article in the *Magazine*. A similar but more concise graphical solution to Instant Insanity appeared in Robin Wilson's *Introduction to Graph Theory*, 4th edition (Addison-Wesley Longman, Ltd., Edinburgh, 1996, pp. 23–25).

• In teaching a mathematics appreciation class it's helpful to have a very simple, yet interesting, game to illustrate the use of a strategy based on symmetry. Such a game is Lose Tic-Tac-Toe, discussed in Daniel I. A. Cohen's "The solution of a simple game," 45 (1972), 213–17. The game is played the same as regular tic-tac-toe, except that the first player to get three on a line loses. An optimal strategy, which guarantees at least a draw, is to go first and play in the center (note that at first glance this is counterintuitive), then play symmetrically to what the opponent plays. The author uses four pages in order to show that this is the only optimal strategy.

• "Seven game series in sports," by Richard A. Groeneveld and Glen Meeden, 48 (1975), 187–92, is an example of probability modeling–here two competing models are discussed–that could be helpful in teaching introductory probability and statistics courses. The seven game series discussed are the best-of-seven championship series played in professional hockey (in the NHL), basketball (in the NBA), and baseball (the World Series) and both models are compared with actual data over many decades. The first model is a simple Bernoulli trial model with a single parameter p, the probability that the stronger team wins any single game. In this model the maximum of the expected values of the length of the series is 5.8125 games, which occurs for $p = 1/2$. The second model includes a second parameter, $\lambda = P$ (series ends in 6|series lasts longer than 5), which is introduced since both baseball and basketball data indicate that the independence assumption in the first model is violated due to "the apparent critical role of the sixth game." For these two sports (but not hockey) empirical data give estimates of λ that are, surprisingly, less than $1/2$.

• When Joseph A. Gallian wrote his survey article "The search for finite simple groups," 49 (1976), 163–79, the classification of all finite simple groups had not been completed. He presents a masterful historical account, accessible to undergraduates, up to 1975, when the classification was complete for groups with order less than a million. It takes three full pages in the article just to tabulate the finite simple groups known then, including their types and orders. (Here "simple group" means nonabelian simple group, thus excluding cyclic groups of prime order.) Some of these groups, like the Monster and the Baby Monster, have orders that are larger than 10^{30}. Otto Hölder had an easier task in 1892 when he initiated what Gallian calls the range problem. Hölder proved that the only simple groups with orders not exceeding 200 were the alternating group A_5 of order 60 and the projective special linear group PSL(2,7) of order 168. For his detailed and scholarly contribution to what was then known about finite simple groups, Gallian was awarded one of the first two Allendoerfer Awards.

• In "Ramanujan's notebooks," 51 (1978), 147–64, Bruce C. Berndt briefly surveys the life of Srinivasa Ramanujan, including the events involving the survival of his famous notebooks, sometimes against all odds. Before their eventual publication, handwritten and photostat copies were made, but getting them safely back and forth between India and England was not guaranteed. S. S. Pillai carried one of three photostat copies when he traveled to the International Congress at Harvard in 1950. The plane crashed in Egypt; Pillai was killed and the notebooks were lost. But by some miracle copies did survive until publication and over recent decades they have been much studied. Berndt provides a generous sampler of formulas from the notebooks. How Ramanujan came up with some of these remains a mystery. In his formal manipulation of series he is compared to Euler and Jacobi. Klein is quoted: "In a certain sense, mathematics has been advanced most by those who are distinguished more for intuition than for rigorous methods of proof."

• Where $f(z) = z + a_2z^2 + a_3z^3 + \cdots$ is a power series in z with complex coefficients $a_2, a_3, \ldots$, we assume that $f(z)$ is one-to-one on the domain of convergence of the series. Ludwig Bieberbach conjectured in 1916 that for such functions, $|a_n| \leq n$, $n = 2, 3, \ldots$ and if for any n, $|a_n| = n$, then f is a "rotation" of the Koebe function. His conjecture on these so-called schlicht (univalent) functions became one of the most celebrated unsolved problems in analysis until it was proved by Louis de Branges in 1984. For classical analysts this was the 1980s equivalent of Wiles' proof of the Fermat conjecture in the 1990s. A beautiful article by Paul Zorn, giving the history of the problem and commenting on the proof, was published in the *Magazine*, 59 (1986), 131–48. The eventual proof was all the more dramatic because it accomplished for all coefficients what had occupied the time and attention of various first-rate mathematicians who had attacked the problem one coefficient at a time—Karl Loewner, Paul Garabedian, M. M. Schiffer, and Donald Spencer, among others. And it turns out that Loewner's very early work on the problem (1923) provided a key part of de Branges' proof.

• In "A curious mixture of maps, dates, and names," 60 (1987), 151–58, J. M. Sachs reflects on the curious fact that we assume that the Mercator projection of 1569 involves logarithms, but Napier did not publish his *Mirifici Logarithmorum Canonis Descriptio* until 1614. Further, careful construction of the projection requires some calculus, even differential geometry. Edward Wright, a Cambridge mathematician and consultant to the East India Company, was the connection between Mercator and Napier and this story is largely about him and his contributions. The author explains how Mercator solved some of the projection problems without using calculus, which didn't come along till 1666 or 1684, whichever winner you choose in the Newton-Leibniz controversy, or differential geometry which came even later with the work of Gauss.

• J. I. Katz in "How to approach a traffic light," 63 (1990), 226–30, raises the question of the "lazy bicyclist, or thrifty motorist": if a red traffic light is spotted in the distance, how should one proceed? Should one maintain constant speed in the hope that the light will turn green before one reaches the light, or should one brake in order to improve the chances that the light will change to avoid having to stop." A similar problem exists when one sees a green light. The solution: "Do not slow down in anticipation of a sighted red light or race to meet a green one. Instead, roll until the light... is reached, or until its color is observed to change." Of course, in real life things are complicated by friction and traffic conditions.

• Following a popular article "The Strong Law of Small Numbers," that appeared in 1988 in the *Monthly*, Richard K. Guy in 63 (1990), 3–20, published his "The Second Strong Law of Small Numbers." The first law states: "There aren't enough small numbers to meet the many demands made of them." The second law says, "When two numbers look equal, it ain't necessarily so!" The author provides 45 examples to support his second law.

• In "Indeterminate forms revisited," 63 (1990), 155–59, Ralph Boas, Jr., explores the first calculus text, *L'Analyse des Infiniments Petits*, published anonymously by the Marquis de l'Hôpital in 1696. He points out that this is not easy reading and gives an example, "One can substitute, one for the other, two quantities which differ only by an infinitely small quantity; or (what amounts to the same thing) a quantity that is increased or decreased only by another quantity infinitely less than it, can be considered as remaining the same." Boas remarks that this "sort of presentation gave calculus a reputation, which has survived to modern times, of being unintelligible." It's small wonder that J. J. Sylvester

wrote that when he was young (about 1830), "a boy of sixteen or seventeen who knew his infinitesimal calculus would have been almost pointed out in the streets as a prodigy like Dante, who had seen hell."

• Inspired by an exhibit in San Francisco's Exploratorium (a science museum), Leon Hall and Stan Wagon, in "Roads and wheels" 65 (1992), 283–301, examined the question of designing a roadway on which a bicycle with square wheels would provide a smooth ride. They actually produced a roadway and a bicycle to demonstrate this curious phenomenon. In the article they go on to generalize this to a wide variety of wheels and roads. Would everyone guess that a limaçon rolling along a trochoid would yield a cycloidal locus?

A similar photograph that appeared in the *American Mathematical Monthly* 105 (1998), 787, makes it clear that the bicycle is technically a tricycle. More pictures of Stan Wagon "taking a ride," and mention of his entry in *Ripley's Believe It or Not!* are in his later article, "The ultimately flat tire," *Math Horizons*, February 1999, pp. 14–17.

The Problem Section

Problems have been posed in the *Magazine* almost from its inception. A separate problems section was initiated in volume 6, issue 6 (1930/31). Problem editors have been: T. A. Bickerstaff, C. W. Trigg, Robert E. Horton, Dan Eustice, Leroy F. Meyers, Loren C. Larson, George T. Gilbert, and Elgin H. Johnston.

Skimming over the Problem Sections of the *Magazine*, one finds lots of the expected names from the "problems community": Leon Bankoff, Leonard Carlitz, Nathan Altshiller Court, Howard Eves, Richard K. Guy, Murray Klamkin, Leo Moser, Charles W. Trigg, and Victor Thébault, names one would probably also find in the *Monthly* in this period. But we also find some names we might not expect: David Blackwell, Eugenio Calabi, George B. Dantzig, Jr., Louis de Branges, and Mark Kac.

It is interesting to look over some of the problems to see how they sometimes get recycled over and over again. One cannot help but wonder where some of them came from originally. For example, in 1939, while a student in Lvov, Mark Kac, 13 (1938/39), 293, proposed the problem of showing that if one trisects the three sides of a triangle and then chooses the first trisection point on each side as one moves around the boundary, connecting them in turn to the respective opposite vertices, the triangle inside determined by these three cevians has one-seventh the area of the original triangle. Kac or the editor pointed out that this problem had previously appeared in a St. Petersburg examination in 1912. And after its publication in the *Magazine* it appeared again (though in slightly more general form) as problem A-3 in the 1962 Putnam Competition (A. M. Gleason, R. E. Greenwood, and L. M. Kelly, *The William Lowell Putnam Mathematical Competition: Problems and Solutions: 1938–1964*. Mathematical Association of America, Washington, DC, 1980, pp. 63, 556–58).

David Blackwell, 10 (1935/36), 67, while still a student at the University of Illinois, solved a diophantine equation resulting from the dyslexic bank teller's confusing dollars and cents when cashing a check (that is, the teller gives the customer as many dollars as there are cents on the check, and vice versa). The customer then pays a bill for \$24.11 after which he finds he has twice as much money as the amount of the original check. What amount was the check made out for? Show that the solution is unique. The problem is now a standard exercise in number theory texts.

In 1954 the inveterate problem poser, Murray Klamkin asked readers to show that

$$F(x, y) = \sum_{n=0}^{\infty} \frac{(-1)^n x^n}{a^{n+1} + y}$$

is symmetric in x and y. George Pólya, by then a professor emeritus at Stanford, sent in a solution (the only one, except for the proposer), 28 (1954/55), 235–36. It is not surprising that the author of *How To Solve It* should break down his solution into three parts: (1) a heuristic consideration in which he outlines a natural approach to the problem, (2) a proof, and (3) a critique in which he discusses values of a, such as $a = 0$, for which the statement is not true.

Twenty-five years later another Klamkin problem had only one non-computer solution submitted, and this time the solver was none other than Paul Erdős! The problem asked if there existed a prime number such that if any digit (in base 10) were changed to any other digit then the resulting number would be composite. Erdős answered yes and proved a slightly stronger result. Further, he proposed several questions of his own. The solution, listed as "Erdős and the computer," 52 (1979), 180–82, included 294,001 as the smallest such prime that was found by a computer search.

At one time solutions to the problems in the *Magazine* were published upside down in order to encourage those reading a problem to try to solve it rather than taking the easy way out and just reading the solution.

Index

Abel, Niels Henrik, 9, 11, 13, 19–20, 61, 81, 86, 87, 88, 92, 205, 218
Académie des Sciences, Paris, 5, 6, 59, 64, 81, 87, 88, 199, 208
Academy of Science, Vienna, 5
Academy of Sciences, Berlin (Prussian Academy), 57, 83, 88, 199, 200
Academy of Sciences, St. Petersburg, 201, 208
Academy of Sciences, U.S.S.R., 199, 201, 207
Acta Mathematica, 90
Ahlfors, Lars Valerian, 11, 13, 16
Albert, A. Adrian, 35
Alexander, James Waddell, 257
Alexander polynomials, 265
al-Khowarizmi, 238
Allendoerfer, Carl Bennett, 109
American Mathematical Society, 5, 23, 45, 161, 257, 275, 276
Andrews, George E., 199
Apollonius of Pergassus, 49
Appell, Paul Émile, 90
Archimedean tilings, 177
Archimedes, 9, 29
Argand, Jean Robert, 144–145
Artin, Emil, 143, 257

Babbage, Charles, 240
Banach, Stefan, 16
Beckenbach, Edwin Ford, 161
Bell, Eric Temple, 9, 10, 51, 79, 223
Bellman, Richard Ernest, 95
Beltrami, Eugenio, 85
Bergman, George, 99
Berndt, Bruce C., 279
Bernoulli, Daniel, 199, 201, 204, 205, 207
Bernoulli, Johann, 202, 205–206
Bernoulli family, 9, 11, 86, 207
Bertrand, Joseph Louis François, 90, 202
Betti, Enrico, 85, 91
Bézout, Étienne, 214
binary quadratic forms, 56
Birkhoff, George David, 16, 35, 89
Bjerknes, Carl Anton, 16
Bjerknes, Vilhelm, 16
Blackwell, David Harold, 281
Blaschke, Wilhelm, 191
Boas, Ralph Philip, Jr., 23, 279
Bohr, Harald, 23
Bolyai, Farkas, 86
Bolyai, János, 86, 202
Bombelli, Rafael, 144
Bonnet, Ossian Pierre, 90
Boole, George, 9, 85, 169, 224
Borel, Émile, 16, 90, 91
Botts, Truman, 12
Bouquet, Jean-Claude, 82
Braden, Bart, 229
Brahe, Tycho, 73
Brahmagupta, 52
Brioschi, Francesco, 85
Briot, Charles Auguste-Albert, 82
Buck, Robert Creighton, 11
Bunyakovskii, Viktor Yakovlevich, 85, 86
Burckhart, Johann Jacob, 199, 201
Burnside, William Snow, 163

Calabi, Eugenio, 281
Cantor, Georg, 9, 226
Carathéodory, Constantin, 89, 91
Cardano, Hieronimo, 77
Carlitz, Leonard, 281
Cartan, Élie Joseph, 13, 14
Cartwright, Dame Mary, 11
Casorati, Felice, 87
Cauchy, Baron Augustin-Louis, 5, 9, 53, 55, 57, 64, 65, 82, 86, 87, 95, 191, 195, 205, 213, 218, 219, 232, 243
Cayley, Arthur, 5, 9, 85, 146, 148, 213, 216, 223, 224, 225, 226
Cayley numbers, 148
Chakerian, Gulbank D., 151
Chasles, Michel, 82
Chebyshev, Pafnuti L'vovich, 85
Chern, Shiing-Shen, 35, 257
Christian IV, King of Norway, 11
Clairaut, Alexis Claude, 207, 208
Clebsch, Rudolf Friedrich Alfred, 83
Clifford, William Kingdon, 57, 242
Cohn-Vossen, Stefan, 192
Collège de France, 82
Commentarii Mathematici Helvetici, 199
commutative ring theory, 137
Comptes Rendus de l'Académie des Sciences, Paris, 5, 7, 82

Condorcet, Marquis de, 239, 243, 244
Connelly, Robert, 195
Copernicus, Nicolaus, 71–79, 240
Coxeter, Harold Scott McDonald, 177
Crelle, August Leopold, 61, 83
Crelle's Journal, 61, 83, 86, 89
Cremona, Antonio Luigi, 85
Curtis, Charles Whittlesey, 299
Curtiss, David Raymond, 161
Curtiss, John Hamilton, 161

d'Alembert, Jean Le Rond, 204, 205, 207, 208
Dantzig, George, 161, 281
Darboux, Jean Gaston, 90
de Branges, Louis, 279, 281
Dedekind, (Julius Wilhelm) Richard, 9, 10, 51, 52, 54, 55, 56, 57, 63, 64, 65, 84, 88, 223, 226
Dedekind inversion formula, 57
Dehn, Max, 191
Delisle, Joseph Nicholas, 207
De Morgan, Augustus, 85
Descartes, René, 9, 78, 109, 214, 237, 239, 243
Dickson, Leonard Eugene, 29, 30
Diophantus of Alexandria, 49, 52
Dirichlet, Peter Gustav Lejeune-, 53, 55, 58, 64, 66, 83, 84, 92, 220
division algebra, 143
Douglas, Jesse, 11, 13
Dowling, Roy J., 115
Dresdner, Arnold, 20
Dudley, Underwood, 133, 199
Dunnington, G. Waldo, 9, 11, 23
Dupin, Pierre François Charles, 82
Dürer, Albrecht, 9

Edwards, Harold M., 199
Eisenhart, Luther Pfahler, 11, 16
Eisenstein, Ferdinand Gotthold, 62, 66, 84
Emch, Arnold, 5
Eneström, Gustaf, 200
Engel, Friedrich, 14
Erdős, Paul, 139, 199
Escher, M. C., 177
Euclid, 55, 79, 238, 248
Euclid's algorithm, 58
Eudoxus, 9
Euler, Leonhard, 9, 52, 54, 57, 58, 59, 83, 86, 87, 89, 144, 148, 171, 199, 201, 203, 205, 207–208, 215, 247

Faraday, Michael, 207
Fehr, Henri, 16
Feit–Thompson theorem, 226
Fellmann, Emil A., 201, 205
Fermat, Pierre de, 9, 59, 201, 205, 237
Fermat's last theorem, 57, 58, 59, 64
Fibonacci numbers, 99–100
Fields medals, 11, 13
Fourier, Jean-Baptiste-Joseph, 9, 82, 84, 208
Frank, Philipp, 84
Fréchet, René Maurice, 16, 90
Frederick the Great, 199, 209
Fredholm, Erik Ivar, 92
Frege, Friedrich Ludwig Gottlob, 170
Frobenius, Ferdinand Georg, 221–222, 225, 226
Fuchs, Immanuel Lazarus, 91
Fulkerson, Delbert Ray, 161
fundamental theorem of algebra, 56
Fuss, Nicholas, 200
Fuss, Paul-Heinrich, 200
Füter, Karl Rudolph, 13

Galileo Galilei, 76–77, 240
Gallian, Joseph A., 278
Galois, Évariste, 9, 55, 60, 63, 66, 82, 216, 218
Galois theory, 143, 214, 217, 218, 219, 222, 223
Garabedian, Paul, 279
Gardner, Martin, 175, 181, 185
Garrett, Zena, 3
Gauss, Carl Friedrich, 9, 10, 51–66, 83, 84, 144, 145, 214, 215, 216, 220, 242
Gaussian integers, 55
Gelfond, Aleksandr Osipovich, 14, 204
George II, King of England, 83
Gergonne, Joseph-Diaz, 82
Gibbs, Josiah Willard, 85, 148, 275
Giornale di Matematiche, 7
Gluck, Herman, 191
Gödel, Kurt, 173–174
Goldbach, Christian, 148, 199, 201
Goldberg, Michael, 191, 275
Goursat, Edouard Jean-Baptiste, 90
Grabiner, Judith V., 237
Grassmann, Hermann Günther, 5, 88, 216
Graves, John T., 145, 148
Green, George, 84
Grünbaum, Branko, 177, 179, 190, 247
Gudermann, Christophe, 88
Guy, Richard K., 279

Haakon VII, King of Norway, 11, 12, 13
Habicht, Walter, 201, 205
Hadamard, Jacques Salomon, 82, 89, 90, 92
Halmos, Paul Richard, 35, 169, 257
Hamilton, Sir William Rowan, 9, 63, 84, 143–147, 222
Hardy, Godfrey Harold, 89, 93, 204, 257
Hasse, Helmut, 16
Hausdorff space, 36
Hecke, Erich, 16
Heesch, Heinrich, 179
Heine, Heinrich Eduard, 88
Helmholtz, Hermann von, 242
Hemenway, David, 255
Hensel, Kurt, 54

Hermann, Jacob, 207
Hermite, Charles, 9, 82, 90
Hesse, Ludwig Otto, 83
Hilbert, David, 16, 51, 66, 83, 89, 92, 93, 95, 192
Hille, Einar, 81
Hirschhorn, Michael, 185, 190
Hölder, Otto, 225, 226
Hölder's inequality, 97, 98, 213
Holmgren, Erik Albert, 92
Hôpital, Marquis de l', 279
Hurwitz, Adolf, 55, 83
Huygens, Christiaan, 201
Hypatia of Alexandria, 45-50

ideal numbers, 65
Institut de France, 7

Jacobi, Carl Gustav Jacob, 9, 10, 60, 61, 66, 83, 86, 87, 88, 200
Jahrbuch über die Fortschritte der Mathematik, 23
James, Richard, 183
Jones, Vaughn Frederick Randal, 258
Jonquières, Ernest de, 5, 7
Jordan, Camille, 90, 213, 219–220, 226
Jordan-Hölder theorem, 51, 90, 213, 225, 226
Journal für die reine und angewandte Mathematik (Crelle's Journal), 5, 60, 83, 86
Julia, Gaston Maurice, 16

Kac, Mark, 257, 281
Kaluza, Theodore, Jr., 83
Kaplansky, Irving, 35
Karl Johann, King of Norway, 13
Kepler, Johannes, 73–76, 208, 240
Kershner, R. B., 175, 179–183, 193
Khintchin, Aleksandr Yakovlevich, 16
Kienzle, Otto, 179
Klamkin, Murray, 159, 281, 282
Klein, Felix, 19, 26, 84, 90, 91, 92, 214, 215, 216, 217, 219, 222, 225
Kleiner, Israel, 213
Kline, Morris, 71, 83
Knopp, Konrad, 83
Knorr, Wilbur, 45
Koch, Helge von, 92
Koebe, Paul, 89
Kovalevskaya, S'ofya, 9
Krantz, Steven G., 35
Kronecker, Leopold, 9, 55, 56, 64, 66, 84, 220, 221
Kummer, Ernst Eduard., 9, 52, 54, 55, 57, 58, 65, 66, 83, 84, 220

Lagrange, Joseph Louis, Comte de, 9, 52, 57, 58, 59, 82, 83, 86, 201, 207, 208, 209, 213, 215, 216, 217, 218, 237, 241, 243
Laguerre, Edmond Nicolas, 90
Lamé, Gabriel, 53, 58, 59, 64
Landau, Edmund Georg Herman, 10, 91
Lange, Lester Henry, 151
Laplace, Pierre Simon, Marquis de, 9, 82, 242
Laugwitz, Detlef, 202
Laurent, Pierre Alphonse, 86
Lebesgue, Henri Léon, 92
Lebesgue integral, 123, 129–131
Lebesgue measure, 129
Legendre, Adrien-Marie, 82, 86, 143, 148
Lehmer, Derrick Henry, 29, 129
Lehmer, Derrick Norman, 29
Leibniz, Gottfried Wilhelm, Freiherr von, 9, 10, 84, 207, 239, 240
Leonardo da Pisa (Fibonacci), 52
Lickorish, William Bernard Raymond, 257
Lie, Marius Sophus, 13–15, 214, 217, 222
Lie groups, 143, 215
Lietzmann, Walter, 14
Lindelöf, Ernst Leonhard, 92
Liouville, Joseph, 64, 67, 82, 92
Littlewood, John Edensor, 93, 205
Lobachevskii, Nikolai Ivanovich, 9, 86, 242
Loewner, Karl, 279
Lucas-Lehmer test, 29

Mac Lane, Saunders, 35
Maclaurin, Colin, 237
Magid, Andy M., 137
Mangoldt, Hans Karl Friedrich von, 89
Mathematical Reviews, 23
Mathematische Annalen, 83
Mathieu, Émile Léonard, 90
Maud, Queen of Norway, 10, 11, 13
Maupertuis, Pierre Louis Moreau de, 207
Maxwell, James Clerk, 79, 84, 207, 242
Mazur, Barry, 257
Méray, Charles, 88
Mersenne, Marin, 3
Mersenne primes, 29
Miller, George Abram, 226
Millett, Kenneth C., 257
Milnor, John Willard, 179
Minkowski, Hermann, 83, 98
Minkowski's inequality, 98
Mississippi Delta State Teachers College, 3
Mittag-Leffler, Gösta Magnus, 90, 92
Möbius, August Ferdinand, 83, 84
Monge, Gaspard, 9, 81
Moore, Robert Lee, 3
Moore method, 3
Mordell, Louis Joel, 16
Morley, Frank, 45

Neugebauer, Otto, 16, 23
Neumann, Carl Gottfried, 91
Nevanlinna, Frithiof, 92
Nevanlinna, Rolf Herman, 92

Newman, Donald J., 159
Newton, Sir Isaac, 9, 84, 240, 243
Niven, Ivan, 190, 277
Noether, (Amelie) Emmy, 143

octonions, 148
Ogilvy, C. Stanley, 125
Øre, Øystein, 16, 65
Oseen, William, 16
Osgood, William Fogg, 89

Painlevé, Paul, 90
Pappus of Alexandria, 45
Pascal, Blaise, 9, 243
Peano, Giuseppe, 170
Peirce, Benjamin Osgood, 85, 242, 275
pentagonal tilings, 179, 183
perfect numbers, 3
Phragmén, Lars Edvard, 92
Picard, Charles Émile, 16, 90, 91, 217
Plücker, Julius, 83, 84
Poincaré, Henri, 9, 56, 65, 90, 91, 215, 216, 217, 222
Poisson, Siméon-Denis, 82
Pólya, George, 109, 151, 161, 166–167, 229, 282
Pólya's enumerationo theorem, 167
Pomerance, Carl, 140, 141, 257
Poncelet, Jean-Victor, 9, 82
Prony, Gaspard François de, 240
Ptolemy, 72
Puiseux, Victor Alexandre, 86
Pythagoreans, 55, 241

quadratic reciprocity, 56
quadratic residues, 55
quaternions, 143
Quêtelet, Adolphe, 241

Ramanujan, Srinivasa, 204
Reid, Constance, 51
Reidemeister move, 263, 266
Rhind papyrus, 23
Rice, Marjorie, 181, 184, 190
Richeson, A. W., 44
Riemann, (Georg Friedrich) Bernhard, 5–6, 9, 10, 16, 84, 88, 89, 90, 92
Riemann hypothesis, 89
Riemann integral, 129, 131
Riemann zeta function, 89
Riesz, Frigyes, 92
Robins, Benjamin, 199
Rosenhain, Johann Georg, 88
Rothman, Tony, 9
Ruffini, Paolo, 85, 218
Russell, Bertrand, 239

Saari, Donald 257
Salmon, George, 85
Schattschneider, Doris, 175, 253
Schiffer, Menahem Max, 279
Schläfli, Ludwig, 5–6
Schouten, Jan Arnoldus, 17
Schwarz, Hermann Amandus, 86, 88
Schweizerische Naturforschende Gesellschaft, 201
Segal, Irving Ezra, 35
Selfridge, John L., 140
Shephard, Geoffrey, 177, 179, 190, 247
Siegel, Carl Ludwig, 14
Singmaster, David Breyer, 133
Smale, Stephen, 257
Smith, Henry John Stephen, 85
Spanier, Edwin Henry, 35
Spencer, Donald Clayton, 279
Staudt, Karl Georg Christian von, 84
Steiner, Jakob, 83, 84
Stickelberger, Ludwig, 221–222
Stieltjes, Thomas Johannes, 90
Stirling approximation formula, 134
Stokes, George Gabriel, 84
Stone, Harlan Fiske, 35
Stone, Marshall Harvey, 35
"Stone age", 35
Størmer, Carl, 13, 17
Straus, Ernst Gábor, 139
Strogatz, Steven H., 211
Sturm, Jacques-Charles-François, 82
Sylvester, James Joseph, 9, 10, 85, 91, 216, 224, 275
Szegő, Gábor, 83
Szekeres, George, 185

Tamarkin, Jacob David, 29
Taylor, Edson H., 89
Thomson, William, Lord Kelvin, 84
tilings, 175–190
Tucker, Alan W., 161
Tucker, Albert William, 161

van der Corput, Johannes Gualtherus, 16
van der Waerden, Bartel Leendert, 143, 205
Veblen, Oswald, 14
Vessiot, Ernest, 217
Viète, François, 214, 239
Vigeland, Gustav, 13, 19–28
Volterra, Vito, 16, 92
von Dyck, Walther, 224–225, 226
von Mises, Richard, 84
von Neumann, John, 169

Waring, Edward, 93
Weber, Heinrich, 84, 224
Weber, Wilhelm Eduard, 84, 207
Weierstrass, Karl Theodor Wilhelm, 9, 84, 88, 89, 237
Weierstrass approximation theorem, 35–44
Weil, André, 35
Weyl, Hermann, 192

Whitehead link, 257
Wiener, Norbert, 16

Zelinsky, David, 137
Zeno, 9
Zentralblatt für Mathematik und ihre Grenzgebiete, 23
Zermelo, Ernst Friedrich Ferdinand, 81,
Zygmund, Antoni, 35

About the Editors

Gerald L. Alexanderson grew up in Northern California and except for undergraduate years at the University of Oregon and short assignments in Illinois and Switzerland, has spent his life there. After graduate school at Stanford University, he joined the faculty at Santa Clara University where he is in his 50th year of teaching. He served as department chair for 35 years and holds the Valeriote Professorship in Science.

From 1986 to 1990 he was editor of *Mathematics Magazine*, having served between 1984 and 1986 as First Vice President of the Mathematical Association of America, later as Secretary and finally, as President in 1997–99. By the most recent count he has served on 64 MAA committees or editorial boards and may hold a near record tenure on the MAA Board of Governors, 24 years. Active in various other professional societies, he was a member of the Phi Beta Kappa Senate for 12 years. In 2005 the MAA awarded him both the Deborah and Franklin Tepper Haimo Award for Distinguished College or University Teaching of Mathematics and the Yueh-Gin Gung and Dr. Charles Y. Hu Award for Distinguished Service to Mathematics.

In addition to articles in professional journals he has written, coauthored or coedited the following books: *Functional Trigonometry, A First Undergraduate Course in Abstract Algebra, Mathematical People, the Santa Clara Silver Anniversary Problem Book, The William Lowell Putnam Mathematical Competitions: Problems and Solutions 1965–1984, International Mathematical Congress: An Illustrated History, Discrete and Combinatorial Mathematics, The Pólya Picture Album, More Mathematical People, Lion Hunting and Other Mathematical Pursuits,* and *The Random Walks of George Pólya.*

Peter Ross was raised in the Midwest (born in Evanston, Illinois and attended high school in Madison, Wisconsin), but attended colleges on both coasts. After getting a BS from MIT and an MA from the University of California at Berkeley, he taught mathematics and physics in India as a Peace Corps Volunteer from 1963 to 1965. With several more interludes of working for the government on "new math," he eventually completed his PhD at Berkeley the day after his 40th birthday, and has taught in universities since then.

Peter has been at Santa Clara University since 1982, teaching mathematics and some computer science. He has been active in the Mathematical Association of America, including writing Media Highlights and book reviews for the *College Mathematics Journal* since 1985. His hobbies include music, as a choral singer, and bicycle-commuting to work in San Jose, the tenth largest city in the country. He has worked with the Sierra Club on eleven "service trips" in locations as diverse as Siberia, the Yukon, Mount Whitney, and Death Valley.